T0137881

Transactions on Engineering Technologies

Sio-Iong Ao · Len Gelman
Haeng Kon Kim
Editors

Transactions on Engineering Technologies

25th World Congress on Engineering (WCE 2017)

Editors
Sio-Iong Ao
International Association of Engineers
Hong Kong
Hong Kong

Len Gelman
School of Computing and Engineering
The University of Huddersfield
Queensgate, Huddersfield
UK

Haeng Kon Kim
Department of Computer and Communication
Engineering College, Catholic University of DaeGu
DaeGu
Korea (Republic of)

ISBN 978-981-13-4490-9 ISBN 978-981-13-0746-1 (eBook)
https://doi.org/10.1007/978-981-13-0746-1

Softcover re-print of the Hardcover 1st edition 2019

This Springer imprint is published by the registered company Springer Nature Singapore Pte Ltd.
The registered company address is: 152 Beach Road, #21-01/04 Gateway East, Singapore 189721, Singapore

Preface

A large international conference on Advances in Engineering Technologies and Physical Science was held in London, UK, July 5–7, 2017, under the World Congress on Engineering 2017 (WCE 2017). The WCE 2017 is organized by the International Association of Engineers (IAENG); the Congress details are available at: http://www.iaeng.org/WCE2017. IAENG is a nonprofit international association for engineers and computer scientists, which was founded originally in 1968. The World Congress on Engineering serves as good platforms for the engineering community to meet with each other and to exchange ideas. The conferences have also struck a balance between theoretical and application development. The conference committees have been formed with over three hundred committee members who are mainly research center heads, faculty deans, department heads, professors, and research scientists from over 30 countries. The congress is truly global international event with a high level of participation from many countries. The response to the Congress has been excellent. There have been more than six hundred manuscript submissions for the WCE 2017. All submitted papers have gone through the peer review process, and the overall acceptance rate is 51%.

This volume contains thirty revised and extended research articles written by prominent researchers participating in the conference. Topics covered include mechanical engineering, engineering mathematics, computer science, knowledge engineering, electrical engineering, wireless networks, and industrial applications. The book offers the state of the art of tremendous advances in engineering technologies and physical science and applications, and also serves as an excellent reference work for researchers and graduate students working on engineering technologies and physical science and applications.

Hong Kong, Hong Kong — Sio-Iong Ao
Queensgate, UK — Len Gelman
DaeGu, Korea (Republic of) — Haeng Kon Kim

Contents

Homogenization of Electromagnetic Fields Propagation in a Composite

Helene Canot and Emmanuel Frenod

Abstract In this paper we study the two-scale behavior of the electromagnetic field in 3D in a composite material. It is the continuation of the paper (Canot and Frenod Method of homogenization for the study of the propagation of electromagnetic waves in a composite 2017) [7] in which we obtain existence and uniqueness results for the problem, we performed an estimate that allows us to approach homogenization. Techniques of asymptotic expansion and two-scale convergence are used to obtain the homogenized problem. We justify the two-scale expansion numerically in the second part of the paper.

Keywords Asymptotic Expansion · Electromagnetism · Finite element Harmonic Maxwell Equations · Homogenization · Simulations · Two-scale Convergence

1 Introduction

We are interested in the time-harmonic Maxwell equations in and near a composite material with boundary conditions modeling electromagnetic field radiated by an electromagnetic pulse (EMP). We recall that composite material is composed by carbon fibers periodically enclosed in an epoxy resin which is charged with nanoparticles. In the first part, we have presented the model and proved the existence of a unique solution of the problem. Our mathematical context is periodic

H. Canot (✉) · E. Frenod
Helene Canot and Emmanuel Frenod are in the Department of Mathematics of University of Bretagne Sud (LMBA), Centre Yves Coppens, Bat. B, 1er et., Campus de Tohannic BP 573, 56017 Vannes, France
e-mail: helene.canot@univ-ubs.fr

E. Frenod
e-mail: emmanuel.frenod@univ-ubs.fr

S.-I. Ao et al. (eds.), *Transactions on Engineering Technologies*,
https://doi.org/10.1007/978-981-13-0746-1_1

homogenization. We consider a microscopic scale ε, which represents the ratio between the diameter of the fiber and thickness of the composite material. So, we are trying to understand how the microscopic structure affects the macroscopic electromagnetic field behavior. Homogenization of Maxwell equations with periodically oscillating coefficients was studied in many papers. N. Wellander homogenized linear and non-linear Maxwell equations with perfect conducting boundary conditions using two-scale convergence in [19, 20]. N. Wellander and B. Kristensson homogenized the full time-harmonic Maxwell equation with penetrable boundary conditions and at fixed frequency in Wellander and Kristensson [21]. The homogenized time-harmonic Maxwell equation for the scattering problem was done in Guenneau et al. [11]. Y. Amirat and V. Shelukhin perform two-scale homogenization time-harmonic Maxwell equations for a periodical structure in Amirat and Shelukhin [4]. They calculate the effective dielectric ε and effective electric conductivity σ. They proved that homogenized Maxwell equations are different in low and high frequencies. In our model, we use the Asymptotic expansion suggested by Bensoussan et al. [6] and we justify rigorously mathematical problem by using the theory of two-scale convergence introduced by Nguetseng [15] and developed by Allaire [2]. The result obtained by two-scale convergence approach takes into account the characteristic sizes of skin thickness and wavelength around the material. Then we compare numerically the theoretical result with the homogenized model. The goal is to validate the homogenized procedure for *epsilon* $= 0.01$. The paper was presented at the World Congress on Engineering in London [13].

2 Homogenization

We recall that our problem is:

$$\nabla\times\nabla\times E^{\varepsilon} - \omega^2\varepsilon^5 k(\varepsilon)E^{\varepsilon} + i\omega[(\mathbf{1}^{\varepsilon}_C(\frac{\mathbf{x}}{\varepsilon}) + \varepsilon^4\mathbf{1}^{\varepsilon}_R(\frac{\mathbf{x}}{\varepsilon}))]E^{\varepsilon} = 0, \text{ in } \Omega. \tag{1}$$

where for a given set $\mathscr{A}$, $\mathbf{1}_{\mathscr{A}}$ stands for the characteristic function of $\mathscr{A}$ and where $\mathbf{1}^{\varepsilon}_{\mathscr{A}}(\mathbf{x}) = \mathbf{1}_{\mathscr{A}}(\frac{\mathbf{x}}{\varepsilon})$, hence $\mathbf{1}^{\varepsilon}_C$ and $\mathbf{1}^{\varepsilon}_R$ are the characteristic functions of the sets filled by carbon fibers and by resin. And where $k(\varepsilon) = (\varepsilon_c\mathbf{1}^{\varepsilon}_C(\mathbf{x}) + \varepsilon_r\mathbf{1}^{\varepsilon}_R(\mathbf{x}))$.

Remark 1 We recall that ε_c and ε_r are respectively the relative permittivity of the carbon fibers and the resin. You should not confused with the microscopic scale ε. Equation (1) is provided with the following boundary conditions:

$$\nabla\times E^{\varepsilon} \times e_2 = -i\omega H_d(x, z) \times e_2 \text{ on } R \times \Gamma_d, \tag{2}$$

and

$$\nabla\times E^{\varepsilon} \times e_2 = 0 \text{ on } R \times \Gamma_L. \tag{3}$$

We start homogenization approach with the two-scale asymptotic expansion. The rigourous mathematical justification of the homogenized problem was made using two-scale convergence. This concept was introduced by Nguetseng [16] and specified by Allaire [3] which studied properties of the two-scale convergence. Neuss-Radu in [14] presented an extension of two-scale convergence method to the periodic surfaces. Many authors applied two-scale convergence approach Cionarescu and Donato [9], Crouseilles et al. [10], Amirat et al. [1] and also Back and Frénod [5]. This mathematical concept were applied to homogenize the time-harmonic Maxwell equations Ouchetto et al. [17], Pak [18].

In our model, the parallel carbon cylinders are periodically distributed in direction x and z, as the material is homogenous in the y direction, we can consider that the material is periodic with a three directional cell of periodicity. In other words, introducing $\mathscr{Z} = [-\frac{1}{2}, \frac{1}{2}] \times [-1, 0]^2$, function Σ^ε given in Canot and Frenod [8] is naturally periodic with respect to (ξ, ζ) with period $[-\frac{1}{2}, \frac{1}{2}] \times [-1, 0]$ but it is also periodic with respect to $\mathbf{y}$ with period $\mathscr{Z}$.

Now, we review some basis definitions and results about two-scale convergence.

2.1 *Two-Scale Convergence*

We first define the function spaces

$$\begin{aligned} \mathbf{H}(\mathrm{curl}, \Omega) &= \{u \in \mathbf{L}^2(\Omega) : \ \nabla \times u \in \mathbf{L}^2(\Omega)\}, \\ \mathbf{H}(\mathrm{div}, \Omega) &= \{u \in \mathbf{L}^2(\Omega) : \ \nabla \cdot u \in L^2(\Omega)\}, \end{aligned} \tag{4}$$

with the usual norms:

$$\begin{aligned} \|u\|^2_{\mathbf{H}(\mathrm{curl},\Omega)} &= \|u\|^2_{\mathbf{L}^2(\Omega)} + \|\nabla \times u\|^2_{\mathbf{L}^2(\Omega)}, \\ \|u\|^2_{\mathbf{H}(\mathrm{div},\Omega)} &= \|u\|^2_{\mathbf{L}^2(\Omega)} + \|\nabla \cdot u\|^2_{L^2(\Omega)}. \end{aligned} \tag{5}$$

They are well known Hilbert spaces.

$$\begin{aligned} \mathbf{H}_\#(\mathrm{curl}, \mathscr{Z}) &= \{u \in \mathbf{H}(\mathrm{curl}, R^3) : u \text{ is } \mathscr{Z}\text{-periodic}\} \\ \mathbf{H}_\#(\mathrm{div}, \mathscr{Z}) &= \{u \in \mathbf{H}(\mathrm{div}, R^3) : u \text{ is } \mathscr{Z}\text{-periodic}\} \end{aligned} \tag{6}$$

We introduce

$$\mathbf{L}^2_\#(\mathscr{Z}) = \{u \in \mathbf{L}^2(R^3), u \text{ is } \mathscr{Z}\text{-periodic}\}, \tag{7}$$

and

$$\mathbf{H}^1_\#(\mathscr{Z}) = \{u \in \mathbf{H}^1(R^3), u \text{ is } \mathscr{Z}\text{-periodic}\}, \tag{8}$$

where $\mathbf{H}^1(R^3)$ is the usual Sobolev space on R^3. First, denoting by $\mathbf{C}^0_\#(\mathscr{Z})$ the space of functions in $\mathbf{C}^0(R^3)$ and $\mathscr{Z}$-periodic, $\mathbf{C}^0_0(R^3)$ the space of continuous functions over R^3 with compact support, we have the following definitions:

Definition 1 A sequence $u^\varepsilon(\mathbf{x})$ in $\mathbf{L}^2(\Omega)$ two-scale converges to $u_0(\mathbf{x},\mathbf{y}) \in \mathbf{L}^2(\Omega, \mathbf{L}^2_\#(\mathscr{Y}))$ if for every $V(\mathbf{x},\mathbf{y}) \in \mathbf{C}^0_0(\Omega, \mathscr{C}^0_\#(\mathscr{Y}))$

$$\lim_{\varepsilon\to 0} \int_\Omega u^\varepsilon(\mathbf{x}) \cdot V(\mathbf{x}, \mathbf{x}/\varepsilon)\, d\mathbf{x} = \int_\Omega \int_{\mathscr{Y}} u_0(\mathbf{x},\mathbf{y}) \cdot V(\mathbf{x},\mathbf{y})\, d\mathbf{x} d\mathbf{y}. \tag{9}$$

Proposition 1 *If $u^\varepsilon(\mathbf{x})$ two-scale converges to $u_0(\mathbf{x},\mathbf{y}) \in \mathbf{L}^2(\Omega, \mathbf{L}^2_\#(\mathscr{Y}))$, we have for all $v(\mathbf{x}) \in C_0(\overline{\Omega})$ and all $w(\mathbf{y}) \in \mathbf{L}^2_\#(\mathscr{Y})$*

$$\lim_{\varepsilon\to 0} \int_\Omega u^\varepsilon(\mathbf{x}) \cdot v(\mathbf{x}) w(\tfrac{\mathbf{x}}{\varepsilon})\, d\mathbf{x} = \int_\Omega \int_{\mathscr{Y}} u_0(\mathbf{x},\mathbf{y}) \cdot v(\mathbf{x}) w(\mathbf{y})\, d\mathbf{x} d\mathbf{y}. \tag{10}$$

Theorem 1 *(Nguetseng). Let $u^\varepsilon(\mathbf{x}) \in \mathbf{L}^2(\Omega)$. Suppose there exists a constant $c > 0$ such that for all ε*

$$\|u^\varepsilon\|_{L^2(\Omega)} \leq c. \tag{11}$$

Then there exists a subsequence of ε (still denoted ε) and $u_0(\mathbf{x},\mathbf{y}) \in \mathbf{L}^2(\Omega, \mathbf{L}^2_\#(\mathscr{Y}))$ such that:

$$u^\varepsilon(\mathbf{x}) \text{ two-scale converges to } u_0(\mathbf{x},\mathbf{y}). \tag{12}$$

Proposition 2 *Let $u^\varepsilon(\mathbf{x})$ be a sequence of functions in $\mathbf{L}^2(\Omega)$, which two-scale converges to a limit $u_0(\mathbf{x},\mathbf{y}) \in \mathbf{L}^2(\Omega, \mathbf{L}^2_\#(\mathscr{Y}))$.*

Then $u^\varepsilon(\mathbf{x})$ converges also to $u(\mathbf{x}) = \int_{\mathscr{Y}} u_0(\mathbf{x},\mathbf{y}) d\mathbf{y}$ in $\mathbf{L}^2(\Omega)$ weakly.

Furthermore, we have

$$\lim_{\varepsilon\to 0} \|u^\varepsilon\|_{\mathbf{L}^2(\Omega)} \geq \|u_0\|_{\mathbf{L}^2(\Omega\times Y)} \geq \|u\|_{\mathbf{L}^2(\Omega)}. \tag{13}$$

Proposition 3 *Let $u^\varepsilon(\mathbf{x})$ be bounded in $\mathbf{L}^2(\Omega)$. Up to a subsequence, $u^\varepsilon(\mathbf{x})$ two-scale converges to $u_0(\mathbf{x},\mathbf{y}) \in \mathbf{L}^2(\Omega, \mathbf{L}^2_\#(\mathscr{Y}))$ such that:*

$$u_0(\mathbf{x},\mathbf{y}) = u(\mathbf{x}) + \widetilde{u_0}(\mathbf{x},\mathbf{y}), \tag{14}$$

where $\widetilde{u_0}(\mathbf{x},\mathbf{y}) \in \mathbf{L}^2(\Omega, \mathbf{L}^2_\#(\mathscr{Y}))$ satisfies

$$\int_{\mathscr{Y}} \widetilde{u_0}(\mathbf{x},\mathbf{y})\, d\mathbf{y} = 0, \tag{15}$$

and $u(\mathbf{x}) = \int_{\mathscr{Y}} u_0(\mathbf{x},\mathbf{y})\, d\mathbf{y}$ is a weak limit in $\mathbf{L}^2(\Omega)$.

Proof Due to the a priori estimates (32), $u^{\varepsilon}(\mathbf{x})$ is bounded in $\mathbf{L}^2(\Omega)$, then by application of Theorem 1, u^{ε} we get the first part of the proposition. Furthermore by defining $\widetilde{u_0}$ as

$$\widetilde{u_0}(\mathbf{x},\mathbf{y}) = u_0(\mathbf{x},\mathbf{y}) - \int_{\mathscr{Y}} u_0(\mathbf{x},\mathbf{y})d\mathbf{y}, \tag{16}$$

we obtain the decomposition (14) of u_0.

Proposition 4 *Let any two-scale limit $u_0(\mathbf{x},\mathbf{y})$, given by Proposition (3), can be decomposed as*

$$u_0(\mathbf{x},\mathbf{y}) = u(\mathbf{x}) + \nabla_{\mathbf{y}}\Phi(\mathbf{x},\mathbf{y}). \tag{17}$$

where $\Phi \in \mathbf{L}^2(\Omega, \mathbf{H}^1_{\#}(\mathscr{Y}))$ is a scalar-valued function and where $u \in \mathbf{L}^2(\Omega)$.

Proof Proof of (17), integrating by parts, for any $V(\mathbf{x},\mathbf{y}) \in \mathbf{C}^1_0(\Omega, \mathbf{C}^1_{\#}(\mathscr{Y}))$, we have

$$\begin{aligned}
&\varepsilon \int_{\Omega} \nabla\times u^{\varepsilon}(\mathbf{x}) \cdot V(\mathbf{x}, \tfrac{\mathbf{x}}{\varepsilon})\, d\mathbf{x} \\
&= \varepsilon \int_{\Omega} u^{\varepsilon}(\mathbf{x}) \cdot \nabla\times V(\mathbf{x}, \tfrac{\mathbf{x}}{\varepsilon})\, d\mathbf{x} \\
&= \int_{\Omega} u^{\varepsilon}(\mathbf{x})\{\varepsilon\nabla_{\mathbf{x}} \times V(\mathbf{x}, \tfrac{\mathbf{x}}{\varepsilon}) + \nabla_{\mathbf{y}} \times V(\mathbf{x}, \tfrac{\mathbf{x}}{\varepsilon})\}\, d\mathbf{x}.
\end{aligned} \tag{18}$$

Taking the two-scale limit as $\varepsilon \to 0$ we obtain

$$0 = \int_{\Omega}\int_{\mathscr{Y}} u_0(\mathbf{x},\mathbf{y}) \cdot \nabla_{\mathbf{y}} \times V(\mathbf{x},\mathbf{y})\, d\mathbf{x}d\mathbf{y}, \tag{19}$$

which implies that $\nabla_{\mathbf{y}} \times u_0(\mathbf{x},\mathbf{y}) = 0$. To end the proof of the proposition we use the following result.

Proposition 5 *If $u_0 \in \mathbf{L}^2(\Omega)$ satisfies*

$$\nabla_{\mathbf{y}} \times u_0(\mathbf{x},\mathbf{y}) = 0, \tag{20}$$

then there exists $u \in \mathbf{L}^2(\Omega)$ and $\Phi \in \mathbf{L}^2(\Omega, \mathbf{H}^1_{\#}(\mathscr{Y}))$ such that $u_0(\mathbf{x},\mathbf{y}) = u(\mathbf{x}) + \nabla_{\mathbf{y}}\Phi(\mathbf{x},\mathbf{y})$.

Applying this proposition we obtain equality (17) ending the proof of Proposition (4).

3 Homogenized Problem

We will explore in this section the behavior of electromagnetic field E^{ε} using the asymptotic expansion and the two-scale convergence to determine the homogenized problem. We place in the context of the case 6 with $\delta > L$ and $\overline{\omega} = 10^6\text{rad.s}^{-1}$, then we have $\eta = 5$ and $\Sigma^{\varepsilon}_a = \varepsilon$, $\Sigma^{\varepsilon}_r = \varepsilon^4$, $\Sigma^{\varepsilon}_c = 1$ which gives the following equation:

$$\nabla\times\nabla\times E^{\varepsilon}-\omega^{2}\varepsilon^{5}k(\varepsilon)E^{\varepsilon}+i\omega[(\mathbf{1}_{C}^{\varepsilon}(\tfrac{\mathbf{x}}{\varepsilon})+\varepsilon^{4}\mathbf{1}_{R}^{\varepsilon}(\tfrac{\mathbf{x}}{\varepsilon}))\mathbf{1}_{\{y<0\}}+\varepsilon\mathbf{1}_{\{y>0\}}]E^{\varepsilon}=0, \tag{21}$$

where for a given set $\mathscr{A}$, $\mathbf{1}_{\mathscr{A}}$ stands for the characteristic function of $\mathscr{A}$ and where $\mathbf{1}_{\mathscr{A}}^{\varepsilon}(\mathbf{x})=\mathbf{1}_{\mathscr{A}}(\frac{\mathbf{x}}{\varepsilon})$, hence $\mathbf{1}_{C}^{\varepsilon}$ and $\mathbf{1}_{R}^{\varepsilon}$ are the characteristic functions of the sets filled by carbon fibers and by resin. And where $k(\varepsilon)=(\varepsilon_{c}\mathbf{1}_{C}^{\varepsilon}(\mathbf{x})+\varepsilon_{r}\mathbf{1}_{R}^{\varepsilon}(\mathbf{x}))\mathbf{1}_{\{y<0\}}+\mathbf{1}_{\{y>0\}}$. First, we will use the classical method of the asymptotic expansion.

3.1 Asymptotic Expansion

We assume that $(E^{\varepsilon}, H^{\varepsilon})$ satisfies the following asymptotic expansion, as $\varepsilon\to 0$:

$$E^{\varepsilon}(\mathbf{x})=E_{0}(\mathbf{x},\frac{\mathbf{x}}{\varepsilon})+\varepsilon E_{1}(\mathbf{x},\frac{\mathbf{x}}{\varepsilon})+\varepsilon^{2}E_{2}(\mathbf{x},\frac{\mathbf{x}}{\varepsilon})+....,\tag{22}$$

where for any $k\in N$ $E_{k}=E_{k}(\mathbf{x},\mathbf{y})$ are considered as $\mathscr{Y}$-periodic functions with respect to $\mathbf{y}$. Applied to functions $E_{k}(\mathbf{x},\frac{\mathbf{x}}{\varepsilon})$ the curl operator becomes $\nabla_{\mathbf{x}}\times E_{k}(\mathbf{x},\frac{\mathbf{x}}{\varepsilon})+\frac{1}{\varepsilon}\nabla_{\mathbf{y}}\times E_{k}(\mathbf{x},\frac{\mathbf{x}}{\varepsilon})$. Plugging (22) in the formulations (21), gathering the coefficients with the same power of ε, we get:

$$\left\{\begin{array}{l}\frac{1}{\varepsilon^{2}}\nabla_{\mathbf{y}}\times\nabla_{\mathbf{y}}\times E_{0}(\mathbf{x},\frac{\mathbf{x}}{\varepsilon})\\ +\frac{1}{\varepsilon}[\nabla_{\mathbf{y}}\times\nabla_{\mathbf{y}}\times E_{1}(\mathbf{x},\frac{\mathbf{x}}{\varepsilon})+\nabla_{\mathbf{y}}\times\nabla_{\mathbf{x}}\times E_{0}(\mathbf{x},\frac{\mathbf{x}}{\varepsilon})+\nabla_{\mathbf{x}}\times\nabla_{\mathbf{y}}\times E_{0}(\mathbf{x},\frac{\mathbf{x}}{\varepsilon})]\\ +\varepsilon^{0}[\nabla_{\mathbf{x}}\times\nabla_{\mathbf{y}}\times E_{1}(\mathbf{x},\frac{\mathbf{x}}{\varepsilon})+\nabla_{\mathbf{x}}\times\nabla_{\mathbf{x}}\times E_{0}(\mathbf{x},\frac{\mathbf{x}}{\varepsilon})\\ +\nabla_{\mathbf{y}}\times\nabla_{\mathbf{y}}\times E_{2}(\mathbf{x},\frac{\mathbf{x}}{\varepsilon})+\nabla_{\mathbf{y}}\times\nabla_{\mathbf{x}}\times E_{1}(\mathbf{x},\frac{\mathbf{x}}{\varepsilon})+i\omega\mathbf{1}_{C}^{\varepsilon}(\mathbf{x})\mathbf{1}_{\{y<0\}}E_{0}(\mathbf{x},\frac{\mathbf{x}}{\varepsilon})]\\ +\varepsilon[\nabla_{\mathbf{x}}\times\nabla_{\mathbf{y}}\times E_{2}(\mathbf{x},\frac{\mathbf{x}}{\varepsilon})+\nabla_{\mathbf{x}}\times\nabla_{\mathbf{x}}\times E_{1}(\mathbf{x},\frac{\mathbf{x}}{\varepsilon}))+\nabla_{\mathbf{y}}\times\nabla_{\mathbf{x}}\times E_{2}(\mathbf{x},\frac{\mathbf{x}}{\varepsilon})\\ +\nabla_{\mathbf{y}}\times\nabla_{\mathbf{y}}\times E_{3}(\mathbf{x},\frac{\mathbf{x}}{\varepsilon})+i\omega(\mathbf{1}_{C}^{\varepsilon}(\mathbf{x})\mathbf{1}_{\{y<0\}}E_{1}(\mathbf{x},\frac{\mathbf{x}}{\varepsilon})\\ +\mathbf{1}_{\{y>0\}})E_{0}(\mathbf{x},\frac{\mathbf{x}}{\varepsilon})]+...)=0.\end{array}\right.\tag{23}$$

In order to write what is in factor of ε in the last equation we used that : $\mathbf{1}_{\{y<0\}}=\mathbf{1}_{\{\frac{y}{\varepsilon}<0\}}$. Since (23) is considered as true for any small ε it gives a cascade of equations, from which we extract the four first equations

$$\left\{\begin{array}{l}\nabla_{\mathbf{y}}\times\nabla_{\mathbf{y}}\times E_{0}(\mathbf{x},\mathbf{y})=0\\ \nabla_{\mathbf{y}}\times\nabla_{\mathbf{y}}\times E_{1}(\mathbf{x},\mathbf{y})+\nabla_{\mathbf{y}}\times\nabla_{\mathbf{x}}\times E_{0}(\mathbf{x},\mathbf{y})+\nabla_{\mathbf{x}}\times\nabla_{\mathbf{y}}\times E_{0}(\mathbf{x},\mathbf{y})=0\\ \nabla_{\mathbf{x}}\times\nabla_{\mathbf{y}}\times E_{1}(\mathbf{x},\mathbf{y})+\nabla_{\mathbf{x}}\times\nabla_{\mathbf{x}}\times E_{0}(\mathbf{x},\mathbf{y})+\nabla_{\mathbf{y}}\times\nabla_{\mathbf{y}}\times E_{2}(\mathbf{x},\mathbf{y})\\ +\nabla_{\mathbf{y}}\times\nabla_{\mathbf{x}}\times E_{1}(\mathbf{x},\mathbf{y})+i\omega\mathbf{1}_{C}(\mathbf{y})\mathbf{1}_{\{\upsilon<0\}}E_{0}(\mathbf{x},\mathbf{y})=0\\ \nabla_{\mathbf{x}}\times\nabla_{\mathbf{y}}\times E_{2}(\mathbf{x},\mathbf{y})+\nabla_{\mathbf{x}}\times\nabla_{\mathbf{x}}\times E_{1}(\mathbf{x},\mathbf{y})\\ +\nabla_{\mathbf{y}}\times\nabla_{\mathbf{x}}\times E_{2}(\mathbf{x},\mathbf{y})+\nabla_{\mathbf{y}}\times\nabla_{\mathbf{y}}\times E_{3}(\mathbf{x},\mathbf{y})\\ +i\omega\big(\mathbf{1}_{C}(\mathbf{y})\mathbf{1}_{\{\upsilon<0\}}E_{1}(\mathbf{x},\mathbf{y})+\mathbf{1}_{\{\upsilon>0\}}\big)E_{0}(\mathbf{x},\mathbf{y})=0\end{array}\right.\tag{24}$$

Applying $\mathrm{div}_{\mathbf{y}}$ in the last two equations in (24), we obtain

$$\nabla_{\mathbf{y}}\cdot\big(i\omega\mathbf{1}_{C}(\mathbf{y})\mathbf{1}_{\{\upsilon<0\}}E_{0}(\mathbf{x},\mathbf{y})\big)=0,\tag{25}$$

and

$$\nabla_{\mathbf{y}} \cdot \big(i\omega(\mathbf{1}_C(\mathbf{y})\mathbf{1}_{\{\nu<0\}}E_1(\mathbf{x},\mathbf{y}) + \mathbf{1}_{\{\nu>0\}})E_0(\mathbf{x},\mathbf{y})\big) = 0. \tag{26}$$

The boundary condition in (2) write:

$$\begin{cases} (\frac{1}{\varepsilon}\nabla_{\mathbf{y}} \times E_0(\mathbf{x},\mathbf{y}) + \nabla_{\mathbf{x}} \times E_0(\mathbf{x},\mathbf{y}) + \nabla_{\mathbf{y}} \times E_1(\mathbf{x},\mathbf{y}+...) \times n \\ = -i\omega H_d \times n,\ \mathbf{x} \in \mathbb{R}^3,\ \mathbf{y} \in \mathscr{Y}. \end{cases} \tag{27}$$

Now we take the first equation of (24) and the equation (25) to obtain:

$$\begin{cases} \nabla_{\mathbf{y}} \times \nabla_{\mathbf{y}} \times E_0(\mathbf{x},\mathbf{y}) = 0, \\ \nabla_{\mathbf{y}} \cdot \{i\omega(\mathbf{1}_C(\mathbf{y})\mathbf{1}_{\{\nu<0\}})E_0(\mathbf{x},\mathbf{y})\} = 0. \end{cases} \tag{28}$$

Multiplying the first equation in (28) by E_0 and integrating by parts over $\mathscr{Y}$ leads to

$$\begin{aligned} &\int_{\mathscr{Y}} \nabla_{\mathbf{y}} \times \nabla_{\mathbf{y}} \times E_0(\mathbf{x},\mathbf{y})E_0(\mathbf{x},\mathbf{y})\ d\mathbf{y} \\ &= \int_{\mathscr{Y}} |\nabla_{\mathbf{y}} \times E_0(\mathbf{x},\mathbf{y})|^2\ d\mathbf{y} \\ &= 0. \end{aligned} \tag{29}$$

We deduce that the equation is equivalent to

$$\nabla_{\mathbf{y}} \times E_0(\mathbf{x},\mathbf{y}) = 0, \tag{30}$$

for any $\mathbf{y} \in \mathscr{Y}$.

Hence from Proposition (5) we conclude that $E_0(\mathbf{x},\mathbf{y})$ can be decomposed as

$$E_0(\mathbf{x},\mathbf{y}) = E(\mathbf{x}) + \nabla_{\mathbf{y}}\Phi_0(\mathbf{x},\mathbf{y}), \tag{31}$$

where $\Phi_0(\mathbf{x},\mathbf{y}) \in \mathbf{L}^2(\Omega; \mathbf{H}^1_\#(\mathscr{Y}))$ and $E(\mathbf{x}) \in \mathbf{L}^2(\Omega)$.

3.2 Mathematical Justification

Now we will show rigorously with two-scale convergence that the solution of problems (1)–(3) converge to the solution of the homogenized problem when ε tends to 0. We recall the following Theorem, we give a proof in Canot and Frenod [8]:

Theorem 2 *For any $\varepsilon > 0$, for any $\eta \geq 0$, there exists a positive constant ω_0 which does not depend on ε and such that for all $\omega \in (0, \omega_0)$, $E^\varepsilon \in X^\varepsilon(\Omega)$ solution of (1)–(3) satisfies*

$$\|E^\varepsilon\|_{\mathbf{X}^\varepsilon(\Omega)} \le C \tag{32}$$

with $C = \frac{C_{\gamma_t} C_{\gamma_T}}{\mathscr{C}_0} \|H_d\|_{H(curl,\Omega)}$.

Theorem 3 *Under assumptions of Theorem 2, sequence* E^ε *solution of (1)–(3) converges to* $E(\mathbf{x}) \in \mathbf{L}^2(\Omega)$ *which is the unique solution of the homogenized problem:*

$$\begin{cases} \theta_1 \nabla_\mathbf{x} \times \nabla_\mathbf{x} \times E(\mathbf{x}) + i\omega\theta_2 E(\mathbf{x}) = 0 \;\; in \; \Omega, \\ \theta_1 \nabla_\mathbf{x} \times E(\mathbf{x}) \times e_2 = -i\omega H_d \times e_2 \;\; on \; \Gamma_d, \\ \nabla_\mathbf{x} \times E(\mathbf{x}) \times e_2 = 0 \;\; on \; \Gamma_L. \end{cases} \tag{33}$$

with $\theta_1 = \int_{\mathscr{Y}} \mathrm{Id} + \nabla_y \chi(\mathbf{y})\, d\mathbf{y}$ *and* $\theta_2 = \int_{\mathscr{Y}} \mathbf{1}_C(\mathbf{y})(\mathrm{Id} + \nabla_y \chi(\mathbf{y}))\, d\mathbf{y}$.

And where the scalar function χ *is the unique solution, up to an additive constant in the Hilbert space of* $\mathscr{Y}$ *periodic functions* $H^1_\#(\mathscr{Y})$*, of the following boundary value problem*

$$\begin{cases} \Delta_\mathbf{y}(\chi(\mathbf{y})) = 0 \;\; in \; \mathscr{Y} \backslash \partial\Omega_C, \\ [\frac{\partial \chi}{\partial n}] = -n_j \;\; on \; \partial\Omega_C, \\ [\chi] = 0 \;\; on \; \partial\Omega_C. \end{cases} \tag{34}$$

where $[f]$ *is the jump across the surface of* $\partial\Omega_C$, n_j, $j = \{1, 2, 3\}$ *is the projection on the axis* e_j *of the normal of* $\partial\Omega_C$.

Proof **Step 1: Two-scale convergence**. Due to the estimate (32), E^ε is bounded in $\mathbf{L}^2(\Omega)$. Hence, up to a subsequence, E^ε two-scale converges to $E_0(\mathbf{x}, \mathbf{y})$ belonging to $\mathbf{L}^2(\Omega, \mathbf{L}^2_\#(\mathscr{Y}))$. That means for any $V(\mathbf{x}, \mathbf{y}) \in \mathbf{C}^1_0(\Omega, \mathbf{C}^1_\#(\mathscr{Y}))$, we have:

$$\lim_{\varepsilon \to 0} \int_\Omega E^\varepsilon(\mathbf{x}) \cdot V(\mathbf{x}, \frac{\mathbf{x}}{\varepsilon})\, d\mathbf{x} = \int_\Omega \int_{\mathscr{Y}} E_0(\mathbf{x}, \mathbf{y}) \cdot V(\mathbf{x}, \mathbf{y})\, d\mathbf{y} d\mathbf{x}. \tag{35}$$

Step 2: Deduction of the constraint equation. We multiply the equation (21) by oscillating test function $V^\varepsilon(\mathbf{x}) = V(\mathbf{x}, \frac{\mathbf{x}}{\varepsilon})$ where $V(\mathbf{x}, \mathbf{y}) \in \mathbf{C}^1_0(\Omega, \mathbf{C}^1_\#(\mathscr{Y}))$:

$$\begin{aligned} &\int_\Omega \nabla \times E^\varepsilon(\mathbf{x}) \cdot (\nabla_\mathbf{x} \times V^\varepsilon(\mathbf{x}, \frac{\mathbf{x}}{\varepsilon}) + \frac{1}{\varepsilon} \nabla_\mathbf{y} \times V^\varepsilon(\mathbf{x}, \frac{\mathbf{x}}{\varepsilon})) + [-\omega^2 \varepsilon^5 k(\varepsilon) \\ &+ i\omega((\mathbf{1}^\varepsilon_C(\frac{\mathbf{x}}{\varepsilon}) + \varepsilon^4 \mathbf{1}^\varepsilon_R(\frac{\mathbf{x}}{\varepsilon}))\mathbf{1}_{\{y<0\}} + \varepsilon \mathbf{1}_{\{y>0\}})] E^\varepsilon \cdot V^\varepsilon(\mathbf{x}, \frac{\mathbf{x}}{\varepsilon})\, d\mathbf{x} \\ &= -i\omega \int_{\Gamma_d} H_d \times e_2 \cdot (e_2 \times V(x, 1, z, \xi, \frac{1}{\varepsilon}, \zeta)) \times e_2\, d\sigma. \end{aligned} \tag{36}$$

Integrating by parts, we get:

$$\begin{aligned}
&\int_{\Omega} E^{\varepsilon}(\mathbf{x}) \cdot (\nabla_{\mathbf{x}} \times \nabla_{\mathbf{x}} \times V^{\varepsilon}(\mathbf{x}, \frac{\mathbf{x}}{\varepsilon}) + \frac{1}{\varepsilon}\nabla_{\mathbf{y}} \times \nabla_{\mathbf{x}} \times V^{\varepsilon}(\mathbf{x}, \frac{\mathbf{x}}{\varepsilon}) \\
&+ \frac{1}{\varepsilon}\nabla_{\mathbf{x}} \times \nabla_{\mathbf{y}} \times V^{\varepsilon}(\mathbf{x}, \frac{\mathbf{x}}{\varepsilon}) + \frac{1}{\varepsilon^2}\nabla_{\mathbf{y}} \times \nabla_{\mathbf{y}} \times V^{\varepsilon}(\mathbf{x}, \frac{\mathbf{x}}{\varepsilon})) + [-\omega^2\varepsilon^5 k(\varepsilon) \\
&+ i\omega\big(\mathbf{1}_C^{\varepsilon}(\frac{\mathbf{x}}{\varepsilon}) + \varepsilon^4 \mathbf{1}_R^{\varepsilon}(\frac{\mathbf{x}}{\varepsilon})\big)\mathbf{1}_{\{y<0\}} + \varepsilon \mathbf{1}_{\{y>0\}}] E^{\varepsilon}(\mathbf{x}) \cdot V^{\varepsilon}(\mathbf{x}, \frac{\mathbf{x}}{\varepsilon})\, d\mathbf{x} \\
&= -i\omega \int_{\Gamma_d} H_d \times e_2 \cdot (e_2 \times V(x, 1, z, \xi, \frac{1}{\varepsilon}, \zeta)) \times e_2 \, d\sigma.
\end{aligned} \tag{37}$$

Now we multiply (37) by ε^2 and we pass to the two-scale limit, applying Theorem 1 we obtain:

$$\int_{\Omega}\int_{\mathcal{Y}} E_0(\mathbf{x}, \mathbf{y})\big(\nabla_{\mathbf{y}} \times \nabla_{\mathbf{y}} \times V(\mathbf{x}, \mathbf{y})\big)\, d\mathbf{y} d\mathbf{x} = 0. \tag{38}$$

We deduce the constraint equation for the profile E_0:

$$\nabla_{\mathbf{y}} \times \nabla_{\mathbf{y}} \times E_0(\mathbf{x}, \mathbf{y}) = 0. \tag{39}$$

Step 3. Looking for the solutions to the constraint equation. Multiplying Equation (39) by E_0 and integrating by parts over $\mathcal{Y}$ leads to

$$\int_{\mathcal{Y}} \nabla_{\mathbf{y}} \times \nabla_{\mathbf{y}} \times E_0(\mathbf{x}, \mathbf{y}) E_0(\mathbf{x}, \mathbf{y}) \; d\mathbf{y} = \int_{\mathcal{Y}} |\nabla_{\mathbf{y}} \times E_0(\mathbf{x}, \mathbf{y})|^2 \; d\mathbf{y} = 0. \tag{40}$$

We deduce that equation (29) is equivalent to

$$\nabla_{\mathbf{y}} \times E_0(\mathbf{x}, \mathbf{y}) = 0, \tag{41}$$

Moreover a solution of (41) is also solution of (39). So (39) and (41) are equivalent.

Hence, from Proposition (17) we conclude that $E_0(\mathbf{x}, \mathbf{y})$ can be decomposed as

$$E_0(\mathbf{x}, \mathbf{y}) = E(\mathbf{x}) + \nabla_{\mathbf{y}} \Phi_0(\mathbf{x}, \mathbf{y}). \tag{42}$$

Step 4. Equations for $E(\mathbf{x})$ **and** $\Phi_0(\mathbf{x}, \mathbf{y})$. divergence equation of (21) is multiplied with $V(\mathbf{x}, \frac{\mathbf{x}}{\varepsilon}) = \varepsilon v(\mathbf{x})\psi(\frac{\mathbf{x}}{\varepsilon})$, where $v \in \mathbf{C}_0^1(\Omega)$ and $\psi \in \mathbf{H}_{\#}^1(\mathcal{Y})$. Theorem 1 and integration by parts yields for all $\psi \in \mathbf{H}_{\#}^1(\mathcal{Y})$ and $v \in \mathbf{C}_0^1(\Omega)$

$$\lim_{\varepsilon\to 0}\int_\Omega \nabla\cdot\{-\omega^2\varepsilon^5 k(\varepsilon)E^\varepsilon(\mathbf{x})+i\omega[(\mathbf{1}^\varepsilon_C(\frac{\mathbf{x}}{\varepsilon})+\varepsilon^4\mathbf{1}^\varepsilon_R(\frac{\mathbf{x}}{\varepsilon}))\mathbf{1}_{\{y<0\}}$$
$$+\varepsilon\mathbf{1}_{\{y>0\}}]E^\varepsilon(\mathbf{x})\}\varepsilon v(\mathbf{x})\psi(\frac{\mathbf{x}}{\varepsilon})\,d\mathbf{x}$$
$$=-\lim_{\varepsilon\to 0}\int_\Omega\{-\omega^2\varepsilon^5 k(\varepsilon)E^\varepsilon(\mathbf{x})+i\omega[\mathbf{1}^\varepsilon_C(\frac{\mathbf{x}}{\varepsilon})+\varepsilon^4\mathbf{1}^\varepsilon_R(\frac{\mathbf{x}}{\varepsilon}))\mathbf{1}_{\{y<0\}} \tag{43}$$
$$+\varepsilon\mathbf{1}_{\{y>0\}}]E^\varepsilon\}\cdot(\varepsilon v(\mathbf{x})\psi(\frac{\mathbf{x}}{\varepsilon})+v(\mathbf{x})\nabla_\mathbf{y}\psi(\frac{\mathbf{x}}{\varepsilon}))\,d\mathbf{x}$$
$$=-\int_\Omega\int_{\mathscr{Y}} v(\mathbf{x})\nabla_\mathbf{y}\psi(\mathbf{y})\cdot[i\omega\mathbf{1}_C(\mathbf{y})E_0(\mathbf{x},\mathbf{y})]\,d\mathbf{y}d\mathbf{x}=0.$$

from which it follows that

$$\nabla_\mathbf{y}\cdot[i\omega\mathbf{1}_C(\mathbf{y})E_0(\mathbf{x},\mathbf{y})]=0. \tag{44}$$

with E_0 given by the decomposition (17). So we obtain the local equation

$$\nabla_\mathbf{y}\cdot[i\omega\mathbf{1}_C(\mathbf{y})\{E(\mathbf{x})+\nabla_\mathbf{y}\Phi_0(\mathbf{x},\mathbf{y})\}]\,d\mathbf{y}=0. \tag{45}$$

The potential Φ_0 may be written on the form

$$\Phi_0(\mathbf{x},\mathbf{y})=\sum_{j=1}^{3}\chi_j(\mathbf{y})e_j\cdot E(\mathbf{x})=\chi(\mathbf{y})\cdot E(\mathbf{x}), \tag{46}$$

From (31) and (46), we get:

$$E_0(\mathbf{x},\mathbf{y})=(\mathrm{Id}+\nabla_\mathbf{y}\chi(\mathbf{y}))E(\mathbf{x}). \tag{47}$$

Inserting E_0 in (26) we obtain

$$\nabla_\mathbf{y}\cdot[i\omega\mathbf{1}_C(\mathbf{y})(\mathrm{Id}+\nabla_\mathbf{y}\chi(\mathbf{y})]=0. \tag{48}$$

Now, we build oscillating test functions satisfying constraint (31) and use them in weak formulation (37). We define test function $V(\mathbf{x},\mathbf{y})=\alpha(\mathbf{x})+\nabla_\mathbf{y}\beta(\mathbf{x},\mathbf{y})$, $V(\mathbf{x},\mathbf{y})\in\mathbf{C}^1_0(\Omega,\mathbf{C}^1_\#(\mathscr{Y}))$ and we inject in (37) test function $V^\varepsilon=V(\mathbf{x},\frac{\mathbf{x}}{\varepsilon})$, which gives:

$$\int_{\Omega} E^{\varepsilon}(\mathbf{x}) \cdot \big(\nabla_{\mathbf{x}} \times \nabla_{\mathbf{x}} \times V(\mathbf{x}, \frac{\mathbf{x}}{\varepsilon}) + \frac{2}{\varepsilon}\nabla_{\mathbf{x}} \times \nabla_{\mathbf{y}} \times V(\mathbf{x}, \frac{\mathbf{x}}{\varepsilon}) \\ + \frac{1}{\varepsilon^2}\nabla_{\mathbf{y}} \times \nabla_{\mathbf{y}} \times V(\mathbf{x}, \frac{\mathbf{x}}{\varepsilon})\big) + [-\omega^2\varepsilon^5 k(\varepsilon) + i\omega\big((\mathbf{1}^{\varepsilon}_{C}(\frac{\mathbf{x}}{\varepsilon}) \\ + \varepsilon^4 \mathbf{1}^{\varepsilon}_{R}(\frac{\mathbf{x}}{\varepsilon})\big)\mathbf{1}_{\{y<0\}} + \varepsilon \mathbf{1}_{\{y>0\}})] E^{\varepsilon}(\mathbf{x}) \cdot V(\mathbf{x}, \frac{\mathbf{x}}{\varepsilon})\, d\mathbf{x} \\ = -i\omega \int_{\Gamma_d} H_d \times e_2 \cdot (e_2 \times V^{\ddagger}(x, 1, z, \xi, \zeta)) \times e_2\, d\sigma, \tag{49}$$

with $V(x, 1, z, \xi, \nu, \zeta) = V^{\ddagger}(x, 1, z, \xi, \zeta)$ the restriction on V which does not depend on ν. The term containing the constraint, the third one, disappears. Passing to the limit $\varepsilon \to 0$ and replacing the expression of V by the term $\alpha(\mathbf{x}) + \nabla_{\mathbf{y}}\beta(\mathbf{x}, \mathbf{y})$, we have

$$\begin{aligned} \nabla_{\mathbf{x}} \times \nabla_{\mathbf{y}} \times V(\mathbf{x}, \mathbf{y}) &= \nabla_{\mathbf{x}} \times \nabla_{\mathbf{y}} \times [\alpha(\mathbf{x}) + \nabla_{\mathbf{y}}\beta(\mathbf{x}, \mathbf{y})] \\ &= \nabla_{\mathbf{x}} \times \nabla_{\mathbf{y}} \times (\alpha(\mathbf{x})) + \nabla_{\mathbf{x}} \times \nabla_{\mathbf{y}} \times (\nabla_{\mathbf{y}}\beta(\mathbf{x}, \mathbf{y})) \\ &= \nabla_{\mathbf{x}} \times \nabla_{\mathbf{y}} \times (\nabla_{\mathbf{y}}\beta(\mathbf{x}, \mathbf{y})). \end{aligned} \tag{50}$$

Since $\nabla_{\mathbf{y}} \times (\nabla_{\mathbf{y}}) = 0$, the term $\frac{2}{\varepsilon}\nabla_{\mathbf{x}} \times \nabla_{\mathbf{y}} \times \nabla_{\mathbf{y}}\beta(\mathbf{x}, \mathbf{y}))$ vanishes. Therefore, (49) becomes:

$$\int_{\Omega}\int_{\mathscr{Y}} E_0(\mathbf{x}, \mathbf{y}) \cdot \nabla_{\mathbf{x}} \times \nabla_{\mathbf{x}} \times (\alpha(\mathbf{x}) + \nabla_{\mathbf{y}}\beta(\mathbf{x}, \mathbf{y})) \\ + i\omega \mathbf{1}_C(\mathbf{y}) E_0(\mathbf{x}, \mathbf{y}) \cdot (\alpha(\mathbf{x}) + \nabla_{\mathbf{y}}\beta(\mathbf{x}, \mathbf{y})\, d\mathbf{y} d\mathbf{x} \\ = -i\omega \int_{\Gamma_d} H_d \times e_2 \cdot (e_2 \times (\alpha(x, 1, z) + \nabla_{\mathbf{y}}\beta(x, 1, z, \xi, \zeta))) \times e_2\, d\sigma. \tag{51}$$

Now in (51) we replace expression E_0 giving by (47). We obtain

$$\int_{\Omega}\int_{\mathscr{Y}} (\mathrm{Id} + \nabla_{\mathbf{y}}\chi(\mathbf{y}))E(\mathbf{x}) \cdot \big(\nabla_{\mathbf{x}} \times \nabla_{\mathbf{x}} \times (\alpha(\mathbf{x}) + \nabla_{\mathbf{y}}\beta(\mathbf{x}, \mathbf{y})) \\ + i\omega \mathbf{1}_C(\mathbf{y})(\mathrm{Id} + \nabla_{\mathbf{y}}\chi(\mathbf{y}))E(\mathbf{x})) \cdot (\alpha(\mathbf{x}) + \nabla_{\mathbf{y}}\beta(\mathbf{x}, \mathbf{y}))\, d\mathbf{y} d\mathbf{x} \\ = -i\omega \int_{\Gamma_d} H_d \times e_2 \cdot (e_2 \times (\alpha(x, 1, z) + \nabla_{\mathbf{y}}\beta(x, 1, z, \xi, \zeta))) \times e_2\, d\sigma. \tag{52}$$

Taking $\alpha(\mathbf{x}) = 0$ in (52), we obtain

$$\int_{\Omega}\int_{\mathscr{Z}} (\mathrm{Id}+\nabla_{\mathbf{y}}\chi(\mathbf{y}))\nabla_{\mathbf{x}}\times\nabla_{\mathbf{x}}\times E(\mathbf{x})\nabla_{\mathbf{y}}\beta(\mathbf{x},\mathbf{y}) + i\omega\mathbf{1}_C(\mathbf{y})(\mathrm{Id}+\nabla_{\mathbf{y}}\chi(\mathbf{y}))E(\mathbf{x})\cdot\nabla_{\mathbf{y}}\beta(\mathbf{x},\mathbf{y})d\mathbf{y}d\mathbf{x} = 0. \tag{53}$$

Integrating by parts

$$\int_{\Omega}\int_{\mathscr{Z}} -\nabla_{\mathbf{y}}\cdot\{(\mathrm{Id}+\nabla_{\mathbf{y}}\chi(\mathbf{y}))\nabla_{\mathbf{x}}\times\nabla_{\mathbf{x}}\times E(\mathbf{x})\}\beta(\mathbf{x},\mathbf{y}) - i\omega\nabla_{\mathbf{y}}\cdot\{\mathbf{1}_C(\mathbf{y})(\mathrm{Id}+\nabla_{\mathbf{y}}\chi(\mathbf{y}))E(\mathbf{x})\}\beta(\mathbf{x},\mathbf{y})\,d\mathbf{y}d\mathbf{x} = 0. \tag{54}$$

And since $\nabla_{\mathbf{y}}\cdot\{\mathbf{1}_C(\mathbf{y})(\mathrm{Id}+\nabla_{\mathbf{y}}\chi(\mathbf{y}))E(\mathbf{x})\} = 0$ we obtain

$$\int_{\Omega}\int_{\mathscr{Z}} -\nabla_{\mathbf{y}}\cdot\{(\mathrm{Id}+\nabla_{\mathbf{y}}\chi(\mathbf{y}))\nabla_{\mathbf{x}}\times\nabla_{\mathbf{x}}\times E(\mathbf{x})\}\beta(\mathbf{x},\mathbf{y})\,d\mathbf{y}d\mathbf{x} = 0. \tag{55}$$

which gives the cell problem

$$\nabla_{\mathbf{y}}\cdot[\mathrm{Id}+\nabla_{\mathbf{y}}\chi(\mathbf{y})] = 0. \tag{56}$$

From (48) and (56), the scalar function χ is the unique solution, thanks to Lax-Milgram Lemma, up to an additive constant in the Hilbert space of $\mathscr{Z}$ periodic function $H^1_\#(\mathscr{Z})$ of the following boundary value problem

$$\begin{cases} \Delta_{\mathbf{y}}(\chi(\mathbf{y})) = 0 \ \text{ in } \mathscr{Z}\backslash\partial\Omega_C, \\ [\dfrac{\partial\chi}{\partial n}] = -n_j \ \text{ on } \partial\Omega_C, \\ [\chi] = 0 \ \text{ on } \partial\Omega_C. \end{cases} \tag{57}$$

where $[f]$ is the jump across the surface of $\partial\Omega_C$, n_j, $j = \{1, 2, 3\}$ is the projection on the axis e_j of the normal of $\partial\Omega_C$.

Remark 2 (34) can be seen as an electrostatic problem. Solving (48) and (56) reduces to look for a potential induced by surface density of charges. Then χ is this potential induced by the charges on the interface of carbon fiber.

Setting $\beta(\mathbf{x},\mathbf{y}) = 0$ in (52) and integrating by parts, we get

$$\int_{\Omega}\int_{\mathscr{Z}} (\mathrm{Id}+\nabla_{\mathbf{y}}\chi(\mathbf{y}))\nabla_{\mathbf{x}}\times\nabla_{\mathbf{x}}\times E(\mathbf{x})\cdot\alpha(\mathbf{x}) + i\omega\mathbf{1}_C(\mathbf{y})(\mathrm{Id}+\nabla_{\mathbf{y}}\chi(\mathbf{y}))E(\mathbf{x})\alpha(\mathbf{x})\,d\mathbf{y}d\mathbf{x} = -i\omega\int_{\Gamma_d} H_d\times e_2\cdot(e_2\times\alpha(x,1,z))\times e_2\,d\sigma. \tag{58}$$

which gives the following well posed problem for $E(\mathbf{x})$

$$\begin{cases} \theta_1 \nabla_{\mathbf{x}} \times \nabla_{\mathbf{x}} \times E(\mathbf{x}) + i\omega\theta_2 E(\mathbf{x}) = 0 \text{ in } \Omega, \\ \theta_1 \nabla_{\mathbf{x}} \times E(\mathbf{x}) \times e_2 = -i\omega H_d \times e_2 \text{ on } \Gamma_d, \\ \nabla_{\mathbf{x}} \times E(\mathbf{x}) \times e_2 = 0 \text{ on } \Gamma_L. \end{cases} \tag{59}$$

with $\theta_1 = \int_{\mathcal{Y}} \text{Id} + \nabla_{\mathbf{y}}\chi(\mathbf{y})\, d\mathbf{y}$ and $\theta_2 = \int_{\mathcal{Y}} \mathbf{1}_C(\mathbf{y})(\text{Id} + \nabla_{\mathbf{y}}\chi(\mathbf{y}))\, d\mathbf{y}$.

This concludes the proof of Theorem (33).

4 Numerical Results

Our goal is to validate the homogenization method by comparing the numerically solution of exact problem (1) with the solution of the homogenized model (33). We solve the problem (1) in Ω with cells of size $\varepsilon = 10^{-2}$. We enter in the software FreeFem ++ [12] the geometry of the problem, the bilinear form and the boundary condition, we perform the computation with Lagrangian P_2 Finite Elements. The numerical results confirm the theoretical study. We start by giving computational parameters that we need in this experiment, for $\varepsilon = 0.01$. The geometry corresponds to the fibers surrounded by resin. The domain is the composite material represented by a colon of carbon fibers in the resin with periodic conditions in the y direction. We compare the direct solution of the adimensioned problem in ε with the homogenized solution. The basic cells in ε periodicity contains a cylinder which the radius is equal to 0.45×10^{-2}m. We take periodic boundary conditions on the right and left sides. In the rescaled system the conductivity began

$$\Sigma^{\varepsilon}(\mathbf{y}) = \Sigma^{\varepsilon}(\xi, \nu, \zeta) = \begin{cases} \Sigma_r^{\varepsilon} = \varepsilon^4 \text{ in } \Omega_r, \\ \Sigma_c^{\varepsilon} = 1 \text{ in } \Omega_c, \end{cases} \tag{60}$$

The values of the permittivity are $\varepsilon_r = 5$ in the resin part and $\varepsilon_c = 2.5$ in the carbon part. To simplify the calculations, we consider the composite illuminated by a Progressive Monochromatic Plane Wave, propagating in the Oy direction electrically polarized according to Oz, with a normal incidence. The electric field with the carbon fibers in the composite are contained in the xOy plane. Then we use 2D system. On the upper frontier, we consider the oscillating source $H_d = -\exp(-ik_y.y)$ with the constant of propagation which depends on ε, $k_y = \sqrt{(-\omega^2\varepsilon^3 + i\omega\varepsilon^4)}$, and $\omega = 1$. The computations are performed in P_2 Finite Elements and the direct and homogenized solutions are projected on a regular meshes and the number of triangles is 600. In the cell problem the number of triangles is 10368.

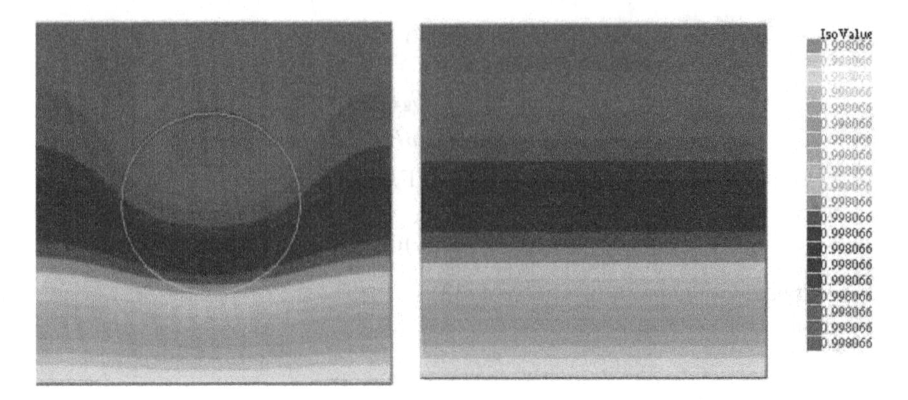

Fig. 1 The z-component of the real part of the electric field E^{ε} (left) and the homogenized solution (right)

Table 1 The relative error according to the number of fibers

n	Err
2 *fibers*	0.012
10 *fibers*	0.00026
26 *fibers*	0.00013
52 *fibers*	0.00005

4.1 Results of Simulation

In Figure 1, we plot the direct solution of (21), the solution of the homogenized problem without fiber, the transmitted electric field is evaluated. We see that we have no difference between the two approximations. The two solutions represent an attenuated wave propagating along the y-axis. We note that fibers do affect the electromagnetic composite response and our homogenized approach is a good agreement with the exact solution. The decay of the amplitude of the electric field is induced by the imaginary part $\sqrt{i\omega}$ in the carbon fiber and the imaginary part, which depends of ε, $\sqrt{i\omega\varepsilon^4}$ in the resin. To obtain a numerical speed of convergence we compute the relative errors, as εgoes to zero, by increasing the number of the cells and the fibers: $E_{rr} = \frac{\|E^{\varepsilon} - E(x)\|_{L^2(\Omega)}}{\|E^{\varepsilon}\|_{L^2(\Omega)}}$ (Table 1).

References

1. Y. Amirat, K. Hamdache , A. Ziani, Homogénéisation d'équations hyperboliques du premier ordre et application aux écoulements missibles en milieux poreux. Ann. Inst. H. Poincaré **6**(5), 397–417 (1989)

2. G. Allaire, Homogenization and two-scale convergence. SIAM J. Math. Anal. **23**(6), 1482–1518 (1992), http://link.aip.org/link/?SJM/23/1482/1. https://doi.org/10.1137/0523084
3. G. Allaire, M. Briand, Multiscale convergence and reiterated homogenization. Proc. Roy. Soc. Edinb. **F126**, 297–342 (1996)
4. Y. Amirat, V. Shelukhin, Homogenization of time-harmonic Maxwell equations and the frequency dispersion effect. J. Maths. Pures. Appl. **95**, 420–443 (2011)
5. A. Back, E. Frenod, Geometric Two-Scale Convergence on Manifold and Applications to the Vlasov Equation. Discrete and Continuous Dynamical Systems - Serie S. Special Issue on Numerical Methods based on Homogenization and Two-Scale Convergence. **8**, 223–241 (2015)
6. A. Bensoussan, J.L. Lions, G. Papanicolaou, Asymptotic analysis for periodic structures, in *Studies in Mathematics and its Applications*, vol. 5 (North Holland, 1978)
7. H. Canot, E. Frenod, Method of homogenization for the study of the propagation of electromagnetic waves in a composite. Part 1: Modeling, Scaling, Existence and Uniqueness Results (2017)
8. H. Canot, E. Frenod, Modeling electromagnetism in and near composite material using two-scale behavior of the time-harmonic Maxwell equations (2016), https://hal.archives-ouvertes.fr/hal-01409522
9. D. Cionarescu, P. Donato, *An Introduction To Homogenization* (Oxford University Press, 1999)
10. N. Crouseilles, E. Frenod, S. Hirstoaga, A. Mouton, Two-Scale Macro-Micro decomposition of the Vlasov equation with a strong magnetic field. Math. Models Methods Appl. Sci. **23**(8), 1527–1559 (2012) (collaboration = CALVI ; IPSO). https://doi.org/10.1142/S0218202513500152, http://hal.archives-ouvertes.fr/hal-00638617/PDF/TSAPSVlas_corr.pdf
11. S. Guenneau, F. Zolla, A. Nicolet. Homogenization of 3D finite photonic crystals with heterogeneous permittivity and permeability. Waves Random Complex Media 653–697 (2007)
12. F. Hecht, O. Pironneau, A. Le Hyaric, FreeFem++ manual (2004)
13. H. Canot, E. Frenod, Method of homogenization for the study of the propagation of electromagnetic waves in a composite part 2: homogenization, in *Lecture Notes in Engineering and Computer Science: Proceedings of The World Congress on Engineering, 5–7 July 2017, London, UK* (2017), pp. 11–15
14. M. Neuss-Radu, Some extensions of two-scale convergence. omptes rendus de l'Academie des sciences. Serie 1 **322**(9), 899–904 (1996)
15. G. Nguetseng, A general convergence result for a functional related to the theory of homogenization. SIAM **20**(3), 608–623 (1989), http://link.aip.org/link/?SJM/20/608/1. https://doi.org/10.1137/0520043?
16. G. Nguetseng, Asymptotic analysis for a stiff variational problem arising in mechanics SIAM J. Math. Anal. **21**(6) 1394–1414 (1990), http://link.aip.org/link/?SJM/21/1394/1. https://doi.org/10.1137/0521078
17. O. Ouchetto, S. Zouhdi, A. Bossavit et al., Effective constitutive parameters of periodic composites, in 2005 European Microwave Conference (IEEE, 2005), p. 2
18. H.E. Pak, Geometric two-scale convergence on forms and its applications to Maxwell's equations, in *2005 European Proceedings of the Royal Society of Edinburgh* vol. 135A, pp. 133–147
19. N. Wellander, Homogenization of the Maxwell equations: case I. Linear Theory Appl. Math. **46**(2), 29–51 (2001)
20. N. Wellander, Homogenization of the Maxwell equations: case II. Nonlinear Cond. Appl. Math. **47**(3), 255–283 (2002)
21. N. Wellander, B. Kristensson, Homogenization of the Maxwell equations at fixed frequency. Technical Report, vol. LUTEDX/TEAT-7103/1-37 (2002)

Statistics of Critical Load in Arrays of Nanopillars on Nonrigid Substrates

Tomasz Derda and Zbigniew Domański

Abstract Multicomponent systems are commonly used in nano-scale technology. Specifically, arrays of nanopillars are encountered in electro-mechanical sense devices. Under a growing load weak pillars crush. When the load exceeds a certain critical value the system fails completely. In this work we explore distributions of such a critical load in overloaded arrays of nanopillars with identically distributed random strength-thresholds (σ_{th}). Applying a Fibre Bundle Model with so-called local load transfer we analyse how statistics of critical load are related to statistics of pillar-strength-thresholds. Based on extensive numerical experiments we show that when the σ_{th} are distributed according to the Weibull distribution, with shape and scale parameters k, and $\lambda = 1$, respectively, then the critical load can be approximated by the same probability distribution. The corresponding, shape and scale, parameters K and Λ are functions of k.

Keywords Array of pillars · Fracture · Load transfer · Scaling · Statistics
Weibull probablity distribution

1 Introduction

Creation and development of new sub-micron scale devices rise questions about reliability of multicomponent systems. One such a question is how performances of individual components combine into a resulting overall performance of the system to which these components belong. This question is important because progressively

T. Derda · Z. Domański (✉)
Institute of Mathematics, Czestochowa University of Technology,
Dabrowskiego 69, PL-42201 Czestochowa, Poland
e-mail: zbigniew.domanski@im.pcz.pl

T. Derda
e-mail: tomasz.derda@im.pcz.pl

S.-I. Ao et al. (eds.), *Transactions on Engineering Technologies*,
https://doi.org/10.1007/978-981-13-0746-1_2

loaded multicomponent systems break when an initial sequence of failures among weakest components develops into an avalanche of failures that may involve all the system components.

Nowadays nanopillar arrays play a crucial role in many areas of technology and science. Photovoltaic devices or grid cells used in experimental biomedicine, to name but a few examples, employ arrays of nanopillars. Fabrication processes of these arrays request a robust transfer of nanopillars between substrates and thus a controllable-fracturing procedure. This is because pillars are detached from the substrate under a suitable-lateral load to ensure a smooth fracturing process.

An effective statistical approach, to study failures in multicomponent systems related to technology, employs Fiber Bundle Models [1–6]. In this work we analyze a set of pillars placed at nodes of a flat-square grid $\mathcal{G}$, and oriented perpendicularly to surface. Pillars imperfections influence strongly the behavior of arrays under load. Due to the imperfections, pillar-strength-thresholds are nonuniform and thus pillars are represented by random load-thresholds (σ_{th}). In our numerical experiments $\{\sigma_{th}^i\}_{i\in\mathcal{G}}$ are quenched random variables distributed according to the Weibull probability distribution function

$$p_{k,\lambda}(\sigma_{th}) = (k/\lambda)(\sigma_{th}/\lambda)^{k-1}\exp[-(\sigma_{th}/\lambda)^k] \quad (1)$$

Parameters $k > 0$ and $\lambda > 0$ define the shape and scale of this distribution. Shape parameter k (so-called Weibull index) controls the amount of disorder in the system. We use the Weibull distribution because this probability distribution is very suitable and frequently employed distribution in the context of engineering systems [7, 8].

2 Loading Process and Statistical Modelling

In our approach we consider pillars located on a nonrigid surface that has a nonvanishing compliance. Within this framework the load redistribution turns out to be localized and thus we employ the local load sharing (LLS). Within a short interval between consecutive fractures the load carried by the broken pillar is transferred only to its nearest intact elements. Such a limited-load-range transfer yields nonhomogeneous distributions of load. As a consequence regions of stress accumulation appear throughout the system. The growing load on the intact pillars leads to other failures, after which each intact pillar undergoes increasing stress. If the load transfer does not trigger further crushes, a stable configuration emerges meaning that this initial value of F is not sufficient to provoke failure of the entire system, and its value has to be increased by an amount δF. In the simulations we applied a quasi-static loading procedure—if the system is in a stable state the external load is uniformly increased on all the intact pillars just to destroy only the weakest intact pillar.

A sequence of increases in the value of the external load gives F_c which induces an avalanche of failures among all still undestroyed pillars. Application of such a quasi-static loading enable us to determine a minimal load F_c, that is necessary for destruction of all the pillars. In order to compare results for systems with different numbers of pillars, we scale the critical loads F_c by the initial number of pillars in the system, i.e. we introduce here an intensive quantity, namely $\sigma_c = F_c/N$.

As already mentioned, the pillar-strength-thresholds σ_{th} are drawn from the Weibull distribution (1). Without loss of generality, we assume $\lambda = 1$ and thus the corresponding probability density reads

$$p_{k,1}(\sigma_{th}) = k\sigma_{th}^{k-1}\exp[-\sigma_{th}^{k}] \tag{2}$$

We address a question how these local critical loads distributed according to (2) combine to create an effective-global critical load F_c. Based on our numerical simulations for systems with $N > 100$, we have found that skewnesses of resulting distributions are negative and they decrease with growing N. For this reason we employ two distributions for fitting our skewed data, namely:

1. three-parameter skew normal distribution (SND) [9, 10] defined by

$$p(\sigma_c) = \frac{\exp[-\frac{(\sigma_c-\xi)^2}{2\omega^2}]\mathrm{erfc}[-\frac{\alpha(\sigma_c-\xi)}{\sqrt{2}\omega}]}{\sqrt{2\pi}\omega} \tag{3}$$

   where ξ, ω, α are location, scale and shape parameters, respectively.
2. the Weibull distribution:

$$p_{K,\Lambda}(\sigma_c) = (K/\Lambda)(\sigma_c/\Lambda)^{K-1}\exp[-(\sigma_c/\Lambda)^K] \tag{4}$$

Studies related to distributions of system's strength in the context of realizations of Fiber Bundle Models are presented in [11–20]. It is worth mentioning that for the GLS rule, σ_c approximately follows the normal distribution, for both the Weibull and uniform distributions of σ_{th}.

3 Results and Discussion

Based on the Fibre Bundle Model and local load sharing rule, we developed a program code for the simulation of the loading process in two-dimensional nanopillar arrays. Intensive numerical simulations are conducted for systems involving $N = L \times L$ pillars, with L ranging from 8 to 128. We have tuned the amount of pillar-strength-threshold disorder by integer values of k ranging from 2 to 20. In order to get reliable statistics, each simulation was repeated 10^5 times.

Figures 1 and 2 show empirical probability density functions of σ_c for chosen systems. In these plots we have also added fitting lines of skew normal (Fig. 1) and Weibull (Fig. 2) probability density functions with parameters computed from the samples. It can be seen that both of these theoretical distributions are in good agreement with empirical distributions of σ_c. Some more precise information concerning

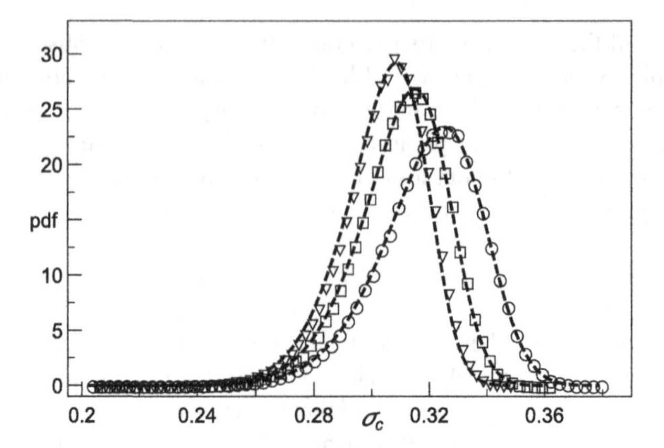

Fig. 1 Empirical probability density functions (pdf) of σ_c for arrays with $L = 64$ (circles), $L = 96$ (squares) and $L = 128$ (triangles). Weibull index $k = 2$ for all presented pdfs. The dashed lines represent skew-normally distributed σ_c with the parameters computed from the simulations

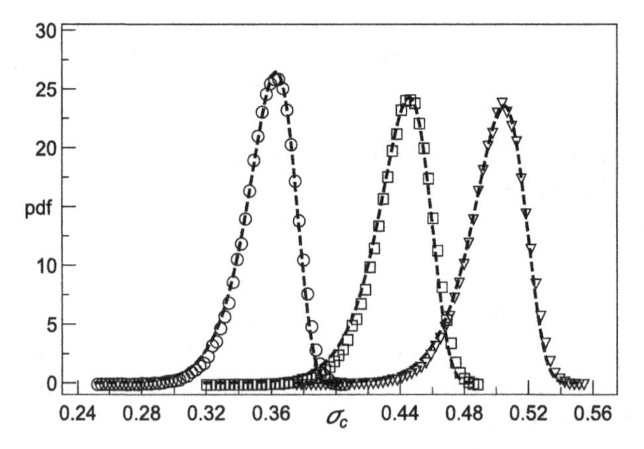

Fig. 2 Empirical probability density functions (pdf) of σ_c for arrays with $L = 128$: $k = 3$ (circles), $k = 5$ (squares) and $k = 7$ (triangles). The dashed lines represent two-parameter Weibull distributed σ_c with the parameters computed from the simulations

the Fig. 2 is gained from a three parameter Weibull distribution, i.e. from an extension of (4) which includes a so-called location parameter. The corresponding cumulative distribution functions reads:

$$P_{K,\Lambda,\mu}(\sigma_c) = 1 - \exp\left[-\left(\frac{\sigma_c - \mu}{\Lambda}\right)^K\right] \tag{5}$$

In Fig. 3 we present distribution (5) for arrays with different number of pillars. We clearly see the support of $\{\sigma_c\}$, with its onset given by the location parameter μ.

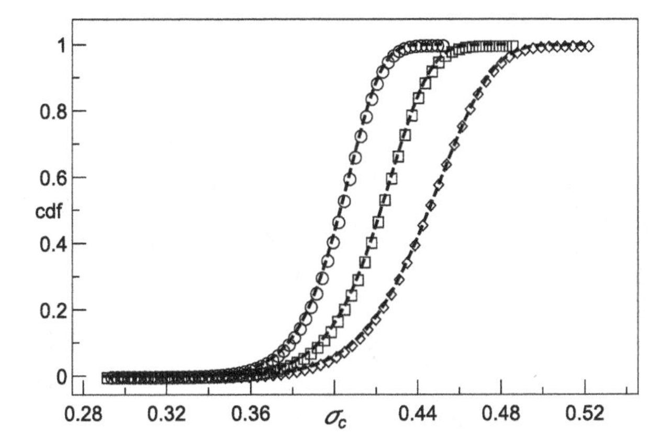

Fig. 3 Empirical cumulative distribution functions (cdf) of σ_c for arrays with $L = 128$ (circles), $L = 64$ (squares) and $L = 32$ (diamonds). Weibull index $k = 4$ for all presented cdfs. The dashed lines represent three-parameter Weibull distributed σ_c with the parameters computed from the simulations

We also present a quantile-quantile plot (Q-Q plot) of the quantiles of the collected data set against the corresponding quantiles given by the SND and Weibull probability distributions. From Figs. 4 and 5, it is seen that the results of fitting by skew normal distribution is slightly better than the Weibull fitting. We also observed that using skew normal distributions give better results than these resulting from two-parameter Weibull fittings. This is true for all analysed systems, especially for the smaller ones. However, it should be noted that the skew normal distribution (3) has one parameter more than two-parameter Weibull distribution (4). Fitting by Weibull distribution allows us to analyse the influence of system properties on the microscopic level (Weibull distributed pillar-strength thresholds) on the macroscopic response (distribution of crical loads) in the framework of one type of distribution. Hence, we focus our attention on the fitting of σ_c distribution by two-parameter Weibull distribution.

In the case of Weibull distribution, values of the fitted parameters K and Λ depend on system size and Weibull index k in the original distribution characterizing the pillar's strength. The plots of the parameters K and Λ are shown in Figs. 6 and 7, respectively. For a fixed value of k, the parameter K is a strictly increasing function of linear system size L. We have found that this relation can be approximated by the following formula

$$K_k(L) = a_1 + a_2\sqrt{L} + a_3 \ln L \tag{6}$$

where a_1, a_2, a_3 are fitted parameters (see page 23). One can also see that fitted curves are (increasingly) ordered according to Weibull index k.

Contrary to K, the parameter Λ is a strictly decreasing function of L, which can be fitted by the formula (see Fig. 7)

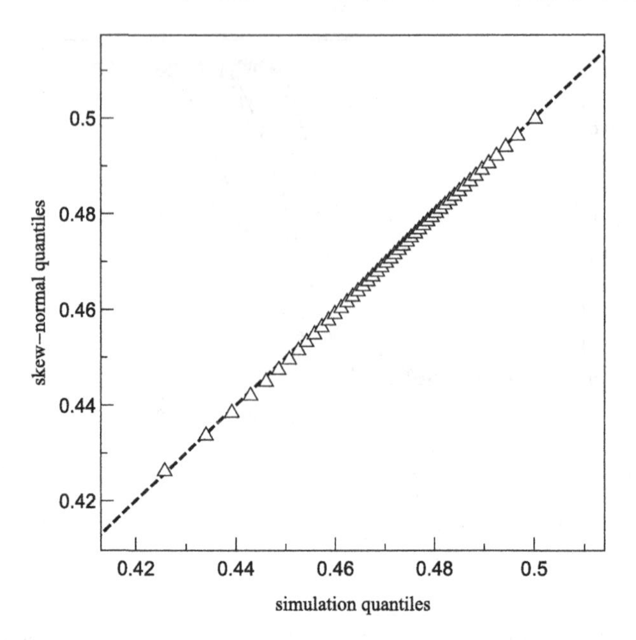

Fig. 4 The Q-Q plot of the quantiles of the set of computed σ_c versus the quantiles of the skew normal distribution. System size $N = 128 \times 128$ and $k = 6$

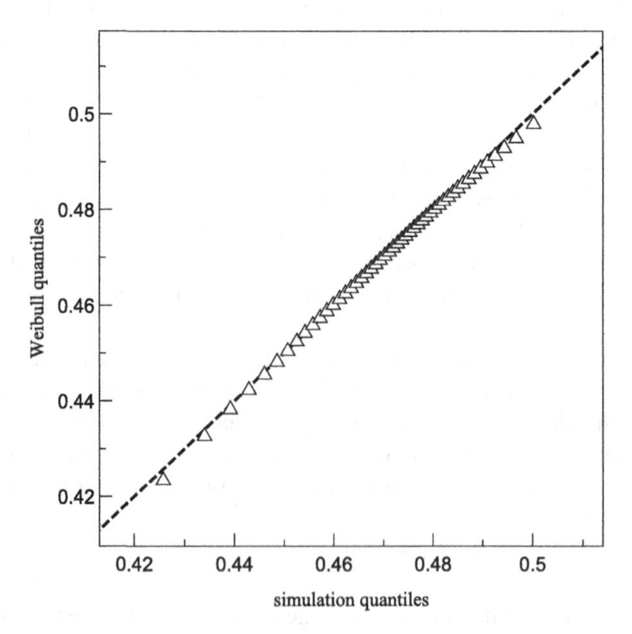

Fig. 5 The Q-Q plot of the quantiles of the set of computed σ_c versus the quantiles of the Weibull distribution. System size $N = 128 \times 128$ and $k = 6$

Weibull index	Fitted parameter		
	a_1	a_2	a_3
$k = 2$	−3.272	−0.200	6.516
$k = 4$	−1.018	−0.454	7.264
$k = 7$	1.742	−0.330	7.446
$k = 10$	3.102	−0.416	8.086
$k = 15$	1.832	−0.927	10.751
$k = 20$	3.268	−0.251	10.218

$$\Lambda_k(L) = b_1 + \frac{b_2}{\sqrt{L}} \tag{7}$$

Weibull index	Fitted parameter	
	b_1	b_2
$k = 2$	0.272	0.440
$k = 4$	0.369	0.470
$k = 7$	0.465	0.490
$k = 10$	0.528	0.492
$k = 15$	0.600	0.482
$k = 20$	0.646	0.484

where b_1, b_2 are matched parameters. The ordering of curves, reported for the previous plot, is preserved.

One of the components of the formula (6) is the natural logarithm of L. If the linear system size is logarithmized, the parameter K can be approximated by the linear function—it is reported in Fig. 8. In turn, Fig. 9 presents values of the parameter Λ in the function of $L^{-1/2}$ which is a part of the function (7). In this case we applied a third degree polynomial as an approximative formula.

Taking assumption that F_c/N follows Weibull distribution with the parameters K and Λ, the expected value of this distribution is given by

$$E[F_c/N] =< F_c/N >= \Lambda\Gamma(1 + \frac{1}{K}) \tag{8}$$

where $\Gamma(1 + \frac{1}{K})$ is the gamma function. From the fitting we have obtained $K \in (9.76, 51.69)$. Substituting limits of this interval into the relation

$$\Gamma(1 + \frac{1}{K})/\Gamma(1) \tag{9}$$

we received two values 0.95 and 0.99. As it was previously mentioned, K is a increasing function of the system size, therefore relation (9) tends to unity with the increasing system size. Consequently, the parameter Λ is a key factor of the formula (8) and the mean critical load can by roughly estimated by

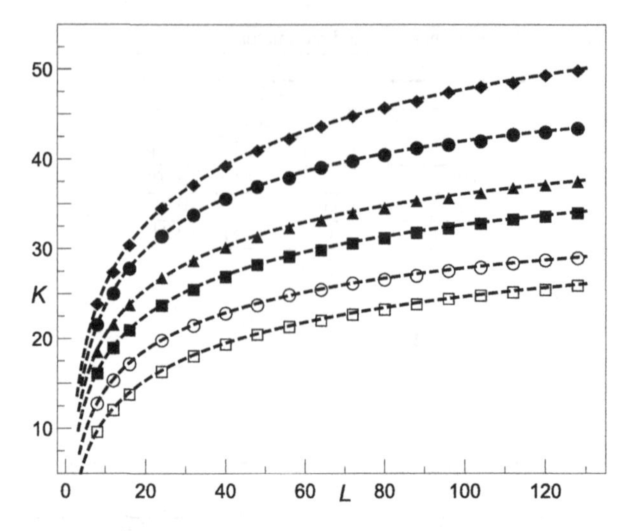

Fig. 6 Parameter K as a function of L, formula (6), for different values of Weibull index: $k = 2$ (squares), $k = 4$ (circles), $k = 7$ (filled squares), $k = 10$ (triangles), $k = 15$ (filled circles), $k = 20$ (diamonds)

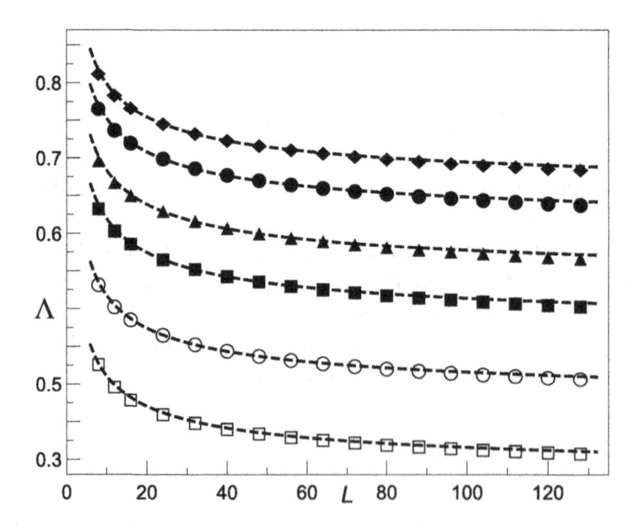

Fig. 7 Parameter Λ as a function of L, formula (7), for different values of Weibull index: $k = 2$ (squares), $k = 4$ (circles), $k = 7$ (filled squares), $k = 10$ (triangles), $k = 15$ (filled circles), $k = 20$ (diamonds)

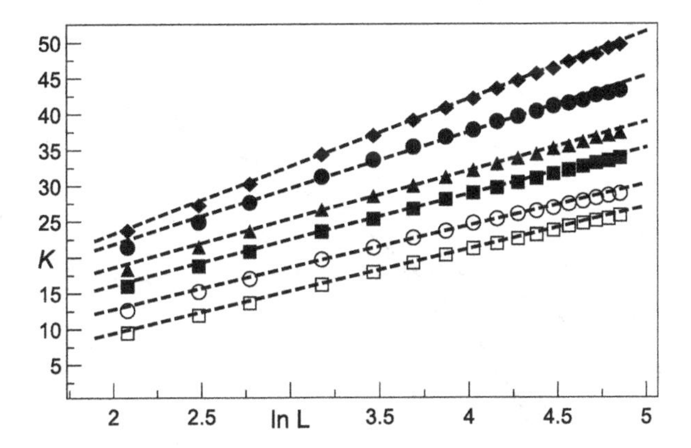

Fig. 8 Same as in Fig. 6, parameter K versus $\ln L$ for different values of Weibull index: $k = 2$ (squares), $k = 4$ (circles), $k = 7$ (filled squares), $k = 10$ (triangles), $k = 15$ (filled circles), $k = 20$ (diamonds). The dashed lines represent a linear function with fitted parameters

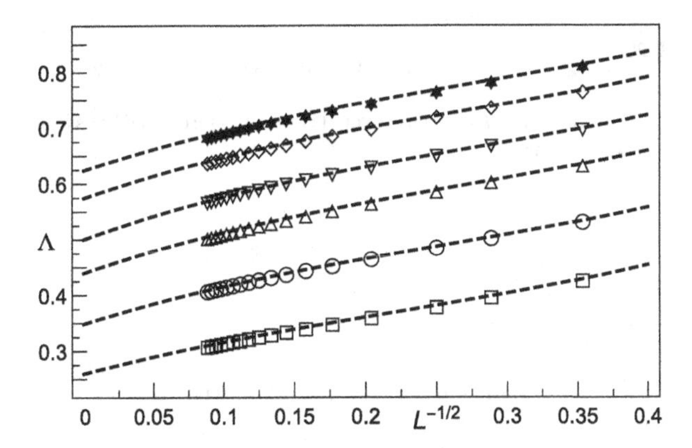

Fig. 9 Same as in Fig. 7, parameter Λ versus $1/\sqrt{L}$ for different values of Weibull index: $k = 2$ (squares), $k = 4$ (circles), $k = 7$ (up triangles), $k = 10$ (down triangles), $k = 15$ (diamonds), $k = 20$ (stars). The dashed lines represent third degree polynomial with fitted parameters

$$< F_c/N > \sim \Lambda\Gamma(1) = \Lambda \tag{10}$$

In the following we propose a universal formula for calculating Λ in dependence of L and k. The function (7) can be rewritten as

$$\Lambda(L, k) = b_1(k) + \frac{b_2(k)}{\sqrt{L}} \tag{11}$$

where parameters b_1, b_2 are replaced by their functions of k. The plot of the parameters b_1 and b_2 with values obtained from simulations is shown in Fig. 10. We have approximated $b_1(k)$ and $b_2(k)$ by

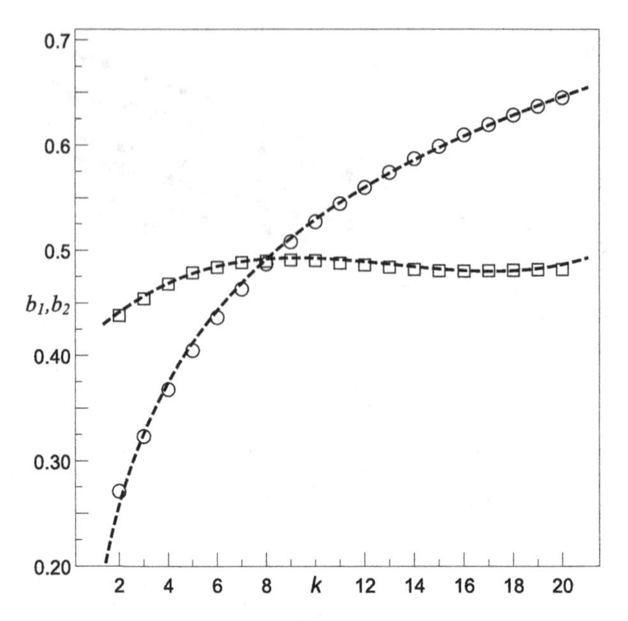

Fig. 10 Parameters b_1 (circles) and b_2 (squares) versus index k. The dashed lines illustrate functions (12) and (13) with fitted parameters

$$b_1(k) = c_1 + c_2 \ln k \tag{12}$$

with $c_1 \approx 0.141$, $c_2 \approx 0.169$ and

$$b_2(k) = d_1 + d_2 k + d_3 k^2 + d_4 k^3 \tag{13}$$

with $d_1 \approx 0.400097$, $d_2 \approx 0.024593$, $d_3 \approx 0.002056$, $d_4 \approx 0.000052$. It can be noticed that $b_2(k)/\sqrt{L} \to 0$ when $L \to \infty$ and so Λ depends only on $b_1(k)$.

Some insight into the strength of the system can be gained by collating, sample by sample, the critical force F_c with the number Δ_c of pillars crushed under this force, i.e., since the load increases in a quasi-static way then $\Delta_c + 1$ pillars bear safely the load $F_c - \delta F$ which means that the average maximal stress $\overline{\sigma}_{\max}$ supported by the system has an upper bound

$$\overline{\sigma}_{\max} < \frac{F_c}{\Delta_c} \tag{14}$$

We employ Pearson correlation coefficient r to measure the relationship between critical loads F_c and sizes of critical avalanches Δ_c. The results concerning r as a function of L are presented in Fig. 11. Following the curves in Fig. 11 an interpolating function is proposed for r:

$$r(L,k) = \frac{h_1(k)}{L^{h_2(k)}} + \frac{h_3(k)}{\ln L} \tag{15}$$

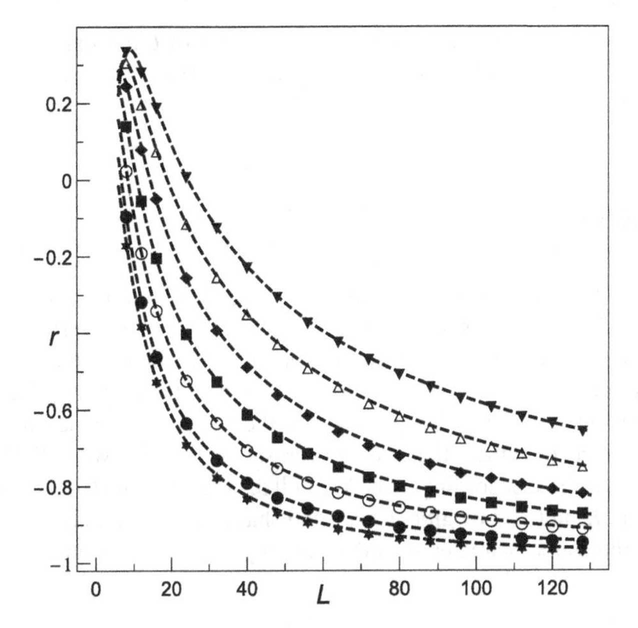

Fig. 11 The Pearson correlation coefficient r between two variables F_c and Δ_c versus the linear system size: $k = 2$ (stars), $k = 5$ (filled circles), $k = 8$ (circles), $k = 11$ (filled squares), $k = 14$ (diamonds), $k = 17$ (triangles), $k = 20$ (down triangles). The dashed lines represent approximative formula (15) with adequate parameters

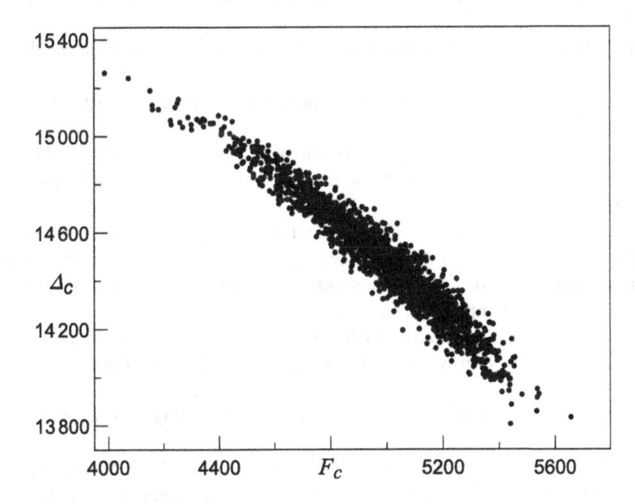

Fig. 12 Size of critical avalanche Δ_c versus critical load F_c for arrays of $N = 128 \times 128$ pillars and $k = 2$ taken from 2×10^3 samples

It is seen from Fig. 11 that r is a decreasing function of L for all illustrated k. For $L > 100$ and $2 \leq k \leq 4$ values of r are in segment $[-0.97, -0.95]$. An exemplary scatter plot of Δ_c versus F_c is shown in Fig. 12.

4 Final Remarks

In conclusion, we have studied numerically effective distributions of critical loads F_c in quasi-statically loaded systems of vertical pillars placed on nonrigid substrates. From large sets of data gathered from numerical experiments we have fitted discrete distributions of critical loads. We have found how the random pillar-strength-thresholds influence the macroscopic yield of the system. We underline two main results of this work: (i) if the pillar-strength-thresholds obey the Weibull distribution, the critical load is fitted according to the Weibull distribution, and (ii) the corresponding parameters K and Λ, i.e., global scale and shape parameters, are functions of k, where the parameter k characterizes the local property of the system.

References

1. A. Hansen, P.C. Hemmer, S. Pradhan, *The Fiber Bundle Model: Modeling Failure in Materials* (Wiley, 2015)
2. S. Pradhan, A. Hansen, B.K. Chakrabarti, Failure processes in elastic fiber bundles. Rev. Mod. Phys. **82**, 499–555 (2010)
3. M.J. Alava, P.K.V.V. Nukala, S. Zapperi, Statistical models of fracture. Adv. Phys. **55**, 349–476 (2006)
4. F. Kun, F. Raischel, R.C. Hidalgo, H.J. Herrmann, Extensions of fibre bundle models, in *Modelling Critical and Catastrophic Phenomena in Geoscience, Lecture Notes in Physics*, vol. 705 (2006) pp. 57–92
5. Z. Domański, T. Derda, and N. Sczygiol, Critical avalanches in fiber bundle models of arrays of nanopillars, in *Lecture Notes in Engineering and Computer Science: Proceedings of The International MultiConference of Engineers and Computer Scientists 2013, IMECS 2013*, 13–15 Mar 2013, Hong Kong, pp. 765-768
6. Z. Domański, T. Derda, Distributions of critical load in arrays of nanopillars, in *Lecture Notes in Engineering and Computer Science: Proceedings of The World Congress on Engineering 2017*, 5–7 July 2017, London, UK, pp. 797–801
7. W. Weibull, A statistical distribution function of wide applicability. J. Appl. Mech. **18**, 293–297 (1951)
8. N.M. Pugno, R.S. Ruoff, Nanoscale Weibull statistics. *J. Appl. Phys.* **99** (2006) (id. 024301)
9. A. Azzalini, A.R. Massih, A class of distributions which includes the normal ones. Scand. J. Statist. **12**, 171–178 (1985)
10. A. Azzalini *The Skew-normal And Related Families* (Cambridge University Press, 2013)
11. H.E. Daniels, The statistical theory of the strength of bundles of threads I. P. Roy. Soc. Lond. A Mat. **183**, 405–435 (1945)
12. R.L. Smith, The asymptotic distribution of the strength of a series-parallel system with equal load sharing. Ann. Probab. **10**, 137171 (1982)
13. L.N. McCartney, R.L. Smith, Statistical theory of the strength of fiber bundles. ASME J. Appl. Mech. **105**, 601608 (1983)

14. D.G. Harlow, S.L. Phoenix, The chain-of-bundles probability model for the strength of fibrous materials I: analysis and conjectures. J. Compos. Mater. **12**, 195–214 (1978)
15. D.G. Harlow, S.L. Phoenix, The chain-of-bundles probability model for the strength of fibrous materials II: a numerical study of convergence. J. Compos. Mater. **12**, 314–334 (1978)
16. S.L. Phoenix, R.L. Smith, A comparison of probabilistic techniques for the strength of fibrous materials under local load-sharing among fibers. Int. J. Sol. Struct. **19**, 479–496 (1983)
17. L.S. Sutherland, C. Guedes, Soares, Review of probabilistic models of the strength of composite materials. Reliab. Eng. Syst. Saf. **56**, 183–196 (1997)
18. M. Ibnabdeljalil, W.A. Curtin, Strength and reliability of fiber-reinforced composites: Localized load-sharing and associated size effects. Int. J. Sol. Struct. **34**, 2649–2668 (1997)
19. S. Mahesh, S.L. Phoenix, I.J. Beyerlein, Strength distributions and size effects for 2D and 3D composites with Weibull fibers in an elastic matrix. In. J. of Fracture **115**, 41–85 (2002)
20. P.K. Porwal, I.J. Beyerlein, S.L. Phoenix, Statistical strength of a twisted fiber bundle: an extension of Daniels equal-load-sharing parallel bundle theory. J. Mech. Mater. Struct. **1**, 1425–1447 (2007)

Quantifying the Impact of External Shocks on Systemic Risks for Russian Companies Using Risk Measure ΔCoVaR

Alexey Lunkov, Sergei Sidorov, Alexey Faizliev, Alexander Inochkin and Elena Korotkovskaya

Abstract One of the biggest recent shocks for Russian economy was the sharp fall of oil prices in 2014. Another big shock was the sanctions imposed by governments of the United States and European Union countries as well as some international organizations. Both sanctions and the sharp fall of oil prices resulted in the weakening of the Russian ruble and led to a sharp slowdown in growth or even to ongoing recession of the Russian economy. Our research examines the impact of the two shocks on systemic risks for some Russian companies using ΔCoVaR, one of the most popular systemic risk measures proposed by M. Brunnermeier and T. Adriany in 2011. The measure provides an opportunity to estimate the mutual influence of certain institutions or the mutual influence of the financial system and a particular institution. The analysis is focused on the static model of the ΔCoVaR estimation. Moreover, this paper uses statistical testing procedures to assess the significance of the findings and interpretations based on this co-risk measure. The results show that the shocks has brought some negative effects of a disintegration of financial intermediation both for banks and some companies.

Keywords CoVaR estimation · Financial risk · Kolmogorov–Smirnov type statistic · Quantile regressions · Risk measures · Systemic risks · Value-at-Risk

---

A. Lunkov · S. Sidorov (✉) · A. Faizliev · A. Inochkin · E. Korotkovskaya
Saratov State University, 83, Astrakhanskaya Str., Saratov 410012, Russia
e-mail: sidorovsp@info.sgu.ru; sidorovsp@yahoo.com

A. Lunkov
e-mail: alunkov@yandex.ru

A. Faizliev
e-mail: faizlievar1983@mail.ru

A. Inochkin
e-mail: InochkinA@info.sgu.ru

E. Korotkovskaya
e-mail: korotkovskaya@yandex.ru

S.-I. Ao et al. (eds.), *Transactions on Engineering Technologies*,
https://doi.org/10.1007/978-981-13-0746-1_3

1 Introduction

Nowadays the entire world economy is a complex system. First of all, that means a high degree of interdependence not only between the economies of different countries but also between separate sectors of national economies. Insufficient identification of these links and their influence on the institutions of these sectors lead to the increasing systemic risk. The crisis of 2008 showed that systemic risk threatens financial economic system as a whole. The possible consequences include the deepest depression due to the so-called knock-on effect of the separate institutions and sectors falling into distress one after another [2, 8, 13]. Thus, problems of early detection, prediction and prevention of the factors contributing to the appearance and development of systemic risk are the priorities of the modern science. One of the most widely spread measures of risk nowadays is the value at risk (VaR) [15] that focuses on the risk of an individual institution in isolation. It shows that with a certain rate of probability potential losses would not exceed the VaR value calculated for the specified period. However, this value does not evaluate risks for the entire financial system.

This value of CoVaR was proposed by American economists Tobias Adriany and Markus K. Brunnermeierz in [3, 4]. To emphasize the systemic nature of the risk measure, its name contains prefix 'Co' which stands for conditional, contagion, or comovement.

The paper [14] presents the analysis of the sanctions impact on systemic risks for four Russian companies using CoVaR and CoVaR measures. In contrast to the paper [14], the main aim of this paper is to examine the effect of the sharp fall of oil prices on some Russian companies using static model of the systemic risk measure CoVaR. Moreover, this paper uses statistical testing procedures proposed in the paper [7] to assess the significance of the findings and interpretations based on this co-risk measure.

Our analysis includes 13 Russian companies from three different sectors:

- Primary sector which involves the retrieval and production of raw materials and includes the largest Russian companies in the field of extraction, production, transportation, and sale of natural gas and oil;
- Finance sector which involves Russian banking and financial services companies;
- Trade and food industry which includes Russia-based holding companies engaged in the retail industry.

The data used for the estimation of ΔCoVaR are weekly stock return data for 13 large Russian companies. The sample covers the period from August 15, 2011 to November 27, 2017, combining in a dataset 323 weekly observations of weekly returns per company. The stock market data for the 13 large Russian companies are taken from Finam.ru.

The first analysis of systemic risks for Russian companies using CoVaR was presented in the paper [6]. As it is pointed out in [6], CoVaR and its derived values are extremely promising from the point of view of financial risk-management, especially for the detection of potential danger for the economic system and companies inside it under systemic risks. The work [6] also demonstrates the capabilities of CoVaR to analyze of the Russian market and shows the adequacy of the obtained values of CoVaR to the real state of economy.

In the empirical part of the paper we use the method of quantile regressions to estimate CoVaR and its derived values. Kolmogorov-Smirnov (KS) type statistic proposed in [7] was used to test a financial institution for the systemic importance. We provide an empirical application of CoVaR measure and KS type statistic on a sample of 13 large Russian companies.

2 Definitions

Given a confidence level $q \in (0, 1)$, Value-at-Risk (VaR) of a random variable $r_{i,t}$ is defined as the solution of the equation

$$\Pr\left(r_{i,t} \leq \text{VaR}^i_{q,t}\right) = q.$$

In other words, $\text{VaR}^i_{q,t}$ is implicitly defined through the q-quantile of the conditional distribution of $r_{i,t}$. Usually (as well as in this paper), $r_{i,t}$ refers to the log return of the financial institution i at time t. Theoretical properties and practical applications of VaR can be found in the book [15]. Nowadays, VaR is one of the most well-known risk measures and is widely used by regulators and banks all over the world.

To start with, CoVaR calculated for the specific institution conditional on the whole system is defined as the Value at Risk of the whole financial sector conditional on that institution being under distress.

Let us now give the formal definition of CoVaR measure as it was proposed in [3] or [4].

$\text{CoVaR}_q^{j|C(r_{i,t})}$ is the value equal to the $\text{VaR}^j_{q,t}$ of institution j (with log return $r_{j,t}$) conditional on some event $C(r_{i,t})$ of institution i. That is, $\text{CoVaR}_{q,t}^{j|C(r_{i,t})}$ is implicitly defined by the q-quantile of the conditional probability distribution:

$$\Pr\left(r_{j,t} \leq \text{CoVaR}_{q,t}^{j|C(r_{i,t})} | C(r_{i,t})\right) = q.$$

In the papers [3, 4], $C(r_{i,t})$ refers to distress of institution i and that event of distress occurs when the return of institution i is equal to its VaR, i.e $r_{i,t} = \text{VaR}^i_{q,t}$. In addition to that, works [3, 4] define the event of median state of an institution as the event when an institution's return is equal to its median, i.e. $r_{i,t} = \text{VaR}^i_{0.5,t}$.

Then we have a legitimate question about the difference between the CoVaR values estimated for institution j conditional on institution i being under distress and in its normal state. That kind of difference is the measure of the contribution of institution i to the risk of institution j and is denoted as $\Delta\text{CoVaR}^{j|i}$.

Thus, $\Delta\text{CoVaR}^{j|i}_{q,t}$ measures the influence of the institution i on the institution j and is defined as follows:

$$\Delta\text{CoVaR}_{q,t}^{j|i} = \text{CoVaR}_{q,t}^{j|r_{i,t}=\text{VaR}_{q,t}^{i}} - \text{CoVaR}_{q,t}^{j|r_{i,t}=\text{VaR}_{0.5,t}^{i}}.$$

The rest of the paper focuses on the conditioning distress event of $r_{i,t} = \text{VaR}_{q,t}^{i}$.

The definition of the $\text{CoVaR}_{q,t}^{j|i}$, namely the VaR of the institution j conditional on the institution i being at its VaR level, allows the study of the spillover effects of the whole process on the financial network. Furthermore, we can obtain value $\text{CoVaR}_{q,t}^{j|\text{system}}$, which can give an answer to the following question: which institutions are most at risk during financial crises due to the fact that it reports the increase of VaR of the institution in the case of a financial crisis in the system.

Papers [3, 4] note that CoVaR and CoVaR are directional, i.e. CoVaR of the system conditional on institution does not equal the CoVaR of institution conditional on the system. While covariance based risk factors are widely used in practice (see e.g. [5]), the applicability of CoVaR measure to portfolio selection is under question.

3 CoVaR Estimation Methods

Estimation of the described value is a nontrivial task, and it can be handled with the help of a great variety of methods, particularly the method of quantile regression, which has been chosen for the empirical part of the study. It is the method of the regression analysis commonly used in statistics and econometric theory [10, 11]. While ordinary least squares (frequently used in Russian studies) are focused on getting estimators approximating conditional mean value of the variable in the case of the defined incoming values, the quantile regression is directed to getting estimation either for 50% or for any other quantile. One more profit of this method is connected with the fact that it is more stable in case of getting the outlying values among incoming data. The case is that this kind of outlying values can be frequently met in practice, especially during the study of financial and economic systems.

It is also the case that there are two models of the CoVaR estimation—the static model and the dynamic model.

3.1 The Static Model

The static model provides an opportunity to calculate CoVaR and ΔCoVaR values that are constant over time and independent of other exogenous factors. According to this model, CoVaR and ΔCoVaR estimation starts with the construction of the quantile regression to find estimated coefficients for institutions i and j.

The q-quantile regression describes the dependance of the predicted value of institution j for q-quantile $\hat{X}_q^{j,i}$ conditional on institution i:

$$\hat{X}_q^{j,i} = \hat{\alpha}_q^i + \hat{\beta}_q^{j|i} X^i, \tag{1}$$

where $\hat{X}_q^{j,i}$ presents the predicted value on the specified quantile and returns of the institution i. We concentrate on the case when $X^i = \mathrm{VaR}_q^i$, which means that the institution i is at its VaR level [3, 4].

Then, after getting the coefficients, we can find the CoVaR and ΔCoVaR values using the following equations:

$$\mathrm{CoVaR}_q^{j|r_i=\mathrm{VaR}_q^i} = \mathrm{VaR}_q^{j|\mathrm{VaR}_q^i} = \hat{\alpha}_q^i + \hat{\beta}_q^{j|i}\mathrm{VaR}_q^i, \tag{2}$$

The value of $\Delta\mathrm{CoVaR}^{j|i}$ is the difference between CoVaR of institution j conditional on the institution i being in distress and CoVaR of institution j conditional on median state of the institution i:

$$\Delta\mathrm{CoVaR}_q^{j|i} = \mathrm{CoVaR}_q^{j|r_i=\mathrm{VaR}_q^i} - \mathrm{CoVaR}_q^{j|r_i=\mathrm{Median}_i} = \hat{\beta}_q^{j|i}\left(\mathrm{VaR}_q^i - \mathrm{VaR}_{0.5}^i\right). \tag{3}$$

However, under the conditions of a real economy not only separate sectors should to be taken into consideration but also macroeconomic indicators which have a strong influence on the estimation results in case of dynamic pattern study. Dynamic model includes this kind of factors and provides an opportunity to capture time variation.

3.2 *Testing for Significance*

Significance can be identified if for an institution $|\Delta\mathrm{CoVaR}|$ exceeds a given threshold level. Following the paper [7] we will assume this threshold level is equal to 0. Thus, a hypothesis test for the identification of a systemically significant institution for institution j is equivalent to the following null hypothesis:

$$H_0 \;:\; \Delta\mathrm{CoVaR}_q^{j|i} = 0 \tag{4}$$

for a given level $q \in (0, 1)$. As it is pointed out in [7], the hypothesis of significance given by (4) is similar to hypothesis test (hypothesis of no effect) of the quantile treatment effects literature. It follows from (4), (2) and (3) that testing the significance of $\mathrm{CoVaR}_q^{j|i}$ reduces to a joint exclusion Wald test of the $\hat{\beta}_q^{j|i}$ for a given q (see. e.g. [12]).

As noted in [7], ΔCoVaR can be interpreted as a quantile treatment effect. For this reason, the paper [7] proposes a test based on the Kolmogorov–Smirnov (KS) type statistic. Since KS type tests are asymptotically distribution free, they are extremely applicable. It is known that the KS test gives a reasonable way to measure the discrepancy between distributions [1]. Furthermore, variants of the two-sample KS test have been widely used for inference based on a quantile process, such as those considered in [7]. The test statistic proposed in [7] is based on the quantile response function. The paper [7] considers a conditional distribution, rather than an unconditional distribution, and in particular the conditional quantile response function of a linear model.

Suppose we have two different values VaR_q^i and $\mathrm{VaR}_{0.5}^i$. Then $\hat{X}_q^{j,i} = \hat{\alpha}_q^i + \widehat{\beta}_q^{j|i}\mathrm{VaR}_q^i$ and $\hat{X}_{0.5}^{j,i} = \hat{\alpha}_q^i + \widehat{\beta}_q^{j|i}\mathrm{VaR}_{0.5}^i$ are the respective empirical quantile response functions evaluated respectively at these two values, where $\widehat{\beta}_q^{j|i}$ correspond to the quantile regression estimate of the parameter of a linear location-scale model. This setup allows us to compare different continuous treatment effects applied to the same population within the framework of a linear location-scale model that relates X^i to the VaR_q^i and $\mathrm{VaR}_{0.5}^i$ covariates. Let T be the amount of observations. Thus, we examine the following parametric empirical process:

$$\nu_T(q) = \sqrt{T}\left(\hat{X}_q^{j,i} - \hat{X}_{0.5}^{j,i}\right) = \sqrt{T}\left(\widehat{\beta}_q^{j|i}\mathrm{VaR}_q^i - \widehat{\beta}_{0.5}^{j|i}\mathrm{VaR}_{0.5}^i\right).$$

In testing the null hypothesis $H_0 : \Delta\mathrm{CoVaR}_q^{j|i} = 0$ it is not difficult to notice that given expression (3), the test is equivalent to a standard significance test $H_0 : \hat{\beta}_q^{j|i} = 0$ for a given $q \in (0, 1)$. The two-sided Kolmogorov–Smirnov type statistic is $K_T = |\nu_T(q)|$.

4 Empirical Results

This part of the paper presents empirical results. Our analysis is based on the original observations of the weekly prices of 13 Russian companies being the part of the RTS index from three different sectors:

- Primary sector which involves the retrieval and production of raw materials and includes such largest Russian companies in the field of extraction, production, transportation, and sale of natural gas and oil as
  - Gazprom (GAZP),
  - Lukoil (LKOH),
  - Rosneft (ROSN),
  - Surgutneftegaz (SNGS),
  - Tatneft (TATN),
  - Transneft (TRNFP);
- Financial sector which involves such Russian banking and financial services companies as
  - Vozrozhdenie Bank (VZRZ),
  - Bank Saint Petersburg (BSPB),
  - Sberbank (SBER),
  - VTB Bank (VTBR);

- Trade and food industry which includes Russia-based holding companies engaged in the retail industry and food production such as
  - Cherkizovo Group PAO (GCHE),
  - Dixy Group PAO (DIXY),
  - Magnit PAO (MGNT).

The data was taken from August 04, 2011 to November 1, 2017. This time interval was divided into two almost equal periods. The first period is from August 04, 2011 to October 6, 2014 (164 weeks). The second period is from October 13, 2014 to November 1, 2017 (159 weeks). The first visible sharp decline in oil prices took place in October of 2014. Despite the fact that October usually sees by a growing demand for fuel, the price decline continued. One of the widely discussed reasons for this fact was the increase in the volume of proposals for shale gas.

In this section we examine the mutual influence of the Russian companies. Moreover, the main question of interest is how it changed after the sharp fall of oil prices in 2014. The oil price was fell by more than 40% from June of 2014 to the end the year, i.e. from $115 to below $70 a barrel. Of course, the negative impact of the sharp fall of oil prices appeared not immediately but quickly enough. Another big shock for the Russian economy was the implementation of sanctions by governments of the United States and European Union countries as well as some international organizations in the same year. Some companies have lost sources of cheap fast loans, some lost their markets or sources of purchase. Later the Russian authorities have also introduced counter sanctions which concerned mostly food producers. The factor of sanctions applied to interactions between companies was discussed in the work [14]. Some nonlinear methods for analysis of time series and data sets are presented in [9, 16].

Based on analysis of the Russian economics data, many researchers conclude that Russian economy is in a recession. Data show that GDP of the first quarter of 2015 is negative with −2.2% with comparison to the first quarter of 2014. Moreover, the cumulative effect of the sanctions and the sharp decline in oil prices in 2014 and 2015 has led to serious downward pressure on the ruble value. The process of flight of capital out of Russia has been enhanced. The sanctions on access to foreign financing have urged Russian government to use part of its fund and reserves to boost the Russian economy. In 2014 the Central Bank of Russian Federation ceased to support the value of the ruble and harshly increased interest rates. On the other hand, the impact of anti-Russian sanctions on the Eurozone economy proved to be negligible. Average economic growth in the eurozone declined slightly. In 2014 negative euro-zone trade balance with Russia decreased only by $3.6 billion.

Accumulated losses from the sanctions of the Russian GDP amounted to 6 percentage points in 2014–2017 compared to the GDP in 2013. Capital flight triggered by sanctions is estimated at $160–170 billion over the same period.

Counter sanctions imposed by Russia in response to Western sanctions affected the inflation rate in the country. Based on the 2014 year data, the Ministry of Economic Development of Russian Federation estimated that the contribution of the counter sanctions to the annual inflation rate (11.4%) was about 1.5 percentage points. At the same time, the food price inflation in 2014 was equal to 15.4%, and 3.8 percentage points of it were due to sanctions.

In this paper we consider how the new economic situation affects on relations between the leading Russian companies within different sectors of economy. Descriptive statistics are presented in Table 1. Log returns did not change significantly after October of 2014. It can be explained by the fact that the inflation and volatility processes start long before October of 2014 (in the summer of 2017) and lie in the first period.

Most of the considered companies were not mentioned in primary sanction list, but these sanctions, one way or another, affected all these companies. Some of them are the leaders in their economy sectors. Cherkizovo is the leader in poultry and pork production, has a unique position in this list due to the absence of universal producers in food industry. Let us note some facts connected with the relationships between these companies and some evident consequences of the sanctions. Sberbank credits Gazprom. Sberbank and Gazprombank credit several companies, including Magnit and Cherkizovo. Creditors are forced to increase credit rates. 56% of large and medium agrarian firms are Sberbank clients. Cherkizovo is oriented to its own production. In 2014 after introduction of the sanctions in view of prohibition for some food products import there was an upturn of Cherkizovo returns connected with meat production. But rouble devaluation can stop that rise because grain and forage are mainly bought abroad. Agrarians do not plan to invest cash in new infrastructures, because their managers do not believe that sanctions will be long-term. Magnit has the following problem: it cannot supply all popular products. The largest Russian retailer changes the product line. It attempts to exclude foreign goods from its list. On the other hand, Magnit, which is interested in product quality, sometimes buys such products by nonstandard ways and pays fines. Some companies suffer the sanctions indirectly, through their affiliated companies. Such structures in Sberbank lose their opportunity for placing their assets long-term period. Gazprom has difficulties in equipment procurement.

We focus on calculating $\Delta\mathrm{CoVar}_q$ with $q = 0.01\%$, i.e we take 1%-quantile for the estimation. The hypothesis of non-significance H_0 for ordered pairs will be checked using the Kolmogorov–Smirnov test as it is described in section "Testing for Significance".

Results for the static model with ΔCoVaR values are given in Tables 2, 3 and 4.

Table 2 (the companies of the primary sector) shows that many values of ΔCoVaR are significant in the first period (before October, 2014), but most of them lost their significance in the second period (after October, 2014). The only significant value of ΔCoVaR with confidence level 0.01 is the impact of Surgutneftegaz on Tatneft. It may be explained by the sharp fall of oil prices and the weakening business activity.

The same picture can be observed in the financial sector. Table 2 shows that the impact on Sberbank has declined in the second period in contrast with the first one. The only significant value of ΔCoVaR corresponds to the impact of Bank Saint Petersburg on Vozrozhdenie Bank which can be explained by the fact that Bank Saint Petersburg was the shareholder of Vozrozhdenie Bank.

Table 1 Descriptive statistics of log returns for 13 Russian companies, %

Company	Weeks from 1 to 323				Weeks from 1 to 164				Weeks from 165 to 323			
	Mean	St.dev.	Skew.	Kurt.	Mean	St.dev.	Skew.	Kurt.	Mean	St.dev.	Skew.	Kurt.
GAZP	0.09	3.53	0.30	0.64	0.02	3.59	0.24	0.75	0.17	3.47	0.37	0.49
LKOH	0.38	3.43	0.39	1.63	0.23	2.69	−0.40	0.97	0.54	4.06	0.53	0.93
ROSN	0.27	3.89	0.43	1.30	0.19	3.41	0.18	2.06	0.34	4.34	0.53	0.65
SNGSP	0.62	4.51	0.51	3.98	0.75	4.22	−0.14	4.19	0.49	4.81	0.98	3.79
TATNP	0.64	3.70	−0.15	4.72	0.51	3.48	−1.08	9.16	0.77	3.91	0.50	1.51
TRNFPP	0.68	4.74	0.22	1.95	0.67	4.66	0.16	3.50	0.70	4.83	0.27	0.54
VZRZ	−0.03	4.92	2.33	20.12	−0.38	4.77	0.55	2.02	0.32	5.06	3.88	34.15
BSPB	−0.07	5.48	0.42	4.46	−0.63	6.38	0.56	4.31	0.51	4.30	0.40	1.07
SBER	0.43	4.36	0.36	2.26	0.06	4.38	0.08	2.87	0.83	4.33	0.66	1.47
VTBR	0.00	4.26	0.78	4.70	−0.24	4.47	−0.12	1.29	0.25	4.03	2.09	9.36
GCHE	0.36	4.58	1.83	18.56	0.09	4.17	0.17	4.21	0.63	4.95	2.80	24.70
DIXY	0.17	4.91	0.35	2.32	0.39	5.47	0.13	1.88	−0.04	4.25	0.72	2.51
MGNT	0.35	4.52	−0.02	0.87	0.83	4.37	−0.28	0.41	−0.14	4.64	0.24	1.45

Table 2 The values of $\Delta\text{CoVaR}_{0.01}^{j|i}$ for static model, the companies of the primary sector, %

$j \setminus i$	GAZP	LKOH	ROSN	SNGSP	TATNP	TRNFP
Weeks from 1 to 164						
GAZP		−3.80	−3.69	−5.01	−10.15***	−4.51
LKOH	−5.36		−3.65	−5.86	−7.60**	−4.45
ROSN	−6.17	−5.56		−4.18	−6.10	−5.75
SNGSP	−10.60***	−5.79	−5.45		−10.66***	−11.00***
TATNP	−8.97***	−6.58*	−12.25***	−12.33***		−9.33***
TRNFP	−9.35***	−11.22***	−13.37***	−7.57**	−10.97***	
Weeks from 165 to 323						
GAZP		−2.47	−5.87	0.91	−2.75	−1.60
LKOH	−4.35		−3.43	−2.61	−3.73	0.88
ROSN	−1.22	−6.29		−1.81	−2.14	−0.86
SNGSP	0.29	−4.60	0.33		−5.64	2.51
TATNP	2.00	−3.34	−3.32	−7.52**		2.27
TRNFP	−2.86	−5.14	−0.37	−3.74	−6.96*	

*,**,***Marks significance at 10%, 5% and 1% respectively

Table 3 The values of $\Delta\text{CoVaR}_{0.01}^{j|i}$ for static model, the companies of the financial sector, %

$j \setminus i$	Weeks from 1 to 164				Weeks from 165 to 323			
	VZRZ	BSPB	SBER	VTBR	VZRZ	BSPB	SBER	VTBR
VZRZ		−6.88*	−2.17	−3.20		−7.18**	−0.82	−4.39
BSPB	−8.55***		−20.57***	−14.74***	5.24		−5.81	3.65
SBER	−8.39***	−10.27***		−11.44***	−1.99	−6.09		−0.55
VTBR	−4.64	−6.05	−8.33***		−1.13	−2.39	−1.45	

*,**,***Marks significance at 10%, 5% and 1% respectively

Table 4 The values of $\Delta\text{CoVaR}_{0.01}^{j|i}$ for static model, the companies from the trade and food industry, %

$j \setminus i$	Weeks from 1 to 164			Weeks from 165 to 323		
	GCHE	DIXY	MGNT	GCHE	DIXY	MGNT
GCHE		−4.05	5.70		−5.86	−10.18***
DIXY	−5.89		−7.45**	−1.02		−0.84
MGNT	0.16	−2.85		−1.80	−1.92	

*,**,***Marks significance at 10%, 5% and 1% respectively

The trade and food industry has only one significant value of ΔCoVaR in the first period (the impact of Magnit PAO on Dixy Group), which became insignificant in the second period. Magnit PAO has recently announced the plan of buying the loss-making Dixy Group. Meanwhile, one can notice the appearance of Magnit's influence on Cherkizovo (the consumer of products affects their producer).

It should be noted that all significant values of ΔCoVaR are positive. Thus, companies of each sector were moving at the same direction.

The Russian economy has been mostly focused on the B2B (business-to-business) model and this resulted (following the fall of the Russian ruble and oil prices) in decreasing of key sectors of the economy. It should be noted that energy and mining sectors give 80% of the total income of the Russian economy. The income of businesses working for the Russian government have sank for at least 30% in 2014. As the result of the ruble weakening, the funding of many government programs was suspended. The B2B companies lost a huge part of their orders. This, in turn, this led to falling consumer demand due to the cuts of real incomes.

Before sanctions many Russian companies took loans in the EU banks at 2–3% per annum. Loans in the Russian banks (17–19% per annum) were not profitable. The sanctions have forced the companies to lend in Russian banks, which can be clearly seen from the data of $\Delta \mathrm{CoVaR}_{0.01}$ in Table 3. For example, the impact of VTB Bank on Sberbank has changed from −11.44 to −0.55.

The main problem in the development of Russian economy is the weak government support of the B2C (business-to-customer) market. The prices of consumer products depend strongly on the US dollar/Ruble rate. As the result of the dollar rate growth and reducing real incomes, the Russian consumer market was dipped significantly. Moreover, companies that produce the products, the component parts for which are bought in dollars, were forced to reduce production. Therefore, sanctions have led to the necessity to create its own production of the component parts and goods for end consumers. The paper [14] shows that the Sberbank effect on some B2C companies became more significant (for example, ΔCoVaR of the pair Sberbank—Magnit has changed from −3.79 to −6.42%).

On the other hand, the counter sanctions and the weak Ruble have led to a sharp rise of the agricultural industry, since agricultural products of Russian companies became competitive in the Russian market. The results in Table 4 confirm this.

5 Conclusion

One of the key features of contemporary economic systems is that they are complex systems consisting of a large number of interdependent parts (agents). The more complex the system, the more dependent its parts are, the more complex behavior it demonstrates. In this case, complex systems can consist of a large number of parts, each of which exhibits a primitive behavior. Processes of evolution and complication of the system should lead to an increase in the co-interdependence of parts, while the processes of disintegration and destruction of systems should lead to a weakening of interdependence. In other words, more primitive systems are characterized by a weaker level of their parts integration. In this regard, it can be noted that the processes of economic development accompanied by the complication of the economic system at the macro level and the emergence or strengthened links between its parts, should objectively lead to a systemic risk increase. Such an increase in systemic risks can be quantified as an increase in ΔCoVaR values. On the other hand, the processes of disintegration and weakening of the economic system can lead to the autonomous functioning of system parts, and consequently, to a systemic risk reduction. The results obtained during the study demonstrate a significant weakening of the links between economic agents (companies, financial institutions, etc.), which are quantitatively expressed in the form of ΔCoVaR. Most ΔCoVaR values are significant in the first period. On the other hand, ΔCoVaR values significantly decreased in the second period, and almost all values are not significant. We interpret such results as the evidence of disintegration processes caused by external shocks for the Russian economy.

Acknowledgements The work was supported by RFBR (grant 18-37-00060).

References

1. A. Abadie, Bootstrap tests for distributional treatment effects in instrumental variable models. J. Am. Stat. Assoc. **97**(457), 284–292 (2002)
2. T. Adrian, M. Brunnermeier, H.L.Q. Nguyen, Hedge fund tail risk, in *Quantifying Systemic Risk* (National Bureau of Economic Research, Inc., 2011), pp. 155–172
3. T. Adrian, M.K. Brunnermeier, CoVaR. Working Paper 17454, NBER (2011)
4. T. Adrian, M.K. Brunnermeier, CoVaR. Am. Econ. Rev. **106**(7), 1705–1741 (2016)
5. F. Alali, A. Cagri Tolga, Covariance based selection of risk and return factors for portfolio optimization, in *Proceedings of the World Congress on Engineering 2016*. Lecture Notes in Engineering and Computer Science, vol. II, 29 June–1 July 2016, London, U.K. (2016), pp. 635–640
6. V. Barabash, S. Sidorov, The analysis of the mutual influence of economic subjects using risk measure CoVaR on the example of some Russian companies. J. Corp. Finance Res. **8**(1), 73–83 (2014)
7. C. Castro, S. Ferrari, Measuring and testing for the systemically important financial institutions. J. Empir. Finance **25**, 1–14 (2014)
8. G. De Nicolo, M. Lucchetta, Systemic risks and the macroeconomy, in *Quantifying Systemic Risk* (National Bureau of Economic Research, Inc., 2011), pp. 113–148
9. I.C. Demetriou, P. Tzitziris, Infant mortality and economic growth: modeling by increasing returns and least squares, in *Proceedings of the World Congress on Engineering 2017*. Lecture Notes in Engineering and Computer Science, vol. II, 5–7 July 2017, London, U.K. (2017), pp. 543–548
10. R. Koenker, *Quantile Regression* (Cambridge University Press, Cambridge, 2005)
11. R. Koenker, K.F. Hallock, Quantile regression. J. Econ. Perspect. **15**(4), 143–156 (2001)
12. R. Koenker, Z. Xiao, Inference on the quantile regression process. Econometrica **70**(4), 1583–1612 (2002)
13. A. Lehar, Measuring systemic risk: a risk management approach. J. Bank. Finance **29**(10), 2577–2603 (2005)
14. A. Lunkov, E. Korotkovskaya, S. Sidorov, V. Barabash, A. Faizliev, Analysis of the impact of sanctions on systemic risks for Russian companies, in *Proceedings of the World Congress on Engineering 2017*. Lecture Notes in Engineering and Computer Science, vol. I, 5–7 July 2017, London, U.K., pp. 380–384 (2017)
15. J. Philippe, *Value at Risk: The New Benchmark for Managing Financial Risk*, 3rd edn. (McGraw-Hill, New York, 2006)
16. K. Tamura, T. Ichimura, Classifying of time series using local sequence alignment and its performance evaluation. IAENG Int. J. Comput. Sci. **44**(4), 462–470 (2017)

An Innovative DSS for the Contingency Reserve Estimation in Stochastic Regime

Fahimeh Allahi, Lucia Cassettari, Marco Mosca and Roberto Mosca

Abstract The problem of sizing and managing contingency reserve is always critical in project management, because of its impact on the project margin. A correct assessment of the contingency reserve to be allocated is, therefore, a main requirement to lead to success the project manager actions. In this research, the Authors propose an innovative Decision Support System to size, starting from an objective phase of risk assessment, the correct contingency reserve. The proposed solution provides the project manager a clear vision of the residual risk of cost overruns to be managed. The Decision Support System uses Failure Mode Effect Analysis and Monte Carlo Simulation.

Keywords Contingency cost · Decision Support System · Monte Carlo simulation · Project management · Risk analysis · Stochastic estimation

1 Introduction

A critical phase in contract of engineering is when the Decision Maker (DM) must determine the contingency reserve, or the extra cost that should be added to the overall project cost to ensure an adequate coverage level against the risk of cost overruns.

F. Allahi · L. Cassettari (✉) · R. Mosca
Department of Mechanical Engineering, Energetics, Management and Transports (DIME), University of Genoa, Genoa, Italy
e-mail: cassettari@dime.unige.it

F. Allahi
e-mail: allahi@dime.unige.it

R. Mosca
e-mail: mosca@dime.unige.it

M. Mosca
Polytechnic School, University of Genoa, Genoa, Italy
e-mail: marcotulliomosca@gmail.com

S.-I. Ao et al. (eds.), *Transactions on Engineering Technologies*,
https://doi.org/10.1007/978-981-13-0746-1_4

A careful contingency assessment is essential to ensure competitiveness in tenders. The higher the contingency reserve is, the higher the coverage level is. Upon the occurrence of a probabilistic risk event, for some years now there has been a movement to replace classic scenario analysis (based on minimum, maximum and most probable values) with the definition of a PDF that describes the probability of possible occurrences of the economic impact values associated with each individual risk. The output of the analysis based on the Monte Carlo simulation, is a PDF curve, capable of linking the reserve amount with the probability of covering expected global risk.

Although this method is, by its nature, more reliable than the traditional methods, it suffers from a strong lack of objectivity. It has also been particularly difficult for uneducated employees to properly use the MCS, despite having the appropriate software. For this reason, the Authors have considered creating a Decision Support System (DSS), which is capable of following tasks:

1. Making the determination of the contingency reserve more objective by focusing on the probability of the occurrence of the risk event that is most familiar to professionals with regard to the PDF of economic impact;
2. Ensuring the scientifically correct use of the MCS, including for users who are uneducated in the methodology;
3. By analyzing the integral curve of the PDF, it can be understand that with allocating a determined budget, a specific level of risks can be covered and vice versa.

2 Literature Review

From the literature analysis, it can be noted that, recently, few Authors have dealt with the issue of allocating contingency reserve despite its relevance in project management.

The most commonly used method for the contingency reserve estimation uses percentages that are predetermined and related to the occurrence probability of the associated risks [1]. This approach however is difficult to be justified or defended because of its unscientific approach [2]. Moreover, it was demonstrated to often lead to an excessive budget allocation [3]. A deeper literature description can be found in [4], where some methods for estimating project cost contingency are shown.

Kuo and Lu [5] focus their work on the risk assessment problem but do not consider the risk impact on the real project development and on the contingency reserve definition.

Yim et al. [6] consider only the risk assessment and applied to engineering design projects. In their research, the possible response strategies to an occurred risk are shown but nor mitigation actions nor contingency reserve allocation are considered.

Hartono et al. [7] aim at identifying stakeholder's perspectives on project risks in Indonesia and their gaps with the rational assumptions of the normative decision theories.

In recent times, three estimation methods have replaced these approaches and have therefore taken on greater importance, namely Monte Carlo Simulation [8], Regression Analysis [9] and Artificial Neural Networks [10]. Regression models and Neural Networks are, however, difficult to use for not particularly skilled users. The approach based on Monte Carlo simulation, instead, starts from the subjective choice by of the impact probability distributions of each risk made by every decision-maker. This choice, however, forces the referent to imagine, in addition to the most likely impact value, a minimum value, a maximum value and a probability function over the whole range of possible impact values. Such approach would be correct only if it is possible to have a rich database and to have the possibility the correct distributions.

Chou et al. [11] emphasize that an inaccurate estimation of project contingency costs may result in a drastic reduction in the total profits. They propose an approach based on heuristic techniques and simulation models in order to provide the Project Manager with valuable input.

Howell et al. [12] review the different approaches adopted by PMs in the contingency costs estimation, rating each strategy according to five drivers: uncertainty, complexity, urgency, team empowerment and criticality.

Idrus et al. [13] criticize the contingency costs allocation as a percentage of the total project cost and propose a non-stochastic calculation model based on risk analysis and fuzzy techniques.

Chou et al. [14], finally, propose a simulation approach using the Monte Carlo method. In particular, it uses a historic set of projects to create stochastic distributions to be associated with the project costs.

Allahi et al. [15, 16] recently have presented a stochastic risk analysis approach to estimate cost contingency in some projects by Monte Carlo Simulation with the aim to determine the probability distribution of the contingency cost and the related level of risk coverage.

Analyzing the literature cited above and industry guides for professionals, the Authors found a lack of an objective base in the used approach. In particular, there is a strong dependence of some key parameters on subjective decisions made by the individual decision-makers. Consequently, the Authors decided to make some substantial changes to the decisional process which lead to the determination of the contingency reserve in order to make the results as objective as possible.

3 Risk Identification

Risk identification is one of the important steps to recognize whole risks, which can affect the project budget. In order to identify these risks, the company uses a checklist guide. It makes possible to identify different kinds of risks: those of a

Table 1 Scale for operational risks

Impact on the schedule parameter	
Number of delayed days	Severity
≤7 days	(1)
7–30 days	(2)
≥30 days	(3)

strictly operational nature and those of a legal/financial nature arising from contract terms. The checklist was filled out with some interviews dedicated to members of the company, directly have involved in the management of the project. At the same time the Work Packages (WP) affected by the risks are identified in order to facilitate both the next step of risk quantification and the contingency allocation phase. Once the most significant risks are identified, qualitative and quantitative assessment is carried out according to Allahi et al. [16].

3.1 Qualitative Analysis

The qualitative survey consists in determining the risks that require special attention in terms of identifying actions to mitigate and/or transfer risk. To do it, an analysis based on a simplification of Failure Mode Effect Analysis (FMEA) methodology is carried out to rate each risk in qualitative terms (high, medium, or low) referred both to the occurrence likelihood and to the severity on project (in terms of schedule, costs and performance).

The indicator used to rank the risks is the so-called "risk factor", which is obtained by the multiplication of risk occurrence probability (ROP) by the level of risk severity (RS).

RS can be qualitatively identify through three main parameters: schedule, cost and performance. For each of these, it is necessary to define a scale to be able to identify a high (3), medium (2) or low (1) impact.

As far as the effect on schedule concerned, there are two types of risks, namely operational and legal/financial. Therefore, it is necessary to define two different scales for the two types of risk. With regard to operational risks, Table 1 presents the number of delayed days on the project completion, caused by the occurrence of the risk. It proposes for the delay more than 30 days, the severity of risk will be high on the cost of the project.

As regards to the legal/financial risks relating to the payment of the penalties for late delivery, the scale of Table 2 is considered.

Table 2 clearly shows that as long as the request subject to a contract penalty has a delay lower than of 15 days, the associated penalty is not actually applied and hence not considered. Obviously, once 15 days are reached, every day of delay accrued will be paid.

Table 2 Scale for legal/financial risks

Impact on the schedule parameter		
Payment	Action	Severity
<15 days	Penalty clauses not applied	(1)
≥15 days	Penalty clauses applied for every late day	(3)

Table 3 Scale for impact on cost

Impact on costs		
Cost	Assessment base on x1 and x2	Severity
<57 k €	Job order K decreases less than x1%	(1)
57 k–1,000 k €	Job order K decreases between x_1% and x_2%	(2)
>1,000 k €	Job order K decreases more than x_2%	(3)

Table 4 Scale for impact on performance

Impact on performance	
Performance	Severity
The system/component does not meet contract specification and operational inefficiency of it exists	(3)
A specific requirement is not satisfied without negative consequences on system's operational performance (Visible risk to the customer)	(2)
Contract requirements not fully satisfied	(1)

For analyzing the impact on costs, a single scale is considered for both types of risk (Table 3). The assessment is closely related to the fact that the risk occurrence would result in a more or less significant decrease in the cost of the project K or in the Gross Margin. Based on several studies, values x_1 and x_2 were identified as discriminants between the various levels of impact (not reported due to corporate intellectual property right).

Besides, the impact on performance, it is not necessary to apply separate analyses for the two types of risk. The scale used in this case is presented in Table 4.

Once the mentioned scales were defined, the scale related to Risk Occurrence Probability (ROP) is determined. In particular, ROP is evaluated according to the followed criteria.

- $ROP < 20\%$: Severity of risk = 1;
- $20\% \leq ROP \leq 50\%$: Severity of risk = 2;
- $ROP > 50\%$: Severity of risk = 3.

Table 5 Risk factors values

Probability of occurrence	3	3(●)	6 (*)	9 (*)
	2	2 (◊)	4(●)	6 (*)
	1	1 (◊)	2 (◊)	3(●)
		1	2	3
		Risk impact		

In order to rank the risks, the risk factor is then determined. It is obtained by combining the qualitative RS assessments and it's ROP in a matrix whose output is a numeric value. The risk factor is obtained by multiplying the probability parameter by the RS one (to quantify the effect, the highest value among those recorded for schedule, cost and performance is considered). The possible risk factor values are summarized in Table 5.

The obtained risk factor makes it possible to rank the identified risks according to priority. In particular, it was decided to proceed with the next step of analysis only for the risks characterized by a risk factor greater than or equal to three (The symbols of (●) and (*) (yellow and red risks) in Table 5).

3.2 Quantitative Analysis

It is now necessary to evaluate, for each risk, the correct value for both the ROP and the economic impact (EI).

In particular, as far as the estimation of the risk impact is concerned, the procedure varies depending on the type of risk under consideration. If the risk is subject to penalties, the maximum impact is estimated as the product of the accrued delay (typically assumed greater than or equal to 15 days) by the sum of the penalties. If the risk is not related to any kind of penalty, the estimated maximum impact is obtained by considering this value equal to a percentage of the costs sum of the WPs affected by that risk. Particularly, in the latter case it is assumed that the risk occurrence leads to a rework of a part of the WPs subject to the risk. At worst, all of the WPs must be reworked. Then, finally, the so called Expected Monetary Value (EMV) can be calculated as the product of the ROP by the EI.

3.3 Mitigation Actions Identifications

Once the EMVs associated to each risk are determined, the possible mitigation actions to be implemented are identified for each risk. This activity is supported by an analysis of the causes and aims at reducing/eliminating the risk impact.

It should be noted that the implementation of a mitigating action allows the reduction of an uncertain cost with a certain incurred cost (the action cost) whose amount is estimated to be lower. In order to ensure the right balance between risk reduction and cost-effectiveness, it is necessary to determine the net benefit of the mitigating actions. This benefit is determined as the difference between the expected value of the risk before and after the risk mitigating action, net of the mitigation cost. Only the actions with a positive net benefit must be implemented, unless there is still an overall benefit for the project and/or the Company.

3.4 Contingency Reserve Estimation

After the reassessment of the residual risks and of their final EMV, the contingency reserve to be allocated has to be defined. In particular it is calculated as the sum of each final EMV. The contingency reserve must make it possible to duly "cover" the project if the risk events occur. After starting the project, the described process must be updated. All the risks that did not occur should be reviewed to recalculate the associated contingency cost. Finally, when checking the condition that makes it possible to declare the risk "closed", the associated contingency reserve can be:

- "Used": if the risk has occurred, causing economic damage to the project. The contingency is considered "used" to cover the damage suffered up to a maximum value equal to the amount previously allocated. Any part of the damage that is not covered by the contingency is to be considered an "extra cost".
- "Releasable" or "Partially Releasable": in case the risk event does not occur and causing fewer damages compared to the allocated contingency.

In order to check the risk occurrences, specific milestones must be assigned to the activities associated with the risk event. When a milestone is reached it indicates that the risk, which has or has not occurred in the past, will not occur in the future.

4 The Contingency Reserve DSS

The complexity of the procedure described in Sect. 3 could make it difficult for the DM to keep control over the process of determining the contingency reserve. Furthermore, the Authors believe that a correct determination of the contingency reserve must be done under a stochastic regime, which is computationally complex. For these reasons, the authors consider it appropriate to construct a DSS that interactively guides the DM through the entire process (deliberately made stochastic), according to the conceptual phases described in section "Conceptual Description". In this way, the DM can concentrate his or her attention solely on determining the most appropriate contingency reserve for a correct ratio between it and the risk coverage.

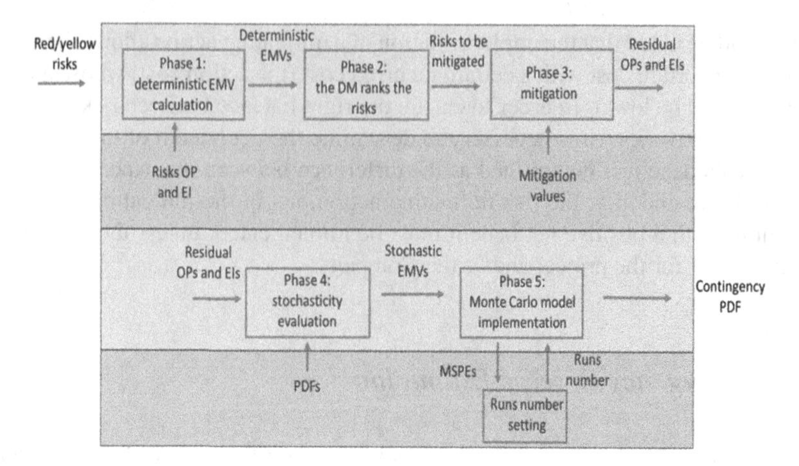

Fig. 1 Conceptual flux diagram

4.1 Conceptual Description

Figure 1 describes the conceptual flux implemented in the DSS. The grey rows contain the main processes while the orange rows represent the user interface activities.

The DSS has in input the (●) and (*) risks (see Table 5) and gives in output the contingency PDF. Between the input and the output, five main phases can be identified.

The first phase uses the risks ROP and EI to calculate the deterministic EMV for each red/yellow risk. The EMV is, here, obtained as the product of the ROP by the EI.

In the second phase, the DM ranks the risks based on their deterministic EMVs and establishes which risks have to be mitigated.

In the third phase, the chosen risks are mitigated by reducing the risks ROP or EI and the residual ROPs and EIs are obtained.

The quantitative analysis described of phase 2 is based on the assumption that the deterministic EMV is enough to define the essential aspects of the overall project risk profile.

However, since the operation regime is stochastic, it is necessary to take into account the fact that any risk can occur with different values in terms of EIs and ROPs.

These deterministic values are transformed into stochastic values in phase 4 thanks to the interaction with the user/users that provide the kind of PDF they want to associate to each element and the main values necessary to define it. Then the EI PDF calculates the stochastic EMVs as the product of the ROP PDF.

These values are, then, used in the Monte Carlo model (phase 5). This model, at the beginning, draws the Mean Square Pure Error of the mean (MSPE Med) and the Mean Square Pure Error of the standard deviation (MSPE St. Dev) that can be,

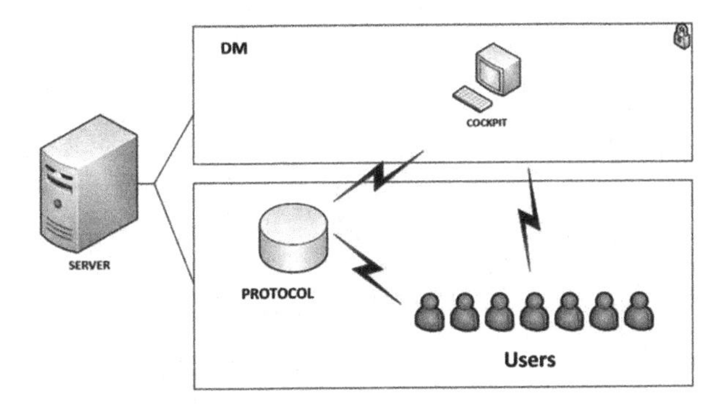

Fig. 2 DSS architecture

here, generally indicated as MSPEs. These curves allow the user to define the correct number of experimental runs [17–19] such that it is possible to obtain an output of statistically reliable results. The MSPE is complex in itself but in the interface with the user, it appears simple and intuitive because its complexity is managed by the DSS.

The chosen runs number is, then, used to runs the Monte Carlo model for the second time and to produce the DSS final output. The curve obtained allows the DM both to identify the "Coverage Level" corresponding to each value of the total allocated contingency reserve and, fixing the "Coverage Level", to determine the contingency reserve able to grant it.

4.2 DSS Architecture and Structure

The DSS developed by the Authors according to the architecture shown in Fig. 2 allows management of the reporting and processing of data in a single database. The functional architecture is composed of three elements:

- User File: interfaces with the person interacting with users or assessors. Consists of a template that is replicated for each user;
- Cockpit: component accessible only to the DM. Gathers all the data from the various User Files, represented as informational dashboards. Individual User Files may be accessed from the Cockpit, but not vice versa;
- Protocol: tool for communication between Cockpit and User File, as well as the DSS Database;

The process described in Fig. 1 is managed through the DSS according to the block scheme shown in Fig. 3.

The DM enters the list of the most critical risks identified in the qualitative analysis phase into the appropriate template. For each risk, the DM will indicate the relevant

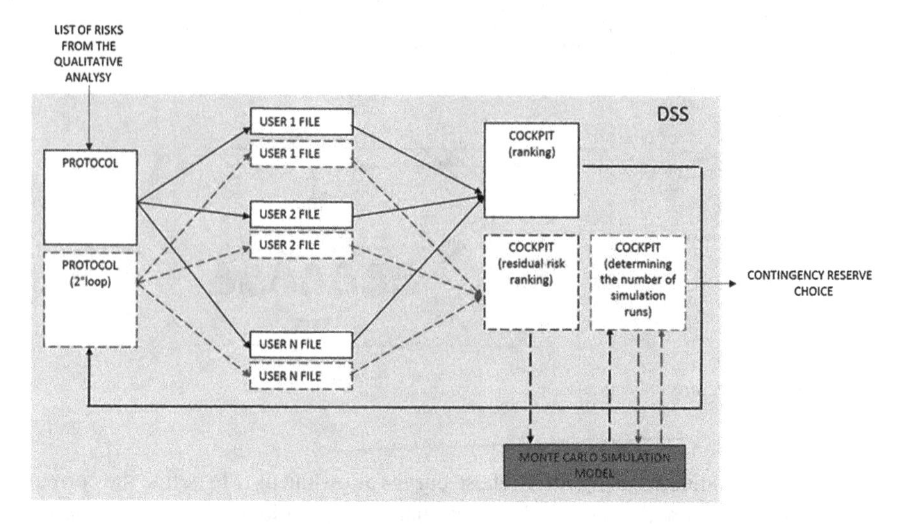

Fig. 3 DSS block scheme

WP and the company department involved (purchasing office, contract engineering, planning office).

The list of risks is copied into N user files, to be sent to the various assessors (department managers), who must carry out a quantitative analysis on the relevant risks, indicating ROP or EI for each risk.

The user files completed by the various department managers are then consolidated and sent to the COCKPIT, which is the display interface for the DM. In particular, the DM will view a single file containing the list of risks organized according to the EMV value assigned by users. It is important to emphasize that, in this phase; the assessors have not yet proposed actions to mitigate risk. In fact, prior to this phase, the DM must identify the most significant risks of the project for which mitigating actions are to be proposed.

The DM therefore selects from the file the risks for which mitigation actions are to be proposed. At this point, the intervention of the assessors is again required (second loop), to propose possible corrective actions for the relevant risks.

In particular, in this new phase of assessment, the assessors will indicate the costs of possible proposed corrective actions and the remaining ROP and PI values for each risk. This time, however, in order to quantify these elements, the assessor must indicate a minimum, maximum, and most probable value, and the most appropriate distribution of probability to be used among those chosen, to be selected from a pre-set pull-down menu.

When the assessors complete the user files, they are consolidated again and sent to the COCKPIT.

The DM will choose the actions to be implemented from among those proposed and provide for the verification/modification of the user assessments.

Table 6 Analyzed risks list

Risk	Occurrence probability (%)	Most likely impact value
Technological maturity	20	€20,872.07
Low quality supply	20	€251,841.41
Reworks	5	€174,311.15
Plant software bugs	10	€176,216.38
Wrong structural calculation	10	€37,733.33
Wrong geotechnical calculation	10	€37,733.33
Not consolidated experience of the subcontractor	10	€226,400.00
Not consolidated experience of the subcontractor	10	€226,400.00
Failure to define supply boundaries among subcontractors	10	€75,466.67
Passing the maximum number of acceptable breakdowns	5	€550,996.39
Possible breakdowns	20	€402,488.89

The file, updated with the DM's modifications, will provide data for the Monte Carlo simulation model, from which emerges the first output to the DM—the development curve of experimental error as a function of the replicated launches. The DM selects the number of simulation runs to be made based on the desired confidence level for the output data, and launches the model again in order to generate the final PDF that forms the basis of the determination, as described in section "Conceptual Description", of the contingency reserve.

4.3 *Application of the DSS to a Railway Construction Project*

The Authors illustrate, in this section, the results obtained from the application of the described DSS to a construction project in the railway sector. Once the critical risks are selected, as illustrated in Sect. 3 (risk factor > 3), each risk (11 risks in total, see Table 6) is set as input of the DSS first phase.

Then phase 1, 2 and 3 are implemented as described in section "Conceptual Description". As far as phase 4 is concerned the project engineers have opted for the use of triangular distributions (Fig. 4).

Then phase 5 and 6 have been implemented. In particular, in this test case, an experimental campaign of 5 simulations, each representing a sample of 20,000 elements, is set. The trend curves of the Mean Square Pure Error of the Mean and the Mean Square Pure Error of the Standard Deviation, shown in Fig. 5, put in evidence

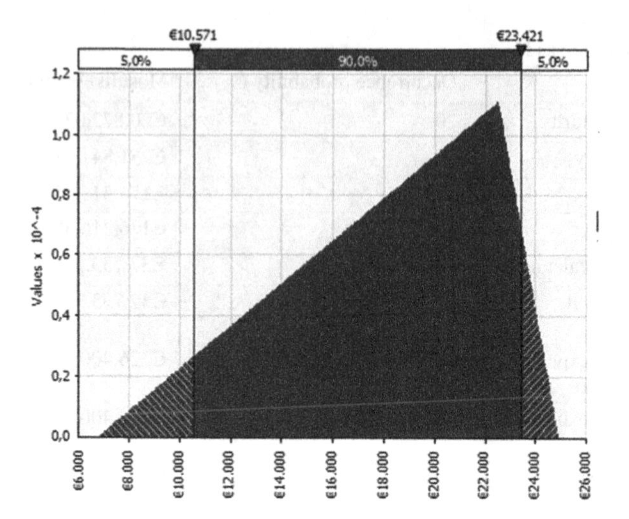

Fig. 4 Triangular distribution used for the "technological maturity" risk

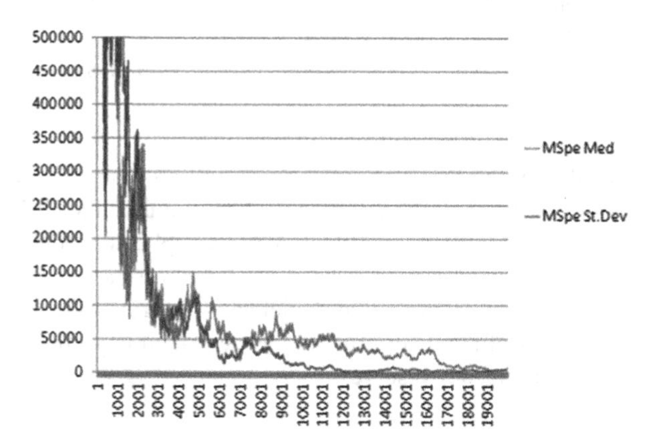

Fig. 5 Trend of the quantities MSPEMED and MSPESTDEV depending on the number of runs

that both curves stabilized at around 20,000 runs. Therefore, for this number of runs the mean value of the output will be very stable and the contribution of these two variables on the output confidence interval will be almost negligible.

The value of 20,000 runs has been set by the user (phase 6) and the DSS gives in output the curve of the contingency probability distribution and its integral curve, shown in Fig. 6, are obtained.

By analyzing the contingency reserve probability curve (Fig. 6), it can be noticed how the distribution mean value is 565,403 €, with a $(1 - \alpha)$ % confidence level $(\alpha = 0.05)$ equal to ±664 €.

The mean value is the value, which has exactly the same probability (50%) to be over or under it.

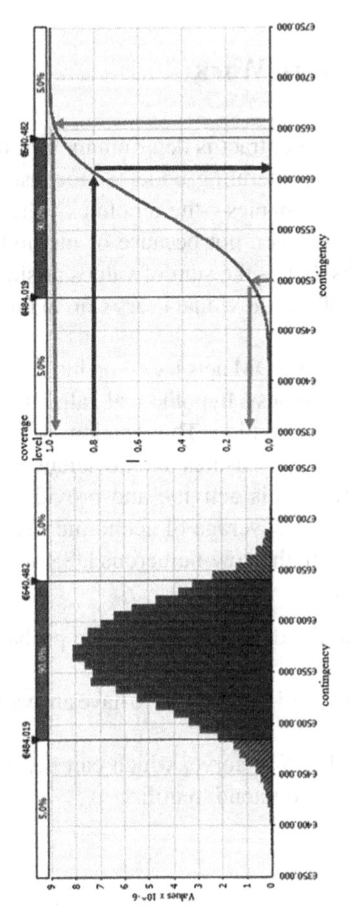

Fig. 6 Probability distribution and cumulative probability curve of contingency with triangular distributions

Focusing on the cumulative curve (Fig. 6), it can be noted that a contingency reserve equal to 653,665.65 € corresponds to a 98% Coverage Level, which is equivalent to a residual risk of incurring extra costs of 2%. With a contingency reserve of 500,000 €, the DSS gives a Coverage Level 7%.

On the contrary, if the Coverage Level is fixed, the contingency reserve to be allocated can be obtained. For example, if the DM needs a Coverage Level of 80%, it is necessary to allocate a contingency reserve equal to 610,000 €.

5 Conclusion and Future Work

The problem of the manager's contract is determination of the contingency reserve that will be accentuated when operating under a stochastic regime. For this reason, often—even in major companies—the amount of the contingency reserve is determined by the project manager, not because of methodology based on appropriate scientific assumptions, but as the sum of values arising from the experiences, impressions, and intuitions of the individual users who manage the various activities that make up the project.

At the time of the decision, the DM acts based on his or her experiences, etc., and so the contingency reserve becomes a hypothetical value that will be in line with the reality of the project under examination. The creation of an appropriately designed DSS, as the one proposed, instead ensures a solid scientific basis for the decision-making process with regard to this activity, and provides the DM an immediate assessment of the percentage of coverage of economic risk incurred with a related value of contingency reserve. In the now-numerous DSS applications that have been implemented, we have found:

1. Satisfactory precision in the determination of the probability curve of the contingency reserve;
2. Increased appreciation by tender managers to have an available scientific methodology base on their decision;
3. The extreme clarity of the PDF curve, which can be, deduced the cost/benefit ratio for the DM strategies to intend to utilize.

References

1. I. Ahmad, Contingency allocation: a computer-aided approach. AACE Trans., 28 June–1 July 1992, Orlando, F.4, pp. 1–7
2. P.A. Thompson, J.G. Perry, *Engineering Construction Risks* (Thomas Telford, London, 1992)
3. F.T. Hartman, *Don't Park Your Brain Outside* (PMI, Upper Darby, PA, 2000)
4. D. Baccarini, Estimating project cost contingency—beyond the 10% syndrome, in *Conference Paper Open Access* (2005)
5. Y.-C. Kuo, S.-T. Lu, Using fuzzy multiple criteria decision making approach to enhance risk assessment for metropolitan construction projects. Int. J. Proj. Manag. **31**, 612–314 (2013)

6. R. Yim, J. Castaneda, T. Doolen, I. Toomer, R. Malak, A study of the impact of project classification on project risk indicators. Int. J. Proj. Manag. **33**(4), 863–876 (2015)
7. B. Hartono, S.R. Sulistyo, P.P. Praftiwi, D. Hasmoro, Project risk: theoretical concepts and stakeholders' perspective. Int. J. Proj. Manag. **32**(3), 400–411 (2014)
8. R.B. Lorance, R.V. Wendling, Basic techniques for analyzing and presenting cost risk analysis. Cost Eng. **43**(6), 25–31 (2001)
9. E.W. Merrow, B.R. Schroeder, Understanding the costs and schedule of hydroelectric projects. AACE Trans. **1**(3), 1–7 (1991)
10. D. Chen, F.T. Hartman, A neural network approach to risk assessment and contingency allocation, in *AACE Transactions*, 24–27 June 2000, Risk, pRIS07
11. J.S. Chou, I.T. Yang, W.K. Chong, Automation in construction, in *Probabilistic Simulation for Developing Likelihood Distribution of Engineering*, vol. 18 (2009), pp. 570–577
12. D. Howell, C. Windahl, R. Seidel, A project contingency framework based on uncertainty and its consequences. Int. J. Proj. Manag. **28**, 256–264 (2010)
13. A. Idrus, M.F. Nuruddin, M.A. Rohman, Development of project cost contingency estimation model using risk analysis and fuzzy expert system. Expert Syst. Appl. **38**, 1501–1508 (2011)
14. J.S. Chou, Cost simulation in an item-based project involving construction engineering and management. Int. J. Proj. Manag. **29**, 706–717 (2011)
15. F. Allahi, L. Cassettari, M. Mosca, A stochastic risk analysis through Monte Carlo simulation to the construction phase of a 600 MW gas turbine plant, in *Proceedings of the MAS 2017*, 18–20 Sept, Barcelona, Spain
16. F. Allahi, L. Cassettari, M. Mosca, Stochastic risk analysis and cost contingency allocation approach for construction projects applying Monte Carlo simulation, in *Proceedings of the World Congress on Engineering*. Lecture Notes in Engineering and Computer Science, 5–7 July 2017, London, U.K., pp. 385–391
17. L. Cassettari, P.G. Giribone, M. Mosca, R. Mosca, The stochastic analysis of investments in industrial plants by simulation models with control of experimental error: theory and application to a real business case. Appl. Math. Sci. **4**(76), 3823–3840 (2010)
18. L. Cassettari, R. Mosca, R. Revetria, Monte Carlo simulation models evolving in replicated runs: a methodology to choose the optimal experimental sample size. Math. Probl. Eng. **2012**, 0–17 (2012)
19. R. Mosca, P. Giribone, R. Revetria, L. Cassettari, S. Cipollina, Theoretical development and applications of the MSPE methodology in discrete and stochastic simulation models evolving in replicated runs. J. Eng. Comput. Archit. **2**(1), 1934–7197 (2008)

A Simulation Study on Indoor Location Estimation Based on the Extreme Radial Weibull Distribution

Kosuke Okusa and Toshinari Kamakura

Abstract In this study, we investigate the possibility of analyzing indoor location estimation under the NLoS environment by radial extreme value distribution model based on the simulation. We assume that the observed distance between the transmitter and receiver is a statistical radial extreme value distribution. The proposed method is based on the marginal likelihoods of radial extreme value distribution generated by positive distribution among several transmitter radio sites placed in a room. Okusa and Kamakura (Lecture notes in engineering and computer science: Proceedings of the world congress on engineering 2017) [16] not discussed the more detail performance of radial distribution based approach. To cope with this, to demonstrate the effectiveness of the proposed method, we carried out a simulation experiments. Results indicate that high accuracy was achieved when the method was implemented for indoor spatial location estimation.

Keywords Extreme value distribution · Indoor location estimation · Positive distribution · Radial distribution · Statistical approach · Weibull distribution

1 Introduction

In this study, we investigated the possibility of analyzing indoor spatial location estimation under the NLoS environment by radial extreme value distribution model. In recent times, the global positioning systems (GPS) are used daily to obtain

---

K. Okusa (✉)
Kyushu University, 4-9-1 Shiobaru, Minami-ku, Fukuoka 815-8540, Japan
e-mail: okusa@design.kyushu-u.ac.jp

T. Kamakura
Chuo University, 1-13-27 Kasuga, Bunkyo-ku,
Tokyo 112-8551, Japan
e-mail: kamakura@indsys.chuo-u.ac.jp

S.-I. Ao et al. (eds.), *Transactions on Engineering Technologies*,
https://doi.org/10.1007/978-981-13-0746-1_5

locations for car navigation. These systems are very convenient, but sometimes we also require location estimation in indoor environments, for instance, to obtain nursing care information in hospitals. Indoor location estimation based on the GPS is very difficult because it is difficult to receive GPS signals.

A study on indoor spatial location estimation is very important in the fields of marketing science and design for public space. For instance, indoor spatial location estimation is an important tool for space planning based on the evacuation model and shop layout planning [4, 5, 9–11]. Recently, indoor spatial location estimation is mostly based on the received signal strength (RSS) method [6, 19, 24], angle of arrival (AoA) method [13, 21], and time of arrival (ToA) method [3, 20, 23].

The RSS is a cost-effective method that uses general radio signals (e.g., Wi-Fi networks). However, the signal strength is affected by signal reflections and attenuation, and hence, it is not robust. Therefore, location estimation accuracy using the RSS method is very low. The AoA is a highly accurate method that uses signal arrival directions and estimated distances. However, this method is very expensive because array signal receivers are required. The ToA method only makes use of the distance between the transmitter and the receiver. The accuracy of this method is higher than that of the RSS method and its cost is also lower than that of the AoA method. For this reason, it has been suggested that the ToA method is the most suitable method for practical indoor location estimation system [18].

In this study, we made use of the ToA data-based measurement system. The location estimation algorithm implemented in previous studies were mostly based on the least-squares method. However, using the least-squares method to process the outlier value is difficult, and such data is frequently encountered in the ToA method.

To address this problem, Okusa and Kamakura [16] propose a method based on the marginal likelihoods of radial extreme value distribution generated by positive distribution among several transmitter radio sites placed within a room. A comparison of the proposed statistical method with other previously implemented methods was carried out to demonstrate its potential for practical use.

However, Okusa and Kamakura [16] not discussed performance of radial distribution based location estimation approach due to calculation performance problem. In this paper, we discuss more detail performance of radial distribution based approach using simulation study to confirm the performance of the proposed method.

The rest of this paper is organized as follows. In Sect. 2, the features and problems of ToA signals are discussed. In Sect. 3, we will present models for indoor location estimation under the NLoS environment based on the radial extreme value distribution. In Sect. 4, we will present some performance results from a simulation study to demonstrate the effectiveness of our model. We will conclude with a summary in Sect. 5.

2 Time of Arrival (ToA) data

ToA is one of the methods used to estimate the distance between the transmitter and receiver. This method is computed from the travel time of radio signals between the transmitter and receiver. When the transmitter' time and the receiver's time have been completely synchronized, the distance d between the transmitter and receiver is calculated as follows:

$$d = C(r_r - r_t), \tag{1}$$

where r_t and r_r are transmitted and received time, respectively. C is the speed of light. In an ideal circumstance, d provides accurate distance between the transmitter and receiver called the Line-of-Sight. In this case, the location of the subject is easily estimated by trilateration (Fig. 1).

However, in many cases, the distance d includes error components called Non Line-of-Sight (NLoS) [1, 22] (Fig. 2).

NLoS conditions are mainly due to obstacles between the transmitter and receiver, i.e., signal reflections. In this case, the observed distance d will be longer than the true distance. Fujita et al. [2] reported that the observed distance of LoS and NLoS are defined as follows:

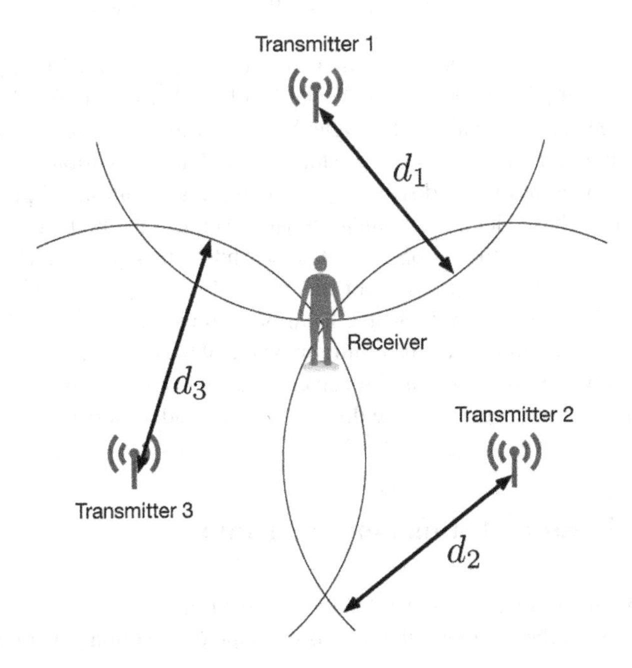

Fig. 1 Location estimation by trilateration (ideal case)

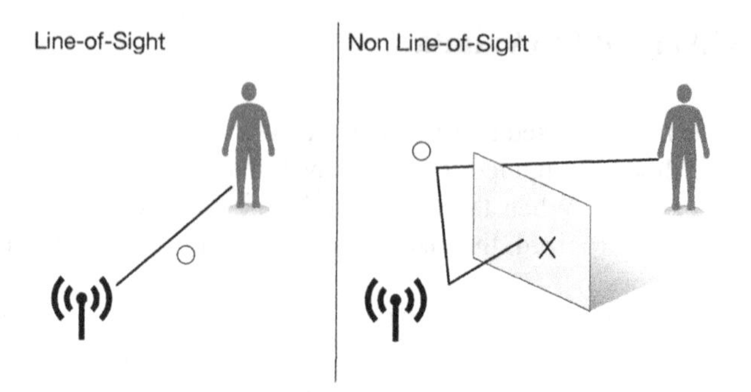

Fig. 2 LoS, NLoS illustration

$$d_{k,LoS} = \sqrt{(x - c_{k1})^2 + (y - c_{k2})^2} + e_k \tag{2}$$

$$d_{k,NLoS} = \sqrt{(x - c_{k1})^2 + (y - c_{k2})^2} + e_k + b_k \tag{3}$$

In the LoS case, the observed value is distributed from the true distance with error term $e_k \sim N(0, \sigma_k^2)$, where $N(\cdot)$ is the normal distribution. However, in the NLoS case, the observed value contains an additional bias term $b_k \sim U(0, B_{\max})$, where $U(\cdot)$ is the uniform distribution. $B_{\max}$ is the possible maximum bias value of the observed value.

To address this, some studies implemented the model-based MLE approach [8, 12, 17]; however, these methods were modelized by 1-D distribution. ToA signals indicate only the distance, and not the angle. We can assume that the observed signal is a 2-D distribution and we propose the statistical radial distribution.

From the viewpoint of 2-D distribution, Kamakura and Okusa [7] proposed the radial distribution-based location estimation. However, this method did not consider the NLoS situation. For the NLoS case, Okusa and Kamakura [14, 15] proposed the NLoS bias correction approach, but this method requires iteration calculation. Therefore, it is not suitable for real-time location estimation.

In this study, we assumed that the minimum value of the observed signal is the true distance between the transmitter and receiver (Fig. 4). We modeled the distribution of the minimum value (extreme value distribution) for indoor location estimation.

3 Indoor Location Estimation Algorithm

In this section, the indoor location estimation algorithm is presented. The proposed method is based on the marginal likelihoods of radial distribution generated by positive distribution.

Considering that the obtained distances were all positive, we propose the following circular distribution based on the 3-parameter Weibull distribution:

$$f(r,\theta)=\frac{1}{2\pi}\Big(\frac{m}{\eta}\Big)\Big(\frac{r-g}{\eta}\Big)^{m-1}\exp\Big\{-\Big(\frac{r-g}{\eta}\Big)^{m}\Big\}$$
$$(r,g,\eta,m>0,\ 0\leq\theta<2\pi). \tag{4}$$

Here, g is the location parameter, η and m are the shape and scale parameters of the Weibull distribution, respectively.

From Eq. 4, we can convert from Polar to Cartesian coordinates:

$$g(x,y)=\frac{\lambda m}{2\pi}(\sqrt{x^2+y^2}-g)^{m-2}\exp\Big\{-\lambda(\sqrt{x^2+y^2}-g)^{m}\Big\}$$
$$(x,y,g,\lambda,m>0). \tag{5}$$

Here, $\lambda=1/\eta^m$ (Fig. 3).

Assuming that each transmitter station observes independent measurements, the likelihood based on the data set is calculated as follows:

$$L(\lambda_1,m_1,g_1,\ldots,\lambda_K,m_K,g_K)=$$
$$\prod_{i=1}^{K}\prod_{j=1}^{n_i}\frac{\lambda_i m_i}{2\pi}(\sqrt{(x_{ij}-c_{i1})^2+(y_{ij}-c_{i2})^2}-g_i)^{m-2}$$
$$\exp\Big[-\lambda_i\{\sqrt{(x_{ij}-c_{i1})^2+(y_{ij}-c_{i2})^2}-g_i\}^m\Big]. \tag{6}$$

Here, K is the number of stations and for each station i, the sample size is n_i. The observed data set for station i is (x_{ij},y_{ij}). The coordinates (c_{i1},c_{i2}) are given transmitter station positions (Fig. 1).

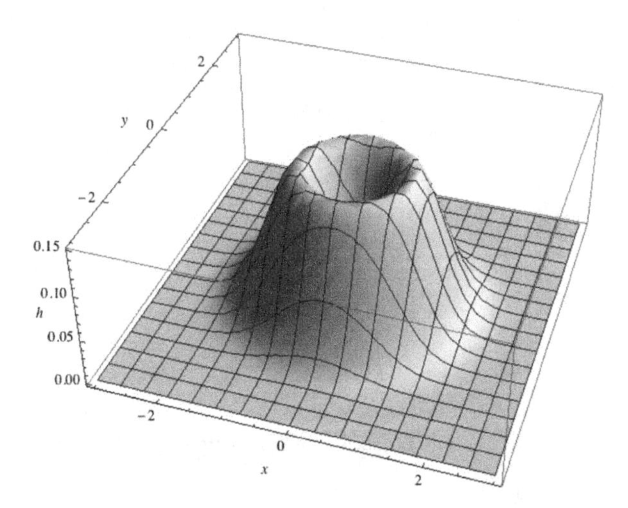

Fig. 3 Radial Weibull distribution ($m=3,\lambda=0.5,g=0$)

3.1 Radial Extreme Value Distribution

In the NLoS case, we can assume that the minimum value of the observed signals is the true distance between the transmitter and receiver (Fig. 4). It is reasonable to assume that extreme value distribution is suitable for the NLoS data case. The extreme value distribution of the 3-parameter Weibull distribution is formulated as follows:

$$\begin{aligned} F_z(z) &= 1 - \{1 - F_x(z)\}^n \\ &= 1 - \left[\exp\left\{-\left(\frac{z-g}{\eta}\right)^m\right\}\right]^n \\ &= 1 - \exp\left\{-\left(\frac{z-g}{\eta/n^{\frac{1}{m}}}\right)^m\right\} \end{aligned} \tag{7}$$

where $z = \min(X_1, X_2, \ldots, X_n)$ and g is the location parameter. The shape and scale parameters of the 3-parameter extreme value Weibull distribution are m and $\eta n^{-\frac{1}{m}}$, respectively. The shape and scale parameters are estimated from the radial Weibull distribution parameters (Eq. 5). Considering the circular extreme value distribution, we can rewrite Eq. 7 as follows:

$$\begin{aligned} F_z(r,\theta) &= \frac{1}{2\pi}\left[1 - \exp\left\{-\left(\frac{r-g}{\eta/n^{\frac{1}{m}}}\right)^m\right\}\right] \\ &\quad (r, g, \eta, m > 0,\ 0 \le \theta < 2\pi). \end{aligned} \tag{8}$$

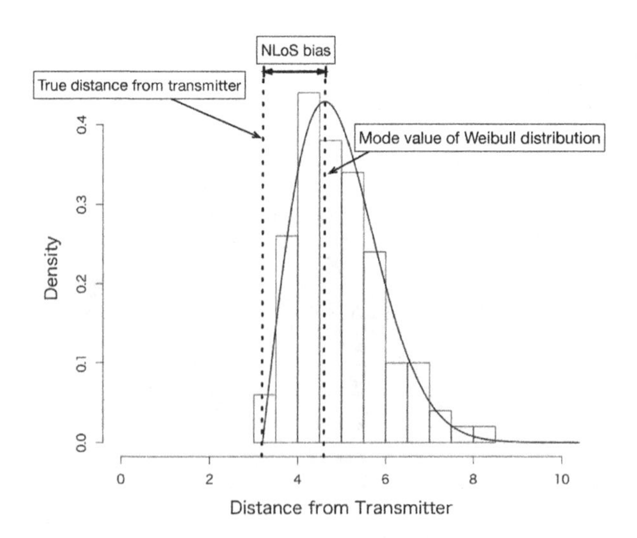

Fig. 4 ToA data NLoS bias

From Eq. 8, we can convert from Polar to Cartesian coordinates, same as in Eq. 5, as follows:

$$h(x, y) = \frac{1}{2\pi\sqrt{x^2+y^2}}\Big[1 - \exp\Big\{-\Big(\frac{\sqrt{x^2+y^2}-g}{\eta/n^{\frac{1}{m}}}\Big)^m\Big\}\Big]$$
$$(x, y, g, \eta, m > 0). \tag{9}$$

Using a similar procedure as in Eq. 6, and assuming that each transmitter station observes independent measurements, the likelihood based on the data set is calculated as follows:

$$L(\eta_1, m_1, g_1, \ldots, \eta_K, m_K, g_K) =$$
$$\prod_{i=1}^{K}\prod_{j=1}^{n_i}\frac{1}{2\pi\sqrt{(x_{ij}-c_{i1})^2+(y_{ij}-c_{i2})^2}}$$
$$\Big[1 - \exp\Big\{-\Big(\frac{\sqrt{(x_{ij}-c_{i1})^2+(y_{ij}-c_{i2})^2}-g_i}{\eta_i/n^{\frac{1}{m_i}}}\Big)^{m_i}\Big\}\Big] \tag{10}$$

It is reasonable to assume that the highest probability location $(\hat{x}, \hat{y})$ is the estimated location of the subject (Fig. 5). The highest probability location from the likelihood function is calculated as follows:

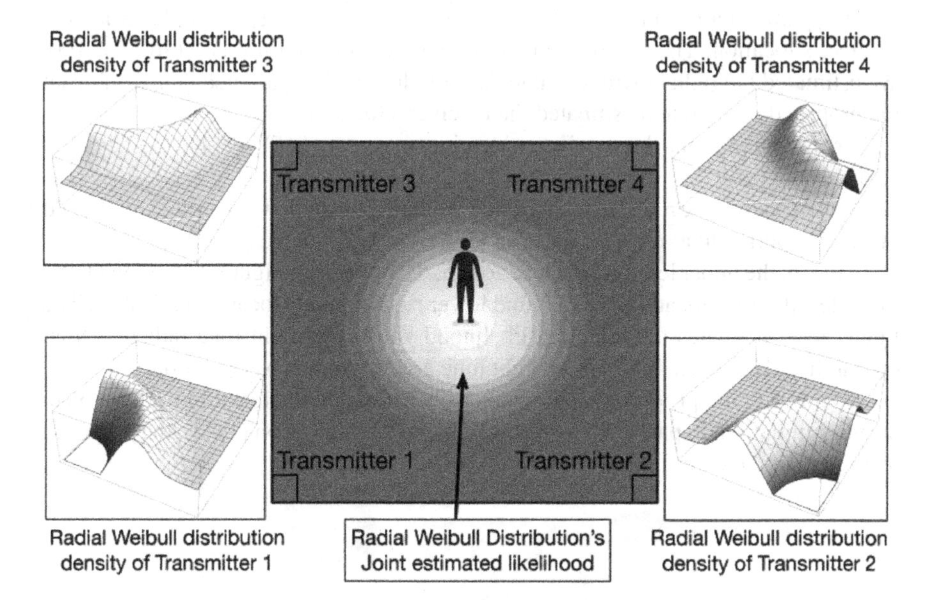

Fig. 5 Location estimation based on radial Weibull distribution

$$
\begin{aligned}
&(\hat{x}, \hat{y}) = \arg\max G(\hat{\eta}_1, \hat{m}_1, \hat{g}_1, \ldots, \hat{\eta}_K, \hat{m}_K, \hat{g}_K) \\
&G(x, y; \eta_1, m_1, g_1, \ldots, \eta_K, m_K, g_K) = \\
&\prod_{i=1}^{K} \frac{1}{2\pi\sqrt{(x-c_{i1})^2+(y-c_{i2})^2}} \\
&\left[1 - \exp\left\{ - \left(\frac{\sqrt{(x-c_{i1})^2+(y-c_{i2})^2} - g_i}{\eta_i / n^{\frac{1}{m_i}}} \right)^{m_i} \right\}\right]
\end{aligned}
\tag{11}
$$

In the next section, a comparison of simulation and experimental results of the proposed method with other methods is presented.

4 Simulation

To demonstrate the effectiveness of the proposed method, we carried out a simulation study to confirm the performance of the proposed method.

In the simulation, we considered the 10.0 m × 10.0 m space chamber, four transmitters at $c_1 = (0, 0)$, $c_2 = (10, 0)$, $c_3 = (0, 10)$, and $c_4 = (10, 10)$. We generated the observation signal from the 3-parameter Weibull distribution random number $\mathbf{X}_i \sim Weibull(m, \lambda, g)$, where $\mathbf{X}_i$ are the random numbers at station i, m is the shape parameter, λ is the scale parameter calculated from $\lambda = g/\Gamma(1 + \frac{1}{m})$, g is the location parameter calculated from $g = \sqrt{(x - c_{i,1})^2 + (y - c_{i,2})^2}$, and x, y are the receiver's location. The number of random number generation was set to $N = 100$. We estimated the radial extreme value Weibull distribution parameters $\hat{\eta}, \hat{m}, \hat{g}$ from the proposed method and estimated the receiver's location.

For the simulation, the receiver's locations were at (5, 5), (1, 1), (5, 2.5), and (2.5, 2.5).

Figure 6 shows the radial extreme value Weibull distribution's marginal likelihood in static experiment at receiver location (5, 5), (1, 1), (5, 2.5), (2.5, 2.5).

In Fig. 6, the upper left figure is location (5, 5), upper right figure is location (1, 1), lower left figure is location (5, 2.5), and lower right figure is location (2.5, 2.5). The figure color indicates the marginal likelihood value, and light tone indicates high probability. The black circle in the light tone area is the receiver location.

From Fig. 6, the proposed method precisely estimated the location of the subject during simulation. In particular, in the nearside of wall (location (1, 1)), the proposed method maintained accuracy.

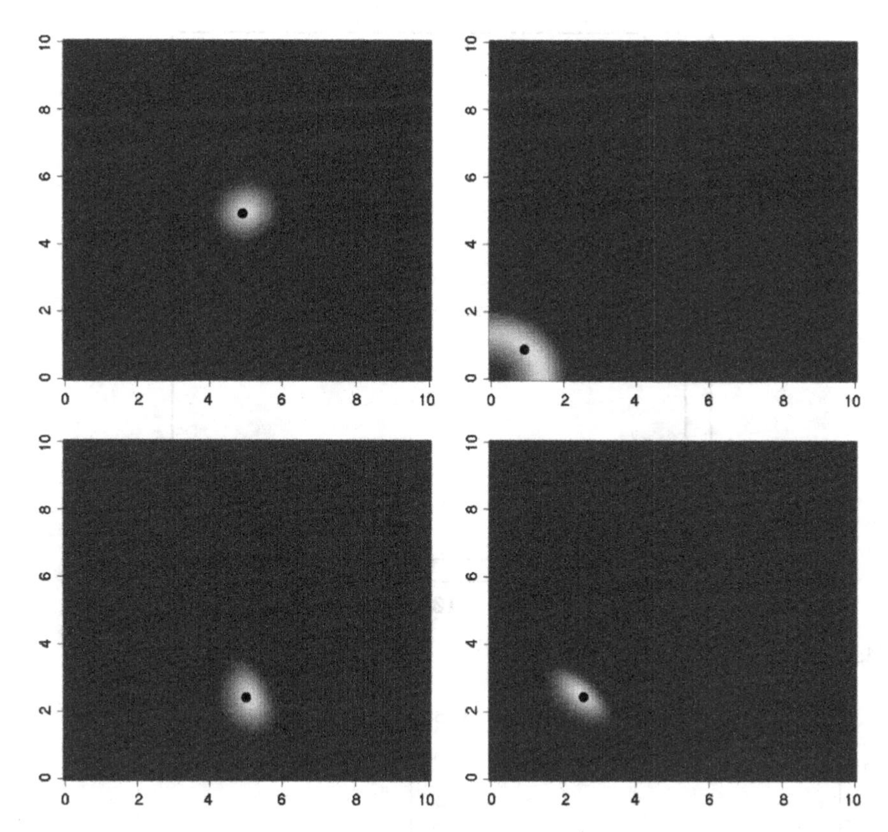

Fig. 6 Radial extreme value Weibull distribution's marginal likelihood in static experiment at receiver location (5, 5) (upper left), (1, 1) (upper right), (5, 2.5) (lower left), and (2.5, 2.5) (lower right)

Next, we conduct the NLoS case simulation. In NLoS case simulation, we consider the obstacle exists in front of transmitter (Fig. 7). We compare the accuracy of LoS and NLoS using proposed method and least squares method.

Figure 8 is error density boxplot of LoS, NLoS case simulation (left side: LoS case, right side: NLoS case). In LoS case, location estimation performance is not different between two methods. On the other hand, in NLoS case, proposed method shows better performance than least squares method. We consider that the proposed method is robust for NLoS case situation.

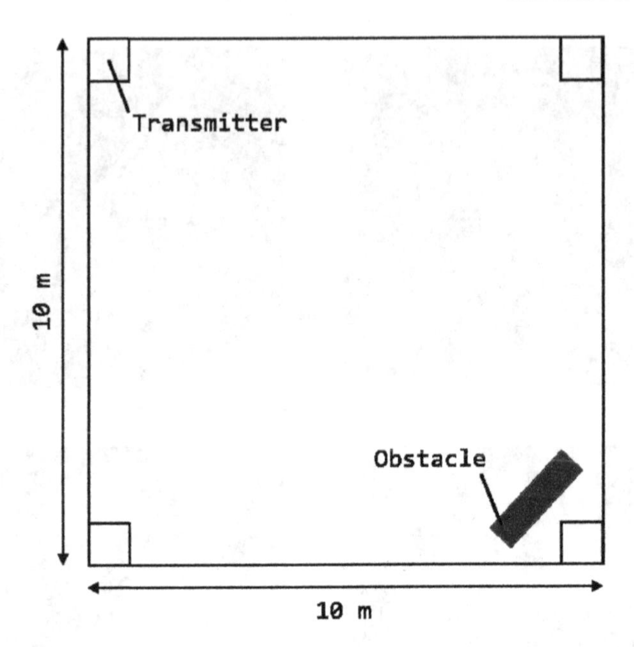

Fig. 7 NLoS case simulation

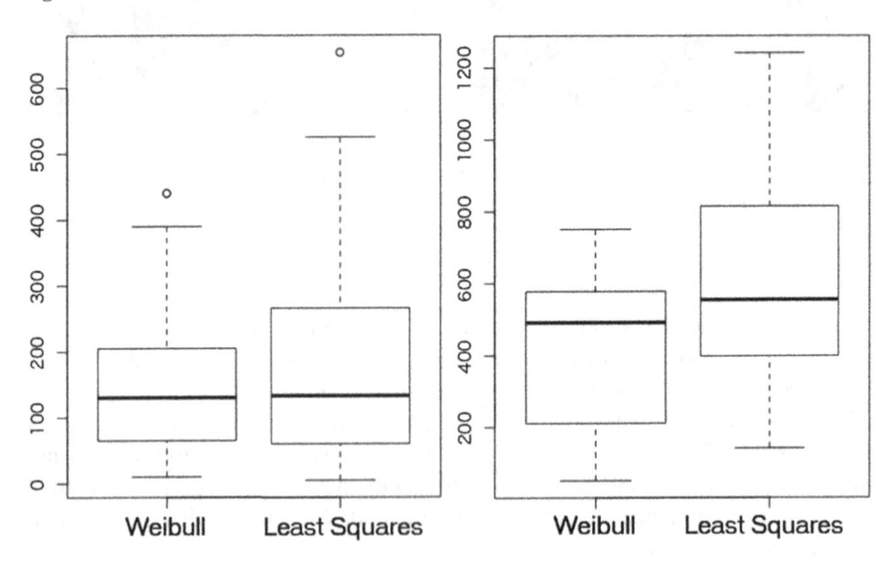

Fig. 8 Error density boxplot of LoS, NLoS case simulation (left side: LoS case, right side: NLoS case)

5 Conclusions

In this article, we proposed an indoor location estimation model based on the radial Weibull distribution. The experimental results suggest that our model can accurately estimate the subject's spatial locations.

In next phase, we will compare our methods with other state-of-the-are indoor location estimation methods, and evaluate its performance. The computational cost of our method is lower than that of Okusa and Kamakura's method [14]; however, its applicability for practical use is limited. Therefore, to address this limitation, we intend to review the location estimation process, especially $(\hat{x}, \hat{y})$ estimation from marginal likelihood of radial extreme value obtained from the Weibull distribution. In addition, we intend to implement the indoor location estimation system based on the proposed method and demonstrate its applicability.

References

1. Y.T. Chan, W.Y. Tsui, H.C. So, P.C. Ching, Time-of-arrival based localization under NLOS conditions. IEEE Trans. Veh. Technol. **55**(1), 17–24 (2006)
2. T. Fujita, T. Ohtsuki, T. Kaneko, Low complexity TOA localization algorithm for NLOS environments. Res. Paper IPS **15**, 69–74 (2007)
3. J. He, Y. Geng, K. Pahlavan, Modeling indoor TOA ranging error for body mounted sensors, in *IEEE International Symposium on Personal Indoor and Mobile Radio Communications (PIMRC)* (2012), pp. 682–686
4. J.C. Herrera, A. Hinkenjann, P.G. Ploger, J. Maiero, Robust indoor localization using optimal fusion filter for sensors and map layout information, in *Proceedings of IEEE Conference on Indoor Positioning and Indoor Navigation* (2013), pp. 1–8
5. T. Ikeda, M. Kawamoto, A. Sashima, J. Tsuji, H. Kawamura, K. Suzuki, K. Kurumatani, An indoor autonomous positioning and navigation service to support activities in large-scale commercial facilities, in *Serviceology for Services* (2014), pp. 191–201
6. K. Kaemarungsi, P. Krishnamurthy, Properties of indoor received signal strength for WLAN location fingerprinting, in *IEEE Conference In Mobile and Ubiquitous Systems: Networking and Services* (2004), pp. 14–23
7. T. Kamakura, K. Okusa, Estimates for the spatial locations of the indoor objects by radial distributions, in *Proceedings of ISI World Statistics Congress 2013 (ISI2013), CPS109* (2013), pp. 3417–3422
8. C.H. Knapp, G.C. Carter, The generalized correlation method for estimation of time delay. IEEE Trans. Acoust. Speech Signal Process. **24**(4), 320–327 (1976)
9. N. Li, B. Becerik-Gerber, Performance-based evaluation of RFID-based indoor location sensing solutions for the built environment. Adv. Eng. Inform. **25**(3), 535–546 (2011)
10. H. Liu, H. Darabi, P. Banerjee, J. Liu, Survey of wireless indoor positioning techniques and systems. IEEE Trans. Syst. Man Cybern. Part C: Appl. Rev. **37**(6), 1067–1080 (2007)
11. W. Martin, *Indoor Location-Based Services—Prerequisites and Foundations* (Springer International Publishing, 2014). ISBN: 978-3-319-10698-4
12. R.L. Moses, D. Krishnamurthy, R.M. Patterson, A self-localization method for wireless sensor networks. EURASIP J. Appl. Signal Process. 348–358 (2003)
13. D. Niculescu, B. Nath, Ad hoc positioning system (APS) using AOA. INFOCOM **3**, 1734–1743 (2003)

14. K. Okusa, T. Kamakura, Indoor location estimation based on the statistical spatial modeling and radial distributions, in *Lecture Notes in Engineering and Computer Science: Proceedings of The World Congress on Engineering and Computer Science 2015*, 21–23 Oct 2015, San Francisco, USA, pp. 835–840
15. K. Okusa, T. Kamakura, Indoor location estimation based on the RSS method using radial log-normal distribution, in *Proceedings of 16th IEEE International Symposium on Computational Intelligence and Informatics (CINTI2015)* (2015), pp. 29–34
16. K. Okusa, T. Kamakura, Statistical indoor location estimation for the NLoS environment using radial extreme value Weibull distribution, in *Lecture Notes in Engineering and Computer Science: Proceedings of The World Congress on Engineering 2017*, 5–7 July 2017, London, U.K., pp. 555–560
17. N. Patwari, A.O. Hero III, M. Perkins, N.S. Correal, R.J. O'dea, Relative location estimation in wireless sensor networks. IEEE Trans. Signal Process. **51**(8), 2137–2148 (2003)
18. N. Patwari, J.N. Ash, S. Kyperountas, A.O. Hero III, R.L. Moses, N.S. Correal, Locating the nodes: cooperative localization in wireless sensor networks. Signal Process. Mag. **22**(4), 54–69 (2005)
19. T. Roos, P. Myllymaki, H. Tirri, P. Misikangas, J. Sievänen, A probabilistic approach to WLAN user location estimation. Int. J. Wirel. Inf. Netw. **9**(3), 155–164 (2002)
20. J. Shen, A.F. Molisch, J. Salmi, Accurate passive location estimation using TOA measurements. IEEE Trans. Wirel. Commun. **11**(6), 2182–2192 (2012)
21. L. Taponecco, A.A. D'Amico, U. Mengali, Joint TOA and AOA estimation for UWB localization applications. IEEE Trans. Wirel. Commun. **10**(7), 2207–2217 (2011)
22. S. Venkatraman, J. Caffery Jr., H.R. You, A novel TOA location algorithm using LOS range estimation for NLOS environments. IEEE Trans. Veh. Technol. **53**(5), 1515–1524 (2004)
23. T. Watabe, T. Kamakura, Localization algorithm in the indoor environment based on the ToA data, in *Australian Statistical Conference 2010 Program and Abstracts* (2010), p. 243
24. I. Yamada, T. Ohtsuki, T. THisanaga, L. Zheng, An indoor position estimation method by maximum likelihood algorithm using received signal strength. SICE J. Control Meas. Syst. Integr. **1**(3), 251–256 (2008)

Infant Mortality and Income per Capita of World Countries for 1998–2016: Analysis of Data and Modeling by Increasing Returns

I. C. Demetriou and P. C. Tzitziris

Abstract Annual cross-country data from the World Bank database during the years 1998 and 2016 demonstrate wide variation in infant mortality rates (IMR) and gross domestic product per capita (GDPpc). The row data, a descriptive analysis of the data and a K-means clustering of the countries into groups depending on the GDPpc and IMR measurements draw attention to the underlying structure of the data. We find that IMR follows a convex descent trend, where there is a range of high infant mortality rates at low GDPpc levels and there is a range of high GDPpc levels with low infant mortality rates. Thus, we assume that GDPpc is subject to increasing returns. This is equivalent to assuming that IMR comes from an unknown underlying convex relationship on the IMR observations at the GDPpc observations. Hence we consider the application of a method that makes least the sum of the squares of the errors to each dataset subject to the constraints from the assumption of non-decreasing returns. Our numerical results show that the estimated IMR values reach a point at which they obtain the lowest value and then increase again while GDPpc is at the highest levels. Thus they suggest that our assumption not only adequately describes reality, but also the ensuing model is able to capture imperceptible features of the underlying process.

Keywords Convexity · Divided difference · Gross domestic product · Increasing returns · Infant mortality rates · K-means clustering · Least squares · Quadratic programming

I. C. Demetriou (✉)
Division of Mathematics and Informatics, Department of Economics,
National and Kapodistrian University of Athens, 1 Sofokleous and Aristidou Street,
10559 Athens, Greece
e-mail: demetri@econ.uoa.gr

P. C. Tzitziris
45A Salaminos Street Nea Chalkidona, 14343 Athens, Greece
e-mail: pantzi87@gmail.com

S.-I. Ao et al. (eds.), *Transactions on Engineering Technologies*,
https://doi.org/10.1007/978-981-13-0746-1_6

1 Introduction

IMR is the number of deaths per 1000 live births of children under one year of age. GDP measures the monetary value of final goods and services produced in a country in a year [1], while the ratio of GDP to the total population of a country is the per capita GDP.

Data for these indicators are maintained in the World Bank [2]. Cross-country data demonstrate wide variation in infant mortality rates as GDPpc varies in its range. They show that infant mortality rates follow a convex descent trend, where there is a range of high infant mortality rates at low GDPpc levels and there is a range of high GDPpc levels with low infant mortality rates. A typical scatter plot of the relationship between IMR and GDPpc is given in the left part of Fig. 1.

In [3] we investigated the claim that a clear link between Infant Mortality Rate and the per capita Gross Domestic Product is provided by assuming that GDPpc is subject to increasing returns. This is equivalent to assuming that IMR comes from an unknown underlying convex relationship on the IMR observations at the GDPpc observations, but convexity has been lost due to errors of measurement [4]. However, due to space restriction it was only possible to present some results by analyzing a subcollection of data for the years 1998, 2002, 2006, 2010 and 2014. Here we extend the results of [3] and provide further evidence for our claim, by analyzing yearly cross-country data for 132 countries for the period of years 1998–2016.

In order to state the problem we address, we need some notation. The data are the plane coordinates (x_i, ϕ_i), for $i = 1, 2, \ldots, n$, where the abscissae x_i are the GDPpc observations in strictly ascending order $x_1 < x_2 < \cdots < x_n$ and ϕ_i is the measurement of some unknown relationship of IMR at x_i. We regard these measurements as components of an n-vector $\underline{\phi}$.

We seek estimates y_i of the ϕ_i that are derived by minimizing the objective function

$$\Phi(\underline{y}) = \sum_{i=1}^{n} (y_i - \phi_i)^2 \tag{1}$$

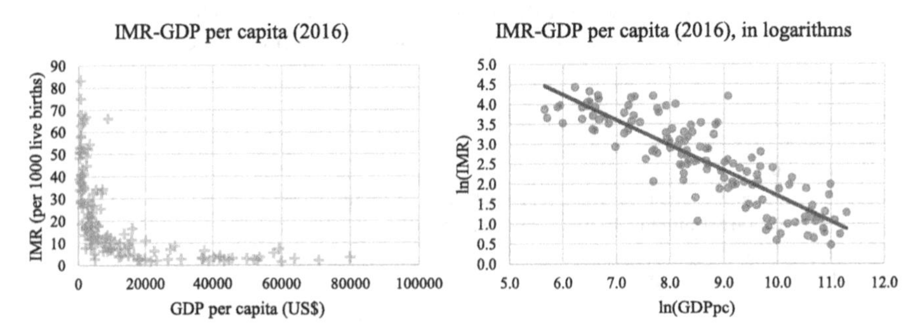

Fig. 1 Left, Infant mortality rate data against gross domestic product per capita for the year 2016. Right, Linear fit to these data in logarithms

subject to the constraints

$$y[x_{i-1}, x_i, x_{i+1}] \geq 0, \; i = 2, 3, \ldots, n-1, \tag{2}$$

where

$$y\left[x_{i-1}, x_i, x_{i+1}\right] = \frac{y_{i-1}}{(x_{i-1} - x_i)(x_{i-1} - x_{i+1})} + \frac{y_i}{(x_i - x_{i-1})(x_i - x_{i+1})} + \frac{y_{i+1}}{(x_{i+1} - x_{i-1})(x_{i+1} - x_i)} \tag{3}$$

is the ith second divided difference on the components of $\underline{y}$. We note that if the data are exact values of a function $f(x)$, which has a continuous second derivative, then $y[x_{i-1}, x_i, x_{i+1}] = \frac{1}{2} f''(\zeta_i)$ for some $\zeta_i \in [x_{i-1}, x_{i+1}]$ [5]. Inequalities (2) allow a convex shape for the smoothed values, as we explain next. Since the ith second divided difference can be expressed as the difference of two consecutive first divided differences divided by the difference between those arguments which are not in common,

$$y[x_{i-1}, x_i, x_{i+1}] = \frac{1}{(x_{i+1} - x_{i-1})} \left(y[x_i, x_{i+1}] - y[x_{i-1}, x_i] \right), \tag{4}$$

the constraints (2) imply the inequalities

$$y[x_i, x_{i+1}] \geq y[x_{i-1}, x_i], \; i = 2, 3, \ldots, n-1, \tag{5}$$

where $y[x_i, x_{i+1}] = (y_{i+1} - y_i)/(x_{i+1} - x_i)$ is the ith first divided difference. Inequalities (5) show non-decreasing rates of change of the smoothed values with respect to changes in x.

In Sect. 2 we present the data for our computation and provide some descriptive analysis. In Sect. 3 we classify the data into groups by applying K-means clustering to the GDPpc measurements for each year. In Sect. 4 we give an outline of the method of Demetriou and Powell [6] for calculating the solution of the optimization problem that has been stated, we apply this method [7] to the datasets of Sect. 2 and demonstrate the suitability of our assumption on the increasing slopes of the IMR curve for each year of the period 1998–2016. The results are typically intended to contribute to decision and policy making.

The data that have been used for this presentation, the results from clustering and the numerical solutions from the application of the ACM TOMS Algorithm 742: L2CXFT [7] to the data are far too many to be presented here. However, throughout the paper, we present summary results from our outputs.

2 Data and Descriptive Analysis

The purpose of this section is to provide some descriptive statistics for the data that we have used in our analysis, to explore the interpretation capability of log-linear transformation to these data and to apply cluster analysis in order to reveal possible structure in the data.

The data have been obtained from the World Bank database, which is freely available on the website [2]. The selection of our sample was based on the availability of IMR and GDPpc indicators for the years 1998, 1999, ..., 2015 and 2016, which resulted to 132 world countries. Therefore in this paper we are able to present results that provide sufficient details of our quantitative modeling.

Some basic descriptive statistics of our sample are displayed on Table 1. They include the mean, quartiles, median, minimum, maximum, range and standard deviation of each data set of the sample. We see in Table 1 that Mean, Median and Quartiles of IMR decreased during the period 1998–2016, while those corresponding to GDPpc increased during the years 1998, ..., 2008 and decreased during 2013, ..., 2016. Since the calculation of the mean is influenced by outliers of the sample, mean is not considered as a very reliable index as to the representation of reality, unlike the median IMR Median which also decreased while GDPpc Median increased. The same situation is observed through quartiles. Moreover, the IMR Range was narrowed almost everywhere during 1998, ..., 2016, while the GDPpc Range was expanded during 1998, ..., 2013 and subsequently was contracted. These figures lead to the general conclusion that IMR has been reduced significantly over the last two decades, but without achieving rates analogous to the growth of GDPpc. The IMR box-and-whisker plots, which are presented at Fig. 2, one for each year 1998 and 2016, quickly reveal the shape of the underlying distribution of data by the five number summary: minimum, 1st quartile, median, 3rd quartile and maximum. Approximately 25% of the 132 IMR data values fall between these five numbers. Specifically, for 1998, the IMR values fall between the numbers 3.5, 10.9, 30, 70.1 and 148.1. Analogously, for 2016 these numbers are 1.6, 6.5, 13.4, 33.7 and 83.3. Indeed, the indicator has been improved, and similarly for GDPpc.

The data are too many to be presented as raw numbers in these pages, but we may easily capture their main features by looking at the scatter plot of Fig. 1 relating to the year 2016, which is quite representative of all the datasets in our analysis. Let us note in advance that the GDPpc and the IMR values for 2016 are presented in the second and third column of Table 4, which we shall meet later on. The scatter plot shows a descending convex pattern of IMR with very high concentration of observations in the left-hand range of GDPpc and spare outliers. Further, the variation of IMR is wider at the low levels than at the middle and higher levels of GDPpc. Most interesting, we see that the rightmost IMR value, namely the one located at the highest GDPpc value (cf. Table 4: $\mathrm{IMR}_{132} = 3.60$ at $\mathrm{GDPpc}_{132} = 79890.52$) is higher than the IMR values in its vicinity. Although, it is not presented here, due to lack of space, this tends to be the rule in all the datasets except for those of the years 2000, 2005 and

Table 1 Descriptive statistics for the indicators Infant Mortality and GDP per Capita by year (Sample size: 132). GDPpc is measured on current US dollars

Year	Indicator	Mean	1st quartile	Median	3rd quartile	Minimum	Maximum	Range	Standard deviation
1998	IMR	42.4	10.9	30.0	70.1	3.5	148.1	144.6	37.2
	GDPpc	6645.8	469.3	1845.4	6630.6	124.7	41487.7	41363.0	9850.8
1999	IMR	40.7	9.9	28.5	66.5	3.2	144.8	141.6	36.4
	GDPpc	6786.4	460.1	1619.9	7037.1	119.1	40581.3	40462.2	10171.6
2000	IMR	39.4	9.5	27.3	64.1	3.0	142.4	139.4	35.4
	GDPpc	6899.0	474.0	1729.9	7071.2	123.9	38532.0	38408.2	10258.8
2001	IMR	38.0	9.1	26.3	61.5	2.8	139.8	137.0	34.2
	GDPpc	6796.9	503.3	1837.0	7066.2	120.2	38549.6	38429.4	10009.3
2002	IMR	36.9	9.3	24.8	61.4	2.6	137.7	135.1	33.2
	GDPpc	7170.9	510.3	2086.6	7290.2	111.5	43061.2	42949.6	10639.7
2003	IMR	35.1	8.0	23.7	57.5	2.5	134.1	131.6	31.7
	GDPpc	8364.9	623.0	2365.8	8870.8	112.8	50111.7	49998.8	12446.9
2004	IMR	33.7	7.9	22.3	55.0	2.3	131.1	128.8	30.4
	GDPpc	9621.5	730.0	2684.2	10386.0	127.4	57570.3	57442.8	14202.1
2005	IMR	32.3	7.8	20.9	52.2	2.3	128.0	125.7	29.3
	GDPpc	10564.5	825.4	3160.3	12336.6	150.5	66775.4	66624.9	15375.3
2006	IMR	31.5	8.1	20.1	50.2	2.3	124.5	122.2	28.8
	GDPpc	11482.1	964.8	3588.3	14265.3	154.9	74114.7	73959.8	16436.1

(continued)

Table 1 (continued)

Year	Indicator	Mean	1st quartile	Median	3rd quartile	Minimum	Maximum	Range	Standard deviation
2007	IMR	29.8	7.5	19.2	46.9	2.2	120.5	118.3	27.0
	GDPpc	13093.0	1131.5	4129.6	16404.2	170.8	85170.9	85000.1	18510.0
2008	IMR	28.6	7.6	18.5	44.5	2.1	116.2	114.1	26.1
	GDPpc	14520.5	1380.2	4686.4	20325.4	196.2	97007.9	96811.7	20065.2
2009	IMR	27.5	7.3	17.4	43.2	2.1	111.5	109.4	25.1
	GDPpc	12662.1	1231.0	4257.2	16199.0	204.9	80067.2	79862.2	17182.5
2010	IMR	27.2	7.1	16.7	42.2	1.9	109.6	107.7	25.4
	GDPpc	13372.0	1447.5	4979.2	16776.4	214.2	87646.3	87432.0	17953.5
2011	IMR	25.5	7.0	16.0	40.3	1.9	101.9	100.0	23.3
	GDPpc	15020.0	1661.3	5546.1	19537.0	260.5	100711.2	100450.7	20323.5
2012	IMR	24.6	6.9	15.3	39.1	1.8	97.3	95.5	22.5
	GDPpc	14903.8	1782.7	5804.7	19296.5	265.3	101668.2	101402.9	20052.1
2013	IMR	23.7	6.9	14.6	37.1	1.7	93.1	91.4	21.7
	GDPpc	15255.4	1893.4	6050.6	19684.8	282.8	103059.2	102776.5	20363.4
2014	IMR	23.5	6.1	14.5	35.8	1.6	98.8	97.2	22.1
	GDPpc	15206.4	1939.3	6058.6	19357.8	286.0	97429.7	97143.7	20343.8
2015	IMR	22.2	6.7	13.8	34.9	1.7	86.2	84.5	20.3
	GDPpc	13520.2	2104.3	5636.9	16759.5	300.7	82016.0	81715.3	17743.4
2016	IMR	21.5	6.5	13.4	33.7	1.6	83.3	81.7	19.7
	GDPpc	13413.6	2169.6	5110.0	16164.4	285.7	79890.5	79604.8	17613.4

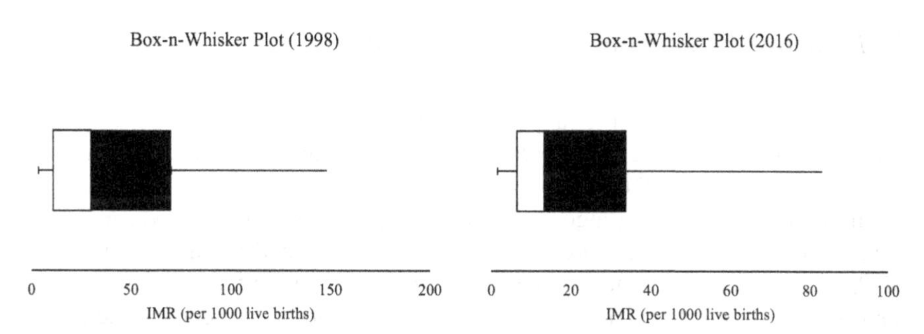

Fig. 2 IMR Box-and-whisker plots for the years 1998 and 2016

2006. Theoretician economists explained this phenomenon by relying upon urban theories [8].

The basic descriptive statistics presented in Table 1 and the ensuing analysis show that the IMR data are skewed to the right. Indeed, the IMR data, as GDPpc increases, seem to run downhill all the way, rapidly at the beginning and more slowly later until they go flatter to the horizontal GDPpc axis.

The data presented in Fig. 1 show large variations in IMR with respect to the GDPpc. This scatter plot displays in general the ranges of IMR at low GDPpc levels and also shows a range of GDPpc at which there are low IMR levels. The analysis so far suggests that it is reasonable to assume that the relationship of IMR and GDPpc has a decreasing slope. Therefore, a clearer link between IMR and GDPpc may be provided by using an exponential fit to the data, as we see in the logarithms plot of Fig. 1. Indeed, a common method for adjusting estimations to observations is by a linear regression equation using ordinary least squares, after transforming the data through the use of logarithms. Therefore, we investigate this regression on the dataset for the year 2016, as regressions for the rest of the datasets are similar. We find that the linear fit to the data is given by the highly statistically significant relationship $\ln(\text{IMR}) = 8.047 - 0.6346 \times \ln(\text{GDPpc})$. We see that the coefficient -0.6346 is the decreasing slope of the line, which is also the elasticity of IMR with respect to GDPpc. Thus, a 1% increase in GDPpc gives a 0.6346% decrease in IMR. A similar calculation for the year 1998 showed elasticity equal to -0.6%.

It is worth mentioning that during the period 1998–2016 the elasticity factor has been improved by 3.46%. However, the log-linear relation is very generic and unable to model the case where at the highest GDPpc value there occurs an IMR value that is higher than the IMR values at the neighboring GDPpc values, as already was mentioned. Therefore, except for the mathematical estimation of the downhill relation between IMR and GDPpc, the specific model presents serious disadvantages in interpreting real conditions and factors leading to them, which may result in implementing incorrect policies for effective solutions to the problem [9]. Thus, further research is needed in order to analyze the relationship of IMR and GDPpc.

3 *K*-means Clustering

The purpose of this section is to apply cluster analysis for each annual dataset separately of the period 1998–2016 in order to form four groups among the 132 countries based on the IMR and GDPpc measurements.

The *K*-means clustering algorithm uses iterative refinement to produce a final result. The algorithm inputs are the number of clusters and the dataset with the GDPpc and IMR measurements. The algorithms starts with initial estimates for the centroids and proceeds to partition the measurements into *K* clusters in which each observation belongs to the cluster with the nearest mean.

Without any preliminary analysis, we fed the data to the *K*-means clustering technique in the R statistical computing environment (see, for example, [10]) by employing the code

```
attach(x)
set.seed(1)
x=data.frame(x[,-1], row.names=x[,1])
km.out=kmeans(x, 4)
km.out
```

where `x` refers to the database. Nineteen different Microsoft Excel files were used as databases, each file containing data for the years 1998, 1999, ..., 2016. Number "4" in the statement `km.out=kmeans(x, 4)` refers to the number of clusters formed. Therefore the 132 countries of the `x` sample were classified into Clusters 1, 2, 3 and 4 demonstrating concentrations of IMR at high, medium-high, low-medium and low GDPpc values respectively.

First we present some results of the clustering with respect to the data of year 1998. The whole clustering vector is rather large to be presented here, but we see in Table 2 some summary results: Cluster 1 contains 15 countries with average of IMR equal to 5.11 and average of GDPpc equal to 29739.11; Cluster 2 contains 11 countries with respective averages equal to 7.48 and 18325.88; Cluster 3 contains 17 countries with averages equal to 16.17 and 7093.25; and, Cluster 4 contains 89 countries with averages equal to 58.08 and 1224.59. The within cluster sum of squares (SS) by cluster is given in the fifth column, while we note that the between SS over total SS was found equal to 95.3%. Further, we see that the mean values of IMR and GDPpc followed two opposite directions, increasing and decreasing.

Secondly, we provide summary results of the clustering analysis for each year in the period 1998–2016 in Table 3 that are similar to Table 2. Now we can see how clustering shifts over the years relating to yearly changes in IMR and GDPpc in the period 1998–2016: for Cluster 4 we see that Mean IMR decreased and Mean GDPpc increased; for Clusters 2 and 3 we see that both Mean IMR and Mean GDPpc fluctuate, which may be due to the fact that some countries are classified in Cluster 2 for one year and in Cluster 3 for another year, because of fluctuation of their economic growth rates; for Cluster 1 we see that Mean GDPpc increased slightly, but Mean IMR varied for the whole period 1998–2016. For example, Mean IMR shows peaks at the years

Table 2 K-means clustering with 4 clusters for the year 1998

j	Size	Mean IMR	Mean GDPpc	Within cluster sum of squares
1	15	5.113333	29739.105	288415970
2	11	7.481818	18325.883	100021927
3	17	16.164706	7093.254	101110971
4	89	58.076404	1224.592	103422101

2002 and 2010 and troughs at the years 1998 and 2006. In general, the behavior of Mean IMR along the years 1998, …, 2016 and the clusters 1, 2, 3, 4 is as follows. In Cluster 1 it varied in the interval [3.5, 5.2], while it has been reduced by 68.6%; in Cluster 2 it varied in the interval [3.9, 14.1], while it has been reduced by 52.0%; in Cluster 3 it varied in the interval [7.2, 19.2], while it has been reduced by 44.4%; and, in Cluster 4 it varied in the interval [30.9, 58.1], while it has been reduced by 53.2%. In all four clusters Mean IMR decreased over the years, while we see that the higher the GDPpc the lower the IMR. This makes sense, but it remains to be investigated the reason that Cluster 1 gave higher IMR than that of Cluster 2 for the years 2009, 2010 and 2013, which is beyond our analytical approach. Finally, we note that the between SS over total SS percentages were found equal to 95.3, 95.50, 94.20, 94.50, 95.50, 95.40, 95.50, 93.60, 92.10, 91.60, 91.10, 94.60, 94.30, 89.20, 94.80, 94.80, 88.80, 94.00, 94.50 for the years 1998, 1999, …, 2016 respectively.

4 Fitting IMR Versus GDPpc Subject to Increasing Returns

As was stated already, we assume that GDPpc is subject to increasing returns. This is equivalent to assuming that IMR comes from an unknown underlying convex relationship on the IMR observations at the GDPpc observations, but convexity has been lost due to errors in the measuring process. For such cases it is desired to make least the sum of squares change to the measurements subject to the given constraints, which is the problem stated in Sect. 1. Therefore we have to solve a strictly convex quadratic programming problem that has a unique solution, $\underline{y}^*$ say, and the method of Demetriou and Powell [6] is highly suitable for calculating the solution. In this section we give a brief description of the output of an implementation of the method [7], which is very instructive to our analysis.

Let $\underline{a}_i$ be the gradient of $y[x_{i-1}, x_i, x_{i+1}]$ with respect to $\underline{y}$. Recalling the definition (3), we see that $\underline{a}_i$ has just three nonzero adjacent components. The Karush-Kuhn-Tucker conditions provide necessary and sufficient conditions for optimality. They state that $\underline{y}^*$ is optimal if and only if the constraints (2) are satisfied and there exist nonnegative Lagrange multipliers $\{\lambda_i^* : i \in \mathscr{A}^*\}$ such that the first order conditions

Table 3 Classification of 132 countries by K-means clustering ($K = 4$)

Year		Cluster		1		2		3		4
1998	Size	Mean IMR	15	5.1	11	7.5	17	16.2	89	58.1
		Mean GDPpc		29739.1		18325.9		7093.3		1224.6
		Sum of squares		288415970.0		100021927.0		101110971.0		103422101.0
1999	Size	Mean IMR	14	5.0	11	6.9	16	12.3	91	55.3
		Mean GDPpc		31060.4		19926.4		7919.6		1264.4
		Sum of squares		267989638.0		99629532.0		124742004.0		117413049.0
2000	Size	Mean IMR	17	5.2	12	7.5	24	18.3	79	58.1
		Mean GDPpc		29793.7		16915.0		5296.1		937.7
		Sum of squares		465656261.0		203799273.0		91405113.0		42386551.0
2001	Size	Mean IMR	16	5.2	13	6.9	21	14.8	82	55.2
		Mean GDPpc		29567.3		17030.5		5572.1		1045.1
		Sum of squares		382842279.0		202137561.0		79252948.0		58672601.0
2002	Size	Mean IMR	12	5.3	13	5.5	13	11.2	94	48.8
		Mean GDPpc		33416.7		21825.0		9389.2		1486.9
		Sum of squares		257875822.0		150854428.0		102210816.0		160201828.0
2003	Size	Mean IMR	16	4.8	13	5.9	19	16.3	84	49.6
		Mean GDPpc		37010.8		20829.6		6825.5		1327.7
		Sum of squares		510907821.0		228693951.0		104576227.0		97129720.0
2004	Size	Mean IMR	14	4.8	11	5.5	13	8.9	94	44.7
		Mean GDPpc		43314.2		28072.4		12794.3		2005.5
		Sum of squares		530115843.0		204644696.0		162693052.0		304493207.0

(continued)

Table 3 (continued)

Year	Cluster			1		2		3		4
2005	Size	Mean IMR	19	4.9	10	5.2	15	19.2	88	43.5
		Mean GDPpc		43639.5		23509.2		10531.6		1957.9
		Sum of squares		1450887115.0		215445306.0		86180348.0		236762923.0
2006	Size	Mean IMR	21	4.6	15	14.1	25	18.2	71	47.8
		Mean GDPpc		45366.4		19076.7		6777.5		1512.1
		Sum of squares		2217505727.0		356699876.0		134770829.0		93364714.0
2007	Size	Mean IMR	20	4.6	17	12.1	22	17.5	73	44.5
		Mean GDPpc		51808.7		22173.9		8099.4		1876.2
		Sum of squares		2798700714.0		690927227.0		134883583.0		145727697.0
2008	Size	Mean IMR	18	4.6	15	11.1	19	13.1	80	41.0
		Mean GDPpc		57515.4		29329.4		12370.5		2580.6
		Sum of squares		3357374188.0		720496805.0		286262656.0		321761917.0
2009	Size	Mean IMR	6	4.3	18	4.1	16	12.3	92	36.3
		Mean GDPpc		62273.4		39726.9		17915.4		3217.7
		Sum of squares		603744883.0		509900071.0		299128604.0		663031798.0
2010	Size	Mean IMR	3	4.7	21	3.9	22	12.2	86	37.5
		Mean GDPpc		77597.9		42885.7		16340.6		3165.4
		Sum of squares		157258922.0		1184755611.0		487500067.0		575240560.0
2011	Size	Mean IMR	17	3.9	16	9.8	28	14.8	71	38.4
		Mean GDPpc		59463.8		29239.9		11703.5		2482.0
		Sum of squares		4264144027.0		990378109.0		345416562.0		225003809.0

(continued)

Table 3 (continued)

Year	Cluster			1		2		3		4
2012	Size	Mean IMR	3	4.6	18	3.8	26	10.3	85	34.1
		Mean GDPpc		91257.1		49119.7		19326.2		3610.6
		Sum of squares		175219715.0		974512694.0		895482544.0		716741976.0
2013	Size	Mean IMR	3	4.5	19	3.7	26	10.4	84	33.1
		Mean GDPpc		92158.9		49367.1		18892.7		3667.2
		Sum of squares		183323333.0		1262321856.0		685899251.0		715241495.0
2014	Size	Mean IMR	19	3.7	17	9.6	13	12.8	70	36.2
		Mean GDPpc		58223.7		24849.0		11207.4		2673.9
		Sum of squares		5004272224.0		611606625.0		208087664.0		242789079.0
2015	Size	Mean IMR	10	3.5	17	4.2	29	9.6	80	32.0
		Mean GDPpc		60623.6		38038.3		15049.6		3093.6
		Sum of squares		1020556130.0		424694064.0		605924461.0		424506998.0
2016	Size	Mean IMR	10	3.5	14	3.9	25	7.2	83	30.9
		Mean GDPpc		59999.3		37309.3		15259.1		3214.4
		Sum of squares		793045133.0		549035798.0		413180913.0		469784539.0

$$\underline{y}^* - \underline{\phi} = \frac{1}{2} \sum_{i \in \mathscr{A}^*} \lambda_i^* \underline{a}_i, \tag{6}$$

hold, where $\mathscr{A}^*$ is a subset of the constraint indices $\{2, 3, \ldots, n-1\}$ with the property

$$y^*[x_{i-1}, x_i, x_{i+1}] = 0, \quad i \in \mathscr{A}^*. \tag{7}$$

We note that the components of $\underline{y}^*$ are the estimated IMR values and we use the descriptive term *convex fit* for $\underline{y}^*$. Equation (7) implies that the points (x_{i-1}, y_{i-1}^*), (x_i, y_i^*) and (x_{i+1}, y_{i+1}^*) are collinear. In practice, it is usual to expect indices $j \notin \mathscr{A}^*$ such that $y^*[x_{j-1}, x_j, x_{j+1}] > 0$. Further, we let $y(x)$ be the function that is defined by linear interpolation to the points $\{(x_i, y_i^*) : i = 1, 2, \ldots, n\}$. Hence and in view of the constraints (2), we see that $y(x)$ is a piecewise linear convex curve. The separate linear pieces are joined at the *knots* x_j, the knots being all in the set $\{x_j : j \in \{2, 3, \ldots, n-1\} \setminus \mathscr{A}^*\}$. It is important to note that the knots are not known in advance bur they are determined automatically by our method.

This method is far faster than a general quadratic programming algorithm, because it takes advantage of the banded matrices that occur during the calculation and makes use of a B-spline representation of the required fit, which allows us to work with much fewer variables than n. For proofs one may consult the reference.

The software package L2CXFT of Demetriou [7] implements this method and includes several useful extensions. It is the main tool of our work. The actual calculations were carried out by supplying the data to L2CXFT, while only a few iterations were needed for termination. We present the best convex fit to the most recent dataset, namely that of year 2016, in Table 4. We have tabulated GDPpc, IMR, convex fit and Lagrange multipliers at $\underline{y}^*$ corresponding to x_i, ϕ_i, y_i^* and λ_i^*. Let us note that the numbers in the first column correspond to the countries of our sample (see [2]). If $\lambda_i^* = 0$ (fifth column) then x_i is a knot of the resultant fit, which implies that $y^*[x_{i-1}, x_i, x_{i+1}] > 0$, except for degeneracy. Moreover, the larger the magnitude of a Lagrange multiplier, the stronger the underlying linearity of the IMR indicator. Despite the existence of outliers, we see that the convex fit follows quite satisfactorily the trend of the data. Fig. 3 presents the relevant convex fit (i.e. to the data of 2016) together with similar fits for the years 2013, 2014 and 2015.

In close association with Table 4 we provide Table 5 that summarizes the results of the run with the data of year 2016. It displays the knot and the endpoint indices $j = 0, \ldots, n-1-\text{card}(\mathscr{A}^*)$, the GDPpc values, the data indices at the knots, the estimated IMR values, i.e. the values y_j^*, and the first differences of the fit (namely, the slopes of the line segments that join two consecutive knots). Further, we see that the best convex fit $y(x)$ contains 10 interior knots at the data indices 22, 33, 55 and so on up to 123, while the value of the best convex fit at a knot is given by the associated estimated IMR value. Therefore, we let B_j, for $j = 1, 2, \ldots, 11$, be a basis of linear B-splines that are defined on the abscissae x_i and satisfy the equations $B_j(\xi_j) = 1$, $B_j(\xi_i) = 0, i \neq j$, where ξ_j is a knot, $B_j(x) = (x - \xi_{j-1})/(\xi_j - \xi_{j-1})$, for $\xi_{j-1} \leq x \leq \xi_j$, $B_j(x) = (\xi_{j+1} - x)/(\xi_{j+1} - \xi_j)$, for $\xi_j \leq x \leq \xi_{j+1}$, and 0 otherwise [5].

Table 4 Infant Mortality and GDP per Capita Data for 2016, and Least Squares Convex Fit [7]

No	GDPpc	IMR	Fit	Lagrange	No	GDPpc	IMR	Fit	Lagrange
1	285.73	48.40	54.49	–	67	5233.47	18.70	18.36	33346.52
2	300.31	38.90	53.71	544.17	68	5284.60	34.20	18.32	32365.90
3	364.17	50.90	53.71	547.27	69	5805.61	13.10	18.18	31203.21
4	382.07	53.10	53.21	2011.36	70	5871.44	17.80	18.10	32966.52
5	401.74	34.00	51.62	11256.11	71	5910.62	10.50	18.09	33436.40
6	505.20	83.30	51.34	12768.91	72	6018.53	17.80	17.98	37882.35
7	578.46	50.70	51.03	14334.98	73	6049.23	11.90	17.66	51077.82
8	580.38	37.70	50.91	14909.51	74	6722.22	25.50	17.60	52640.45
9	627.10	52.70	50.88	15003.41	75	6924.15	32.60	17.14	48939.80
10	641.60	57.80	50.15	13265.42	76	7179.34	34.30	16.74	53786.08
11	661.53	58.30	49.79	12839.98	77	7469.03	6.50	16.06	36190.94
12	664.30	75.20	49.33	13068.34	78	7713.55	10.10	15.40	28225.89
13	702.84	29.20	49.15	13717.87	79	8123.18	8.50	15.37	27636.19
14	706.76	41.00	48.87	14214.05	80	8208.56	12.60	15.15	24226.74
15	729.12	28.40	48.83	14384.22	81	8649.95	13.50	14.97	24100.58
16	739.60	50.90	48.49	14915.51	82	8747.35	66.20	14.86	24468.38
17	779.94	68.00	48.41	14666.56	83	8748.36	6.60	14.79	25001.60
18	789.44	63.10	48.02	12538.55	84	9508.24	7.10	14.47	32578.26
19	795.84	37.10	47.29	8040.63	85	9519.88	7.70	14.33	38127.71
20	879.19	40.30	46.81	5489.03	86	9630.94	12.20	13.61	75418.49
21	952.77	33.60	46.71	5154.52	87	10118.06	7.30	13.40	87196.08
22	1077.04	18.80	44.16	0	88	10862.60	10.90	13.13	73073.23
23	1269.57	43.80	43.64	971.04	89	11824.64	7.70	12.71	52006.38

(continued)

Table 4 (continued)

No	GDPpc	IMR	Fit	Lagrange	No	GDPpc	IMR	Fit	Lagrange
24	1269.91	26.30	43.05	3570.24	90	12160.11	4.00	12.48	42656.82
25	1358.78	28.20	41.76	12080.87	91	12440.32	9.90	11.63	22703.11
26	1374.51	52.80	41.31	14821.93	92	12814.95	4.40	10.98	16505.90
27	1443.63	64.20	41.26	15160.04	93	13680.24	14.10	10.64	10985.18
28	1455.36	35.60	41.17	15585.17	94	13792.93	7.20	10.46	8027.97
29	1513.46	41.20	40.33	19412.62	95	14879.68	4.30	9.86	2291.58
30	1534.97	66.00	40.26	19446.46	96	14982.36	9.20	8.58	4758.97
31	1709.59	34.60	38.24	11416.15	97	15220.57	7.90	7.97	6631.63
32	1900.20	13.70	35.31	691.56	98	15891.63	11.40	7.97	6615.04
33	2151.38	16.80	32.03	0	99	16040.52	16.50	7.47	7400.38
34	2175.67	66.90	31.96	296.18	100	16535.92	4.90	6.94	6309.96
35	2185.73	7.80	31.85	1276.55	101	17727.49	2.30	6.78	7442.42
36	2214.39	17.30	31.64	4859.56	102	17930.16	3.10	6.38	2078.85
37	2353.14	48.90	30.63	15785.23	103	18491.94	2.50	6.03	0
38	2361.16	16.00	29.48	33352.26	104	19839.64	2.90	5.41	0
39	2415.04	44.80	28.87	39514.05	105	20028.65	11.10	5.29	7030.10
40	2770.20	52.40	28.62	42019.95	106	21652.28	1.80	5.27	5447.55
41	2773.55	26.80	27.46	38587.11	107	22579.09	6.50	5.12	4633.82
42	2832.43	23.30	27.27	38240.74	108	23324.20	2.10	5.09	3318.91
43	2951.07	21.50	26.81	38178.52	109	26639.74	2.70	4.86	9760.53
44	2997.75	18.20	26.24	39585.15	110	27359.23	7.20	4.59	31000.49
45	3104.96	29.50	26.06	38458.65	111	28785.48	8.60	4.58	30877.37

(continued)

Table 4 (continued)

No	GDPpc	IMR	Fit	Lagrange	No	GDPpc	IMR	Fit	Lagrange
46	3308.70	54.60	24.48	26121.01	112	30674.84	2.80	4.50	22210.59
47	3477.85	19.40	23.20	16714.16	113	36854.97	3.20	4.13	847.13
48	3570.29	22.20	22.77	14026.81	114	37175.74	2.90	4.03	0
49	3614.69	11.90	22.29	12646.64	115	37622.21	6.60	3.98	1144.28
50	3688.65	11.70	21.88	6660.88	116	38900.57	2.00	3.81	8677.80
51	3694.08	15.40	21.49	2625.62	117	40341.41	3.70	3.66	2331.90
52	3835.39	8.00	21.39	1873.29	118	42069.60	3.20	3.59	0
53	3865.79	9.50	20.91	407.30	119	42157.93	4.30	3.54	945.84
54	3876.94	27.20	20.84	112.56	120	43402.86	1.90	3.49	134.97
55	3916.88	21.60	20.77	0	121	44676.35	2.90	3.48	140.56
56	4077.74	17.00	20.04	0	122	45669.81	3.20	3.46	0
57	4087.94	15.10	20.03	120.56	123	49927.82	3.10	3.16	0
58	4124.98	12.00	20.01	368.06	124	51949.27	2.40	3.16	226.43
59	4146.74	23.90	19.98	1273.91	125	52962.49	2.20	3.24	9975.69
60	4223.58	12.90	19.98	1275.08	126	53549.70	3.70	3.26	12257.97
61	4414.98	32.30	19.55	4785.53	127	57638.16	5.60	3.36	29333.73
62	4529.14	26.90	19.37	11637.29	128	59324.34	7.30	3.36	29446.41
63	4744.74	12.80	19.29	12155.50	129	59976.94	1.60	3.55	38618.89
64	4808.41	5.20	19.21	13471.57	130	63861.92	3.00	3.70	49336.25
65	4878.58	13.20	19.20	13802.89	131	70911.76	2.10	3.98	9651.40
66	4986.50	2.90	18.63	30456.03	132	79890.52	3.60	4.24	–

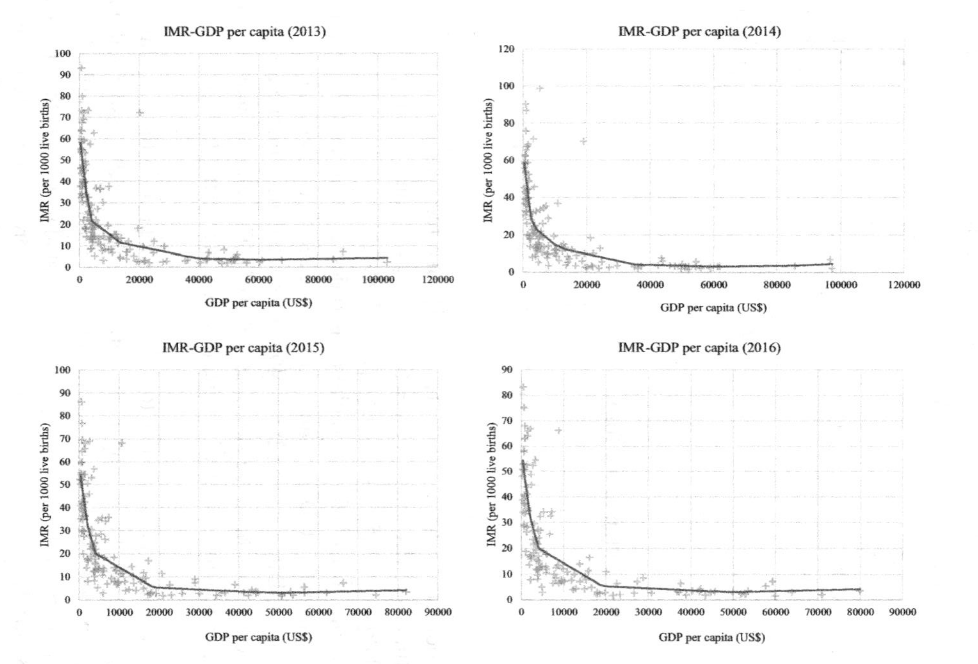

Fig. 3 Least squares convex fit (line) to infant mortality rate data (cross) against gross domestic product per capita for the years 2013, 2014, 2015 and 2016

Table 5 GDPpc and knots of the estimated IMR values for 2016

j	GDP per capita	Knot	Estimated IMR	First difference
0	285.7	1	54.5	
1	1077.0	22	44.2	−1.3E−02
2	2151.4	33	32.0	−1.1E−02
3	3916.9	55	20.8	−6.4E−03
4	4077.7	56	20.0	−4.5E−03
5	18491.9	103	6.0	−9.7E−04
6	19839.6	104	5.4	−4.6E−04
7	37175.7	114	4.0	−7.9E−05
8	42069.6	118	3.6	−9.0E−05
9	45669.8	122	3.5	−3.7E−05
10	49927.8	123	3.2	−7.1E−05
11	79890.5	132	4.2	3.6E−05

Then $y(x)$ may be written in the form

$$y(x) = \sum_{j=0}^{11} \sigma_j B_j(x), \ \ \mathrm{GDPpc_1} \le x \le \mathrm{GDPpc_n}, \tag{8}$$

where the coefficients $\{\sigma_j : j = 0, 1, \ldots, 11\}$ take their values from the column labeled ‘Estimated IMR’, that is $\sigma_0 = 54.5$, $\sigma_1 = 44.2$ and so on up to $\sigma_{11} = 4.2$; $n = 132$, $\mathrm{GDPpc_1} = 285.7$ and $\mathrm{GDPpc_{132}} = 79890.5$.

Hence one may use (8) for interpolation in order to obtain an IMR estimation for x, where x is a tentative GDPpc value. Further, we see that the estimated IMR values decline over the GDPpc range [285.7, 49927.8], reaching a minimum equal to 3.2 and then increase over the range [49927.8, 79890.5]. The rates of change are negative up to 49927.8 and subsequently become positive, increasing from -1.3×10^{-2} up to 3.6×10^{-5}.

Table 6 presents results similar to Table 5 associated with the datasets of the years 1998, …, 2015. In the upper left corner we see results associated with year 1998. The estimated IMR values decline over the GDPpc range [124.7, 30901.1], reaching a minimum value that is approximately equal to 4.7, and then increase over the range [30901.1, 41487.7]. The rates of change are negative up to 30901.1 and subsequently become positive, increasing from −0.14 up to 0.000038. In the lower right corner we see results associated with year 2015. The estimated IMR values decline over the GDPpc range [300.7, 50734.4], reaching a minimum value that is approximately equal to 3.2, and then increase over the range [50734.4, 82016.0]. The rates of change are negative up to 50734.4 and subsequently become positive, increasing from −0.013 up to 0.000035. It is interesting to note the similarity of the year 2015 values with those of 2016 (indeed, they have the same ‘Knots’, while the

Table 6 GDPpc and knots of the estimated IMR values for the years 1998, 1999, …, 2015

j	GDPpc	Knot	Est IMR	1st diff	GDPpc	Knot	Est IMR	1st diff	GDPpc	Knot	Est IMR	1st diff
	1998				**1999**				**2000**			
0	124.7	1	115.0		119.1	1	108.6		123.9	1	102.6	
1	360.6	24	81.1	−1.4E−01	216.0	9	94.0	−1.5E−01	231.4	10	88.9	−1.3E−01
2	869.1	49	42.9	−7.5E−02	597.4	38	57.0	−9.7E−02	401.5	28	73.6	−9.0E−02
3	951.9	50	39.5	−4.1E−02	628.9	39	54.1	−9.4E−02	875.4	45	41.1	−6.8E−02
4	1834.8	66	32.8	−7.6E−03	1246.5	56	34.2	−3.2E−02	946.1	46	38.1	−4.2E−02
5	5650.3	97	16.7	−4.2E−03	5135.5	94	16.7	−4.5E−03	1088.8	51	35.2	−2.1E−02
6	12202.7	106	9.2	−1.1E−03	12474.8	106	8.3	−1.1E−03	5101.4	94	16.4	−4.7E−03
7	22252.4	117	6.1	−3.1E−04	13245.2	107	7.7	−8.4E−04	12043.0	106	9.2	−1.0E−03
8	25101.4	119	5.6	−1.8E−04	32198.9	127	4.7	−1.5E−04	14672.9	108	7.5	−6.2E−04
9	30901.0	127	4.7	−1.7E−04	36371.4	131	4.5	−6.2E−05	24253.3	119	5.4	−2.2E−04
10	41487.7	132	5.1	3.8E−05	40581.3	132	4.7	5.2E−05	38532.04	132	5.4	−3.6E−07
	2001				**2002**				**2003**			
0	120.2	1	99.1		111.5	1	94.0		112.8	1	89.3	
1	407.7	29	71.6	−9.6E−02	477.1	31	66.7	−7.5E−02	436.7	26	70.2	−5.9E−02
2	837.7	45	40.2	−7.3E−02	1000.8	48	38.5	−5.4E−02	1157.3	51	31.3	−5.4E−02
3	1053.1	50	34.4	−2.7E−02	1453.6	57	30.2	−1.8E−02	1174.8	52	30.6	−3.9E−02
4	5283.0	94	14.8	−4.6E−03	6053.7	95	14.8	−3.3E−03	7805.9	96	13.5	−2.6E−03
5	12538.2	106	8.7	−8.3E−04	14110.3	107	6.9	−9.8E−04	8423.4	97	12.8	−1.2E−03
6	15062.9	108	6.7	−7.8E−04	15988.3	108	6.4	−2.9E−04	15772.7	106	6.1	−9.0E−04
7	24537.5	119	5.2	−1.6E−04	26351.4	119	5.1	−1.2E−04	29691.2	117	5.3	−6.2E−05
8	38549.6	132	5.3	6.1E−06	43061.2	132	5.1	−1.2E−06	50111.7	132	4.1	−5.7E−05

(continued)

Table 6 (continued)

j	1998				1999				2000			
	GDPpc	Knot	Est IMR	1st diff	GDPpc	Knot	Est IMR	1st diff	GDPpc	Knot	Est IMR	1st diff
	2004				**2005**				**2006**			
0	127.4	1	84.1		150.5	1	79.7		154.9	1	78.1	
1	1367.4	51	28.3	−4.5E−02	1259.8	44	38.4	−3.7E−02	1448.8	43	36.4	−3.2E−02
2	1408.5	52	28.0	−8.5E−03	1828.7	53	26.7	−2.1E−02	3394.4	65	23.5	−6.6E−03
3	11685.9	102	12.1	−1.5E−03	3126.1	66	24.2	−1.9E−03	4428.5	73	21.8	−1.7E−03
4	17260.9	105	6.1	−1.1E−03	3933.3	76	22.7	−1.9E−03	26455.1	110	5.9	−7.2E−04
5	31174.6	115	5.2	−6.3E−05	20557.1	107	7.3	−9.2E−04	28482.6	111	4.6	−6.4E−04
6	57570.3	132	4.0	−4.6E−05	26510.7	111	5.1	−3.8E−04	33410.7	112	4.4	−4.7E−05
7	–	–	–	–	29869.9	112	4.8	−7.6E−05	74114.7	132	5.1	1.8E−05
8	–	–	–	–	66775.4	132	4.4	−1.2E−05	–	–	–	–
	2007				**2008**				**2009**			
0	170.8	1	75.8		196.2	1	73.0		204.9	1	71.5	
1	1644.8	42	34.1	−2.8E−02	2054.5	43	31.0	−2.3E−02	1825.3	42	31.4	−2.5E−02
2	4324.9	69	22.2	−4.4E−03	4378.7	65	24.2	−2.9E−03	2106.7	43	27.4	−1.4E−02
3	4620.5	70	21.0	−4.0E−03	7261.8	80	17.0	−2.5E−03	6969.6	78	18.0	−1.9E−03
4	4735.5	73	20.8	−2.4E−03	39721.0	113	4.6	−3.8E−04	8562.8	89	15.6	−1.5E−03
5	28827.3	109	5.6	−6.3E−04	61257.9	128	4.4	−1.3E−05	29711.0	109	6.0	−4.5E−04
6	31386.6	110	4.4	−4.7E−04	97007.9	132	4.7	1.0E−05	32105.8	110	5.0	−4.3E−04
7	85170.9	132	4.4	2.8E−07	–	–	–	–	40855.2	119	3.9	−1.2E−04
8	–	–	–	–	–	–	–	–	80067.2	132	4.2	7.5E−06
	2010				**2011**				**2012**			

(continued)

Table 6 (continued)

j	1998				1999				2000			
	GDPpc	Knot	Est IMR	1st diff	GDPpc	Knot	Est IMR	1st diff	GDPpc	Knot	Est IMR	1st diff
0	214.2	1	68.5		265.3	1	61.0		265.3	1	61.0	
1	1631.5	36	39.7	−2.0E−02	2046.5	36	35.5	−1.4E−02	2046.5	36	35.5	−1.4E−02
2	2819.7	45	28.2	−9.7E−03	3408.5	46	24.3	−8.2E−03	3408.5	46	24.3	−8.2E−03
3	4514.9	64	22.1	−3.6E−03	3684.8	47	23.3	−3.6E−03	3684.8	47	23.3	−3.6E−03
4	11938.3	91	14.3	−1.0E−03	4137.5	55	22.5	−1.7E−03	4137.5	55	22.5	−1.7E−03
5	30736.4	110	5.8	−4.5E−04	13236.0	91	11.7	−1.2E−03	13236.0	91	11.7	−1.2E−03
6	42935.3	119	3.6	−1.8E−04	44333.9	117	4.0	−2.5E−04	44333.9	117	4.0	−2.5E−04
7	87646.3	132	4.5	1.9E−05	47415.6	118	3.8	−4.2E−05	47415.6	118	3.8	−4.2E−05
8	–	–	–	–	101668.2	132	4.4	9.9E−06	101668.2	132	4.4	9.9E−06
9	–	–	–	–	59593.3	127	3.6	1.8E−05	–	–	–	–
10					100711.2	132	4.5	2.1E−05	–	–	–	–
	2013				**2014**				**2015**			
0	282.8	1	57.9		286.0	1	58.5		300.7	1	54.5	
1	2244.0	37	34.6	−1.2E−02	2434.3	37	29.3	−1.4E−02	1121.1	22	44.2	−1.3E−02
2	4029.7	53	21.5	−7.3E−03	4102.1	52	23.0	−3.8E−03	2096.0	33	32.0	−1.2E−02
3	13574.7	89	11.6	−1.0E−03	10011.8	80	14.8	−1.4E−03	3947.0	55	20.8	−6.1E−03
4	36291.2	112	5.0	−2.9E−04	13480.7	90	12.2	−7.6E−04	4096.1	56	20.0	−4.9E−03
5	40454.4	113	4.1	−2.3E−04	35179.7	111	4.2	−3.7E−04	18070.8	103	6.0	−1.0E−03
6	60283.2	127	3.5	−2.9E−05	58900.0	127	3.1	−4.7E−05	19252.6	104	5.4	−5.2E−04
7	103059.2	132	4.3	1.9E−05	61330.9	128	3.1	2.8E−06	35691.3	114	4.0	−8.4E−05
8	–	–	–	–	85610.8	130	3.7	2.3E−05	42419.6	118	3.6	−6.6E−05
9	–	–	–	–	97429.7	132	4.6	7.6E−05	44746.3	122	3.5	−5.7E−05
10	–	–	–	–	–	–	–	–	50734.4	123	3.2	−5.1E−05
11	–	–	–	–	–	–	–	–	82016.0	132	4.2	3.5E−05

'Estimated IMR' values are identical due to rounding). The results suggest that it is unlikely to be achieved more than marginal IMR improvements in the coming years.

All the results in Table 6 showed increasing rates of change of the IMR estimates (cf. columns labeled '1st diff') that started from a negative value and the majority of them reached a small positive value. For the years 2003, 2005 and 2006, these rates increased up to a small negative value.

5 Conclusions

We investigated empirically the validity of our claim that a relation between Infant Mortality Rate and the per capita Gross Domestic Product is provided by assuming that GDPpc is subject to increasing returns. We extended the results of [3] and provided further evidence for our claim, by analyzing yearly cross-country data for 132 countries for the period of years 1998–2016.

We started by calculating some descriptive statistics for the data and we applied cluster analysis. The raw data and the descriptive statistics show that there is a downhill non-linear relation between IMR and GDPpc, while a log-linear relationship was proved inadequate to reveal subtle features of IMR and GDPpc changes. The K-means method of clustering classified the data into four groups demonstrating low, low-medium, medium-high and high GDPpc. The main finding is that the IMR slope went downwards in all four clusters for each year, while the GDP slope went upwards. It is interesting to note that for the years 2009, 2010, 2012 and 2013, Cluster 1 gave mean IMR that is higher than the mean IMR of Cluster 2.

Fitting the data subject to increasing slopes of the IMR curve by the L2CXFT software package has been the main consideration of our work. All the fits stated clearly a relationship of IMR and GDPpc and provided a quantitative explanation to what economists have observed: that at high GDPpc levels, IMR increases after bottoming out. This is highly important, because it shows that the convexity method we have employed provides a clear quantitative model for confirmation of assumptions which have so far only qualitative support. In order to use the method, one has only to solve a quadratic programming calculation that gives a convex shape to the IMR process with respect to GDPpc. This is the subject of our software [7].

The main result of this analysis is that the assumption that Infant Mortality Rates and Gross Domestic Product per capita follow increasing returns not only adequately describes reality, but also it is capable to capture imperceptible features of the underlying process. It seems, therefore, that it would be useful to design a system that will monitor the convexity method so as to provide a prescriptive analytics tool for policy actions.

References

1. T. Callen, Gross domestic product: an economy's all. IMF. http://www.imf.org/external/pubs/ft/fandd/basics/gdp.htm. Accessed 10 Nov 2016
2. Mortality Rate, Infant, The World Bank. http://data.worldbank.org/indicator/sp.dy.imrt.in. Accessed 20 Oct 2017
3. I.C. Demetriou, P. Tzitziris, Infant mortality and economic growth: modeling by increasing returns and least squares. In: *Lecture Notes in Engineering and Computer Science: Proceedings of The World Congress on Engineering 2017, WCE 2017*, 5–7 July 2017, London, U.K., ed. by S.I. Ao, L. Gelman, D.W.L. Hukins, A. Hunter, A.M. Korsunsky, vol. II, pp. 543–548
4. C. Hildreth, Point estimates of ordinates of concave functions. J. Am. Stat. Assoc. **49**, 598–619 (1954)
5. C. de Boor, *A Practical Guide to Splines*, Revised Edition (Springer, New York, 2001)
6. I.C. Demetriou, M.J.D. Powell, The minimum sum of squares change to univariate data that gives convexity. IMA J. Numer. Anal. **11**, 433–448 (1991)
7. I.C. Demetriou, Algorithm 742: L2CXFT, a Fortran 77 subroutine for least squares data fitting with non-negative second divided differences. ACM Trans. Math. Softw. **21**(1), 98–110 (1995)
8. L. Bertinelli, *Urbanization and Growth* (Duncan Black, London, 2000)
9. I. Chakraborty, Living standards and economic growth: a fresh look at the relationship through the nonparametric approach. Centre for Development Studies. https://opendocs.ids.ac.uk/opendocs/handle/123456789/2942. Accessed 10 Nov 2016
10. G. James, D. Witten, T. Hastie, R. Tibshirani, *An Introduction to Statistical Learning with Applications in R* (Springer, Heidelberg, 2013)

Mathematical Models for the Study of Resource Systems Based on Functional Operators with Shift

Oleksandr Karelin, Anna Tarasenko, Viktor Zolotov and Manuel Gonzalez-Hernandez

Abstract In previous works we proposed a method for the study of systems with one renewable resource. The separation of the individual and the group parameters and the discretization of time led us to scalar linear functional equations with shift. Cyclic models, in which the initial state of the system coincides with the final state, were considered. In this work, we present models for systems with two renewable resources. In modelling, the interactions and the reciprocal influences between these two resources are taken into account. Analysis of the models is carried out in weighted Holder spaces. For cyclic models a method for the solution of the balance equations is proposed. The equilibrium state of the system is found. Some problems for the optimal exploitation of resources of open systems are formulated.

Keywords Degenerate kernel · Equilibrium state · Exploitation of resources
Holder space · Invertibility · Renewable resources

O. Karelin · M. Gonzalez-Hernandez
Department of Industrial Engineering, Institute of Basic Sciences and Engineering,
Hidalgo State Autonomous University, 42184 Pachuca, Hgo, Mexico
e-mail: karelin@uaeh.edu.mx

M. Gonzalez-Hernandez
e-mail: mghdez@uaeh.edu.mx

A. Tarasenko (✉)
Department of Mathematics, Institute of Basic Sciences and Engineering,
Hidalgo State University, 42184 Pachuca, Hgo, Mexico
e-mail: anataras@uaeh.edu.mx; tarasenko@uaeh.edu.mx

V. Zolotov
Department of Economics of Nature-Resource Management,
Institute for Market Problems and Economic and Ecological Research
of the National Academy of Sciences of Ukraine, Odessa 65044, Ukraine
e-mail: odessaaa48@gmail.com

S.-I. Ao et al. (eds.), *Transactions on Engineering Technologies*,
https://doi.org/10.1007/978-981-13-0746-1_7

1 Introduction

Systems whose state depends on time and whose resources are renewable form an important class of general systems. A great number of works has been dedicated to systems with renewable resources [1, 2]. The core of the mathematical apparatus used for the study of such systems consists of differential equations in which the sought for function is dependent on time [3–5].

Our approach presupposes discretization of the processes with respect to time. We move away from tracking the changes in the system continuously to tracking the changes at fixed time points. This discretization and the identification of the individual parameter and the group parameter lead us to functional equations with shift.

Section 5 is devoted to a discussion of the problems on the exploitation of open resource systems.

This paper is a continuation and expansion of article [6].

2 Cyclic Model of a System with Two Renewable Resources

Let S be a system with two resources λ_1, λ_2 and let T be a time interval. The choice of T is related to periodic processes taking place in the system and to human interferences.

Let these resources λ_1, λ_2 have the same individual parameter but scales of measurement of values of the individual parameter may be different:

$$x_{min} = x_1 < x_2 < \cdots < x_{n_1} = x_{max},$$

$$y_{min} = y_1 < y_2 < \cdots < y_{n_2} = y_{max}.$$

We introduce the group parameters by functions $v(x_i, t)$, $w(y_i, t)$ which express a quantitative estimate of the elements of resources λ_1, λ_2 with the individual parameter $x_i, i = 1, 2, \ldots, n_1$ and $y_i, i = 1, 2, \ldots, n_2$ at the time t.

Let t_0 be the initial time and S the system under consideration.

As in our previous works [6–8] on modelling the system, we will hold the following principles:

I. The description of changes that occur on the interval $(t_0, t_0 + T)$ will be substituted by the fixing of the final results at the moment $t_0 + T$;
II. The separation of parameters into individual parameters, group parameters and the study of dependence of group parameters from individual parameters.

The initial state of the system S at time t_0 is represented as density functions of a distribution of the group parameter by the individual parameter for each resource

$$v(x, t_0) = v(x), 0 < x < x_{max},$$

$$w(y, t_0) = w(y), 0 < y < y_{max}.$$

We will now analyze the system's evolution. In the course of time, the elements of the system can change their individual parameter—e.g. fish can change their weight and length.

Modifications in the distribution of the group parameters by the individual parameters is represented by a displacement. The state of the system S at the time $t = t_0 + T$ is:

$$v(x, t_0 + T) = \frac{d}{dx}\alpha(x) \cdot v(\alpha(x)), \tag{1}$$

$$w(y, t_0 + T) = \frac{d}{dy}\beta(y) \cdot w(\beta(y)). \tag{2}$$

In the article [7], the appearance of derivatives in (1), (2) was explained.

Over the period $j_0 = [t_0, t_0 + T]$, extractions might be taken from the system as a result of human economic activity; these are represented by summands $\rho(x)$, $\delta(y)$. If an artificial entrance of elements into the system has taken place, it shall be accounted for by adding terms $\zeta(x)$, $\xi(y)$.

We take natural mortality into account with the coefficients $c(x)$, $d(y)$.

The process of reproduction will be represented by

$$\sum_{i=1}^{n} P_i p_i(x),$$

where

$$P_1 = \int_{\nu_0}^{\nu_1} v(x)dx, \; P_2 = \int_{\nu_1}^{\nu_2} v(x)dx, \ldots, P_n = \int_{\nu_{n-1}}^{\nu_n} v(x)dx,$$

$$0 = \nu_0 < \nu_1 < \cdots < \nu_n = x_{max},$$

and

$$\sum_{i=1}^{m} Q_i q_i(y),$$

where

$$Q_1 = \int_{\mu_0}^{\mu_1} w(y)dy, \; Q_2 = \int_{\mu_1}^{\mu_2} w(y)dy, \ldots, Q_m = \int_{\mu_{m-1}}^{\mu_m} w(y)dy,$$

$$0 = \mu_0 < \mu_1 < \cdots < \mu_m = y_{max}.$$

We obtain

$$v(x, t_0 + T) = c(x)\frac{d}{dx}\alpha(x)v(\alpha(x)) + \rho(x) + \zeta(x) + \sum_{i=1}^{n} P_i p_i(x)$$

and

$$w(y, t_0 + T) = d(y)\frac{d}{dy}\beta(y)w(\beta(y)) + \delta(y) + \xi(y) + \sum_{i=1}^{m} Q_i q_i(y).$$

Resources λ_1 and λ_2 are not independent. We will account for reciprocal influence by terms

$$\sum_{i=1}^{k} R_i r_i(x),$$

where

$$R_1 = \int_{\gamma_0}^{\gamma_1} w(y)dy, R_2 = \int_{\gamma_1}^{\gamma_2} w(y)dy, \ldots, R_k = \int_{\gamma_{k-1}}^{\gamma_k} w(y)dy,$$

$$0 = \gamma_0 < \gamma_1 < \cdots < \gamma_k = y_{max},$$

and

$$\sum_{i=1}^{l} F_i f_i(y),$$

where

$$F_1 = \int_{\varepsilon_0}^{\varepsilon_1} v(x)dx, F_2 = \int_{\varepsilon_1}^{\varepsilon_2} v(x)dx, \ldots, F_l = \int_{\varepsilon_{l-1}}^{\varepsilon_l} v(x)dx,$$

$$0 = \varepsilon_0 < \varepsilon_1 < \cdots < \varepsilon_l = x_{max}.$$

Thereby, the final state of the system at the moment $[t_0 + T]$ is described as follows:

$$v(x, t_0 + T) = c(x)\frac{d}{dx}\alpha(x)v(\alpha(x)) + \rho(x) + \zeta(x) + \sum_{i=1}^{n} P_i p_i(x) + \sum_{i=1}^{k} R_i r_i(x), \tag{3}$$

$$w(y, t_0 + T) = d(y)\frac{d}{dy}\beta(y)w(\beta(y)) + \delta(y) + \xi(y) + \sum_{i=1}^{m} Q_i q_i(y) + \sum_{i=1}^{l} F_i f_i(y). \tag{4}$$

Let our goal be to find the equilibrium state of system S, that is, to find such an initial distribution of group parameters by the individual parameter $v(x, t_0)$, $w(x, t_0)$, that after all transformations during the time interval $(t_0, t_0 + T))$, it would coincide with the final distribution:

$$v(x) = v(x, t_0 + T), \tag{5}$$

$$w(y) = w(y, t_0 + T). \tag{6}$$

From here, substituting relations (3) and (4) into (5), (6), it follows that

$$v(x) = c(x)\frac{d}{dx}\alpha(x)v(\alpha(x)) + \rho(x) + \zeta(x) + \sum_{i=1}^{n} P_i p_i(x) + \sum_{i=1}^{k} R_i r_i(x), \tag{7}$$

$$w(y) = d(y)\frac{d}{dy}\beta(y)w(\beta(y)) + \delta(y) + \xi(y) + \sum_{i=1}^{m} Q_i q_i(y) + \sum_{i=1}^{l} F_i f_i(y). \tag{8}$$

Equations (7), (8) are called equilibrium proportions or balance equations. A model is called cyclic if the state of system S at the initial time t_0 coincides with the state of system S at the final time $t_0 + T$.

The application of principles I and II leads us to functional operators with shift.

We recall the definition of spaces of Holder functions with weight and the conditions of invertibility for scalar linear functional operators with shift.

3 Conditions of Invertibility in the Space of Holder Functions with Weight

The norm in weighted Holder spaces is defined as follows [9]. A function $\varphi(x)$ that satisfies the condition on $J = [0, x_{max}]$:

$$| \varphi(x_1) - \varphi(x_2) | \leq C \mid x_1 - x_2 |^{\varsigma}, \; x_1 \in J, x_2 \in J, \; \varsigma \in (0, \; 1),$$

is called a Holder function with exponent ς and constant C on J.

Let ρ be a function which has zeros at the endpoints $x = 0$, $x = x_{max}$:

$$\rho(x) = x^{\varsigma_0}(x_{max} - x)^{\varsigma_1}, \quad \varsigma < \varsigma_0 < 1 + \varsigma, \; \varsigma < \varsigma_1 < 1 + \varsigma.$$

The functions that become Holder functions and turn into zero at the points $x = 0$, $x = x_{max}$, after being multiplied by $\rho(x)$, form a Banach space. Functions of this space $H^0_{\varsigma}(J, \rho)$, are called Holder functions with weight ρ.

The norm in the space $H_\varsigma^0(J,\rho)$ is defined by

$$\| f(x) \|_{H_\varsigma^0(J,\rho)} = \| \rho(x) f(x) \|_{H_\varsigma(J)},$$

where

$$\| \rho(x) f(x) \|_{H_\varsigma(J)} = \|\rho(x) f(x)\|_C + \|\rho(x) f(x)\|_\varsigma,$$

$$\|\rho(x) f(x)\|_C = \max_{x\in J} |\rho(x) f(x)|,$$

$$\|\rho(x) f(x)\|_\varsigma = \sup_{x_1,x_2\in J, x_1\neq x_2} |\rho(x) f(x)|_\varsigma,$$

$$|\rho(x) f(x)|_\varsigma = \frac{| \rho(x_1) f(x_1) - \rho(x_2) f(x_2) |}{| x_1 - x_2 |^\varsigma}.$$

Let $\beta(x)$ be a bijective orientation-preserving displacement on J:
if $x_1 < x_2$ then $\beta((x_1) < \beta(x_2)$ for any $x_1 \in J,\ x_2 \in J$;
and let $\beta(x)$ have only two fixed points: $\beta(0) = 0$,
$\beta(x_{max}) = x_{max},\ \beta(x) \neq x,\ when\ \ x \neq 0,\ x \neq x_{max}$.

In addition, let $\beta(x)$ be a differentiable function and $\frac{d}{dx}\beta(x) \neq 0,\ x \in J$.

We consider the equation

$$(A\nu)(x) = f(x),$$

$$(A\nu)(x) \equiv a(x)(I\nu)(x) - b(x)\left(\Gamma_\beta \nu\right)(x), \quad x \in [0, x_{max}] \tag{9}$$

where I is the identity operator and Γ_β is the shift operator:

$$(I\nu)(x) = \nu(x), \quad \left(\Gamma_\beta \nu\right)(x) = \nu[\beta(x)].$$

Let functions $a(x),\ b(x)$ from the operator A belong to $H_\varsigma(J)$.

We will now formulate conditions of invertibility for the operator A from (9) in the space of Holder class functions with weight [7].

Theorem 1 *Operator A, acting in the Banach space $H_\varsigma^0(J,\rho)$, is invertible if the following condition is fulfilled:*

$$\theta_\beta[a(x), b(x), H_\varsigma^0(J,\rho)] \neq 0, \quad x \in J,$$

where the function σ_β is defined by:

$$\theta_\beta[a(x), b(x), H_\varsigma^0(J,\rho)] =$$

$$\begin{cases} a(x), \; when \mid a(0) \mid > \left[\beta'(0)\right]^{-\varsigma_0+\varsigma} \mid b(0) \mid; \\ \quad and, \mid a(x_{max}) \mid > \left[\beta'(x_{max})\right]^{-\varsigma_1+\varsigma} \mid b(x_{max}) \mid; \\ b(x), \; when \mid a(0) \mid < \left[\beta'(0)\right]^{-\varsigma_0+\varsigma} \mid b(0) \mid; \\ \quad and, \mid a(x_{max}) \mid < \left[\beta'(x_{max})\right]^{-\varsigma_1+\varsigma} \mid b(x_{max}) \mid; \\ 0 \quad in\ other\ cases. \end{cases}$$

Corollary 1 *If the following condition is fulfilled:*

$$\theta_\beta[a(x), b(x), H^0_\varsigma(J, \rho)] \neq 0, \quad x \in J,$$

then the operator

$$U = I - u\Gamma_\beta$$

is invertible in the space $H^o_\varsigma(J, \rho)$ and its inverse operator is

$$U^{-1} = \left(I + u\Gamma_\beta + \cdots + \left(\prod_{j=0}^{n-2} u[\beta_j(x)]\right) \Gamma_\beta^{n-1}\right) \cdot \left(I - \left(\prod_{j=0}^{n-1} u[\beta_j(x)]\right) \Gamma_\beta^n\right)^{-1}.$$

where

$$\beta_j(x) = (\Gamma_\beta^j x)(x)$$

and the number n is selected so that

$$\left\| \left(\prod_{j=0}^{n-1} u[\beta_j(x)]\right) \Gamma_\beta^n \right\|_{H^o_\varsigma(J,\rho)} < 1.$$

4 Analysis of Solvability of the Balance Equations and Finding of the Equilibrium State of S

Let S be a system with two resources, considered in Sect. 2. We find the equilibrium state of the system in which the initial distribution of the group parameters by the individual parameters $v(x)$, $w(y)$, $x \in (0, x_{max})$ coincide with the final distribution, after all transformations during the time interval T.

Rewrite the balance equations of the cyclic model (7), (8) for system S

$$(Vv)(x) = \sum_{i=1}^{n} P_i p_i(x) + \sum_{i=1}^{k} R_i r_i(x) + g(x), \tag{10}$$

$$(Ww)(y) = \sum_{i=1}^{m} Q_i q_i(y) + \sum_{i=1}^{l} F_i f_i(y) + h(y), \tag{11}$$

where

$$(Vv)(x) = v(x) - c_\alpha(x)v(\alpha(x)), g(x) = \rho(x) + \zeta(x), \quad x \in (0, x_{max}),$$

$$(Ww)(y) = w(x) - d_\beta(y)w(\beta(y)), \quad h(y) = \delta(y) + \xi(y), \quad y \in (0, y_{max})$$

and

$$c_\alpha(x) = c(x)\frac{d}{dx}\alpha(x), \quad d_\beta(y) = d(y)\frac{d}{dy}\beta(y).$$

Let us study the model in the space of Holder class functions with weight:

$$H_\varsigma^o(J, \rho), \ J = 0, x_{max}], \ \rho(x) = x^{\varsigma_0}(x_{max} - x)^{\varsigma_1},$$

$$\varsigma < \varsigma_0 < 1 + \varsigma, \ \varsigma < \varsigma_1 < 1 + \varsigma.$$

$$H_\vartheta^0(L, \sigma), \ L = [0, y_{max}], \ \sigma(y) = y^{\vartheta_0}(y_{max} - y)^{\vartheta_1},$$

$$\vartheta < \vartheta_0 < 1 + \vartheta, \ \vartheta < \vartheta_1 < 1 + \vartheta,$$

considering conditions of invertibility of operators V and W fulfilled

$$\theta_\alpha[1, c_\alpha(x), H_\varsigma^0(J, \rho)] \neq 0, \quad x \in J,$$

$$\theta_\beta[1, d_\beta(y), H_\vartheta^0(L, \sigma)] \neq 0, \quad y \in L.$$

Additionally, let us consider as known the integer positive constants N, M, for which the following inequalities are fulfilled:

$$\left\| \left(\prod_{j=0}^{N-1} c_\alpha(x)[\alpha_j(x)] \right) \Gamma_\alpha^N \right\|_{H_\varsigma^o(J,\rho)} < 1,$$

$$\left\| \left(\prod_{j=0}^{M-1} d_\beta[\beta_j(y)] \right) \Gamma_\beta^M \right\|_{H_\vartheta^0(L,\sigma)} < 1,$$

where

$$(\Gamma_\alpha \varphi)(x) = \varphi[\alpha(x)], \quad \alpha_j(x) = (\Gamma_\alpha^j x)(x),$$

$$(\Gamma_\beta \varphi)(y) = \varphi[\beta(y)], \quad \beta_j(y) = (\Gamma_\beta^j y)(y)$$

From Theorem 1 and Corollary 1 from Sect. 3, operators inverse to operators V and W are:

$$V^{-1} = \left(I + c_\alpha \Gamma_\alpha + \cdots + \left(\prod_{j=0}^{N-2} c_\alpha[\alpha_j(x)] \right) \Gamma_\alpha^{N-1} \right) \cdot \left(I - \left(\prod_{j=0}^{N-1} c_\alpha[\alpha_j(x)] \right) \Gamma_\alpha^{N} \right)^{-1},$$

$$W^{-1} = \left(I + d_\beta \Gamma_\beta + \cdots + \left(\prod_{j=0}^{M-2} d_\beta[\beta_j(y)] \right) \Gamma_\beta^{M-1} \right) \cdot \left(I - \left(\prod_{j=0}^{M-1} d_\beta[\beta_j(y)] \right) \Gamma_\beta^{M} \right)^{-1}.$$

First, let us apply on the left side operators V^{-1}, W^{-1} to Eqs. (10), (11); we have obtained a system of linear equations:

$$v(x) = \sum_{i=1}^{n} P_i (V^{-1} p_i)(x) + \sum_{i=1}^{k} R_i (V^{-1} r_i)(x) + (V^{-1} g)(x),$$

$$w(y) = \sum_{i=1}^{m} Q_i (W^{-1} q_i)(y) + \sum_{i=1}^{l} F_i (W^{-1} f_i)(y) + (W^{-1} h)(y).$$

For solving the system of equations, we use the idea for solution of integral equations of Fredholm of the second type with degenerate kernel [10, 11].

Having integrated the first equation of system over intervals $[\nu_{j-1}, \nu_j]$, $j = 1, 2, \ldots, n$ corresponding to constants

$$P_j = \int_{\nu_{j-1}}^{\nu_j} v(x)dx,$$

and over intervals $[\varepsilon_{j-1}, \varepsilon_j]$, $j = 1, 2, \ldots, l$ corresponding to constants

$$F_j = \int_{\varepsilon_{j-1}}^{\varepsilon_j} v(x)dx,$$

and having subsequently integrated the second equation of system over intervals $[\mu_{j-1}, \mu_j]$, $j = 1, 2, \ldots, m$ corresponding to constants

$$Q_j = \int\limits_{\mu_{j-1}}^{\mu_j} w(y)dy,$$

and over intervals $[\gamma_{j-1}, \gamma_j]$, $j = 1, 2, \ldots, k$ corresponding to constants

$$R_j = \int\limits_{\gamma_{j-1}}^{\gamma_j} w(y)dy$$

we have

$$P_j = \sum_{i=1}^{n} P_i \int\limits_{\nu_{j-1}}^{\nu_j} (V^{-1}p_i)(x)dx + \sum_{i=1}^{k} R_i \int\limits_{\nu_{j-1}}^{\nu_j} (V^{-1}r_i)(x)dx$$
$$+ \int\limits_{\nu_{j-1}}^{\nu_j} (V^{-1}g)(x)dx, \ \ j = 1, 2, \ldots, n,$$

$$F_j = \sum_{i=1}^{n} P_i \int\limits_{\varepsilon_{j-1}}^{\varepsilon_j} (V^{-1}p_i)(x)dx + \sum_{i=1}^{k} R_i \int\limits_{\varepsilon_{j-1}}^{\varepsilon_j} (V^{-1}r_i)(x)dx$$
$$+ \int\limits_{\varepsilon_{j-1}}^{\varepsilon_j} (V^{-1}g)(x)dx, \ \ j = 1, 2, \ldots, l,$$

$$Q_j = \sum_{i=1}^{m} Q_i \int\limits_{\mu_{j-1}}^{\mu_j} (W^{-1}q_i)(y)dy + \sum_{i=1}^{l} F_i \int\limits_{\mu_{j-1}}^{\mu_j} (W^{-1}f_i)(y)dy$$
$$+ \int\limits_{\mu_{j-1}}^{\mu_j} (W^{-1}h)(y)dy, \ \ j = 1, 2, \ldots, m,$$

$$R_j = \sum_{i=1}^{m} Q_i \int_{\gamma_{j-1}}^{\gamma_j} (W^{-1}q_i)(y)dy + \sum_{i=1}^{l} F_i \int_{\gamma_{j-1}}^{\gamma_j} (W^{-1}f_i)(y)dy$$
$$+ \int_{\gamma_{j-1}}^{\gamma_j} (W^{-1}h)(y)dy, \; j = 1, 2, \ldots, k.$$

5 Exploitation Problems for Resources of Open Systems

We assume that the individual and the group parameters for both resources are the same, and the ranges of variation of the individual parameters are different. Resource λ_1 has a range of variation equal to $I_1 = (0, x^1_{max})$ and resource λ_2 has a range of variation equal to $I_2 = (0, x^2_{max})$.

Time interval $[t_0, t_0 + T]$ is divided into subintervals T_k, $[t_0, t_0 + T] = \bigcup T_k$. The subintervals are chosen taking into account the changes taking place in system S and human activity.

To avoid introducing new notation, symbols in this section are independent and are not related to symbols in previous sections.

Let β_{1k}(x) be a shift which describes the change of the individual parameter x of resource λ_1 during $T_k = [t_{k-1}, t_k]$. For resource λ_2, we denote the corresponding shift by β_{2k}. Let $\beta_{1k}(x)$, $\beta_{2k}(x)$ be differentiable functions and their inverse functions be $\alpha_{1k}(x)$, $\alpha_{2k}(x)$.

Applying the approach proposed in Sect. 2, we will obtain a system of balance relations with non-Carleman shifts $\alpha_{1k}(x)$, $\alpha_{2k}(x)$

$$\nu_{1k+1}(x) = (A_{1k}\nu_{1k})(x), \quad \nu_{2k+1}(x) = (A_{2k}\nu_{2k})(x), \tag{12}$$

where

$$(A_{1k}\nu_{1k})(x) = (B_{1k}\nu_{1k})(x) + (B_{1k}q_{1k})(x) + P_{1k}(x) + R_{1k}(x) - g_{1k}(x),$$
$$(B_{1k}\nu_{1k})(x) = d_{1k}(x)\alpha'{}_{1k}(x)\nu_{1k}[\alpha'{}_{1k}, (x)],$$

$$(A_{2k}\nu_{2k})(x) = (B_{2k}\nu_{2k})(x) + (B_{2k}q_{2k})(x) + P_{2k}(x) + R_{2k}(x)p_{2k}(x) - g_{2k}(x),$$
$$(B_{2k}\nu_{2k})(x) = d_{2k}(x)\alpha'{}_{2k}(x)\nu_{2k}[\alpha'{}_{2k}(x)].$$

Here, the processes of the reduction of the group parameters (natural mortality) is described by d_{ik}.

Terms P_{1k}, P_{2k} are responsible for the processes of natural increase of the group parameters (reproduction) and terms R_{1k}, R_{2k} are responsible for the reciprocal influence between resources. These terms have the same form as their corresponding terms in balance equations (3) and (4), namely, integrals with degenerate kernels.

The extractions from the system as a result of human economic activity are represented by $g_{ik}(x)$.

The artificial input into the system is accounted for q_{ik}.

We will assume that the actions related to q_{ik} take place at the beginning of period T_k and the actions related to g_{ik} take place at the end of period T_k.

The state $\nu_{1k}(x)$, $\nu_{2k}(x)$ of the system at time t_k is described by Eq. (12).

Note that in this section we do not require that the final state and the initial state of system S should coincide. Thus, we have an open system

$$\nu_{1k+1}(x) \neq \nu_{1k}(x), \quad \nu_{2k+1}(x) \neq \nu_{2k}(x).$$

Let us formulate problems related to the exploitation of system S.

We introduce into consideration a functional that estimates the system for the period $[t_0, t_0 + T]$

$$E = E^g - E^q.$$

Let C^g_{ikj} be some known constants

$$G^g_{ikj} = \int_{x_{ij}}^{x_{ij+1}} g_{ik}(\tau) d\tau,$$

where $g_{ik}(\tau)$ describes the extraction of resource λ_i from the system during the period $T_k = [t_{k-1}, t_k]$.

Here, interval $[0, x^i_{max}]$ is divided into sections $0 = x_{i0} < x_{i1} < x_{i2} < \cdots < x^i_{max}$.

Let us consider the functional

$$E^g_{ik} = \sum C^g_{ikj} G^g_{ikj}, \tag{13}$$

where C^g_{ikj} are some known constants (product prices). It economically estimates the cost of $g_{ik}(\tau)$. The term E^q has a similar structure.

Based on the presented model, it is possible to formulate some problems for the rational use of system S. What should the strategy of selection of q_{ik} and g_{ik} be on each interval T_k in order to reach the maximum value of E at the final moment $t_0 + T$? Various restrictions should, of course, be complied with. Some are listed below.

Restrictions on functions $q_{ik}(x)$ are:

$$Q_{ikj} \le Q_{ikj}^{max}, \quad Q_{ikj} = \int\limits_{I_{ij}} q_{ik}(\tau)d\tau, \tag{14}$$

and

$$E_{ikj}^{qmin} \le E_{ikj}^{q} \le E_{ikj}^{qmax}, \quad E_{ikj}^{q} = C_{ikj}^{q} Q_{ikj},$$

$$E_{ik}^{qmin} \le \sum_{j} E_{ikj}^{q} \le E_{ikj}^{qmax}, F_{i0}^{qmin} \le \sum_{k}\sum_{j} E_{ikj}^{q} \le F_{i0}^{qmax},$$

$$F_{0k}^{qmin} \le \sum_{i}\sum_{j} E_{ikj}^{q} \le F_{0k}^{qmax}, \quad E^{qmin} \le E^{q} \le F^{qmax}.$$

Restrictions on functions $g_{ik}(x)$ are:

$$G_{ikj}^{gmin} \le G_{ikj}^{g} \le G_{ikj}^{gmax}, \quad G_{ikj}^{g} = \int\limits_{I_{ij}} g_{ik}(\tau)d\tau,$$

$$G_{ik}^{qmin} \le \sum_{j} G_{ikj}^{g} \le G_{ikj}^{gmax}, K_{i0}^{qmin} \le \sum_{k}\sum_{j} E_{ikj}^{q} \le K_{i0}^{qmax},$$

$$K_{0k}^{gmin} \le \sum_{i}\sum_{j} G_{ikj}^{g} \le K_{0k}^{gmax}, \quad G^{gmin} \le G^{g} \le F^{gmax},$$

$$G_{00}^{gmin} \le G_{00}^{g} \le G_{00}^{gmin}, \quad G_{00}^{g} = \sum_{ik} G_{ik}.$$

The state of the system can not be worse than the permissible state and ecologically acceptable norms must be met.

Restrictions on functions $v_{ik}(x)$ are:

$$N_{ikj}^{min} \le N_{ikj}, \quad N_{ikj} = \int\limits_{I_{ij}} v_{ik}(\tau)d\tau, \quad N_{ik0}^{min} \le \sum_{j} N_{ikj}, \quad N_{0k0}^{min} \le \sum_{i} N_{ik}.$$

The functions $v_{ik}(x)$, $q_{ik}(x)$, $g_{ik}(x)$ must be non-negative.

Simplification of the form of functions, functionals and algorithms for solving particular problems, the authors plan to submit to WCE 2018.

6 Conclusions

On modelling systems with renewable resources, equations with shift appear [7, 8]. The theory of linear functional operators with shift is the adequate mathematical instrument for the investigation of such systems. In this work, we study systems with two renewable resources and our approach is based on functional operators with shift. We constructed the inverse operators with shift acting in the weighted Holder spaces and used these operators to find the equilibrium state of the considered systems. Based on the presented model, it is possible to formulate some problems for the rational use of resources of open system S.

References

1. C.W. Clark, *Mathematical Bioeconomics: The Optimal Management of Renewable Resources*, 2nd edn. (Wiley, New York, 1990)
2. Y. Xiao, D. Cheng, H. Qin, Optimal Impulsive control in periodic ecosystem. Syst. Control Lett. **55**(2), 558–565 (2006)
3. M. Bohner, A. Peterson, *Dynamic Equations on Time Scales: An Introduction with Applications* (Birkhauser, Boston, 2001)
4. L. Wang, M. Hu, Nonzero periodic solutions in shifts Delta(+/−) for a higher-dimensional nabla dynamic equation on time scales. IAENG Int. J. Appl. Math. **47**(1), 37–42 (2017)
5. L. Yang, Y. Yang, Y. Li et al., Almost periodic solution for a Lotka-Volterra recurrent networks with harvesting terms on time scales. Eng. Lett. **24**(4), 455–460 (2016)
6. O. Karelin, A. Tarasenko, V. Zolotov et al., Mathematical model for the study of the equilibrium state of renewable systems based on functional operators with shift, in *Lecture Notes in Engineering and Computer Science: Proceedings of the World Congress on Engineering 2017*, 5–7 July 2017, London, U.K. (2017), pp. 40–44
7. A. Tarasenko, O. Karelin, G.P. Lechuga et al., Modelling systems with renewable resources based on functional operators with shift. Appl. Math. Comput. **216**(7), 1938–1944 (2010)
8. O. Karelin, A. Tarasenko, M.G. Hernández, Application of functional operators with shift to the study of renewable systems when the reproductive processed is describedby integrals with degenerate kernels. Appl. Math. (AM) **4**, 1376–1380 (2013)
9. R.V. Duduchava, Convolution integral equations with discontinuous presymbols, singular integral equations with fixed singularities and their applications to problem in mechanics. Trudy Tbilisskogo Mat. Inst. Acad. Nauk Gruz. SSR **60**(4), 2–136 (1979)
10. K.E. Atkinson, *The Numerical Solution of Integral Equations of the Second Kind* (Cambridge University Press, 1997)
11. A.J. Jerry, *Introduction to Integral Equations with Application*, 2nd edn. (Wiley, 1999)

A Useful Extension of the Inverse Exponential Distribution

Pelumi E. Oguntunde, Adebowale O. Adejumo, Mundher A. Khaleel, Enahoro A. Owoloko, Hilary I. Okagbue and Abiodun A. Opanuga

Abstract This chapter explores the three-parameter Weibull Inverse Exponential distribution. The various and basic structural properties of the distribution are defined and established. Applications to real life datasets were provided and the unknown model parameters were estimated using the maximum likelihood estimation method. The results show that the Weibull Inverse Exponential distribution is a viable alternative to its counterpart distribution(s) based on the selection criteria used.

Keywords Distribution · Generalized model · Inverse exponential · Mathematical statistics · Statistical properties · Weibull generalized family of distributions

---

P. E. Oguntunde (✉) · A. O. Adejumo · E. A. Owoloko · H. I. Okagbue · A. A. Opanuga
Department of Mathematics, Covenant University, Ota, Nigeria
e-mail: pelumi.oguntunde@covenantuniversity.edu.ng

A. O. Adejumo
e-mail: aodejumo@unilorin.edu.ng

E. A. Owoloko
e-mail: alfred.owoloko@covenantuniversity.edu.ng

H. I. Okagbue
e-mail: hilary.okagbue@covenantuniversity.edu.ng

A. A. Opanuga
e-mail: abiodun.opanuga@covenantuniversity.edu.ng

A. O. Adejumo
Department of Statistics, University of Ilorin, Ilorin, Kwara State, Nigeria

M. A. Khaleel
Faculty of Computer Science and Mathematics, Department of Mathematics,
University of Tikrit, Tikrit, Iraq
e-mail: mun880088@gmail.com

S.-I. Ao et al. (eds.), *Transactions on Engineering Technologies*,
https://doi.org/10.1007/978-981-13-0746-1_8

1 Introduction

The Weibull Inverse Exponential (WIE) distribution is a three-parameter distribution obtained by generalizing the Inverse Exponential distribution using the Weibull generalized family of distributions due to [1]. The Inverse Exponential distribution itself was introduced by [2] as a modification of the well-known Exponential distribution, its introduction caters for some of the short comings of the Exponential distribution (the constant failure rate and memoryless property).

Developing new compound distributions have become the order of the day in Statistics, particularly in probability distribution theory. These compound distributions involve extra parameters which are known as shape parameters and their role is to induce skewness into the parent (or baseline) distribution. On application to real life datasets, they also appear better and flexible than the parent distribution. Examples of well-known family of distributions include Beta generalized family of distribution [3], Kumaraswamy family of distributions [4] and several others which are mentioned in [5–7] and the references therein.

Examples of various authors that have as well used the Weibull generalized family of distributions include [8–12] and the references therein. In this research however, the work of [8] is extended to include more statistical properties and applications.

2 Weibull Inverse Exponential (WIE) Distribution

According to [1], the Weibull generalized family of distributions was derived from:

$$F(x) = \int_{0}^{\frac{G(x)}{1-G(x)}} abt^{b-1}e^{-at^{b}} \tag{1}$$

In clear terms, its cumulative distribution function (cdf) and probability density function (pdf) are:

$$F(x) = 1 - \exp\left\{-a\left[\frac{G(x)}{1-G(x)}\right]^{b}\right\} \tag{2}$$

and

$$f(x) = abg(x)\frac{G(x)^{b-1}}{[1-G(x)]^{b+1}}\exp\left\{-a\left[\frac{G(x)}{1-G(x)}\right]^{b}\right\} \tag{3}$$

respectively. Where $a > 0$ and $b > 0$ are additional shape parameters.

Also, the cdf and the pdf of the Inverse Exponential distribution which is the parent distribution are:

$$G(x) = \exp\left(-\frac{c}{x}\right); \quad x > 0, c > 0 \tag{4}$$

and

$$g(x) = \frac{c}{x^2}\exp\left(-\frac{c}{x}\right); \quad x > 0, c > 0 \tag{5}$$

respectively. Where c is a scale parameter.

Now, the cdf of the Weibull Inverse Exponential (WIE) distribution is obtained by substituting Eq. (4) into Eq. (2) as:

$$F(x) = 1 - \exp\left\{-a\left[\frac{\exp(-\frac{c}{x})}{1-\exp(-\frac{c}{x})}\right]^b\right\} \quad ; x > 0, a, b, c > 0 \tag{6}$$

Similarly, the corresponding pdf is obtained by substituting both Eqs. (4) and (5) into Eq. (3) and simplified as:

$$f(x) = ab\frac{c}{x^2}\frac{\{\exp(-\frac{c}{x})\}^b}{[1-\exp(-\frac{c}{x})]^{b+1}}\exp\left\{-a\left[\frac{\exp(-\frac{c}{x})}{1-\exp(-\frac{c}{x})}\right]^b\right\} \tag{7}$$

where a and b are shape parameters and c is the scale parameter.

2.1 Special Case

When parameter $b = 1$, the WIE distribution would reduce to give the Exponential Inverse Exponential (EIE) distribution proposed by [13].

Figure 1 shows the graphical representation of the pdf of the WIE distribution at different parameter values.

From Fig. 1, the WIE distribution shows the following shapes; unimodal and decreasing.

Figure 2 shows a possible plot for the cdf of the WIE distribution.

3 Basic Statistical Properties of the WIE Distribution

The reliability analysis which include the survival, hazard, reversed hazard and odd functions are derived, the quantile function, median, distribution of order statistics and estimation of parameters are also provided in this section.

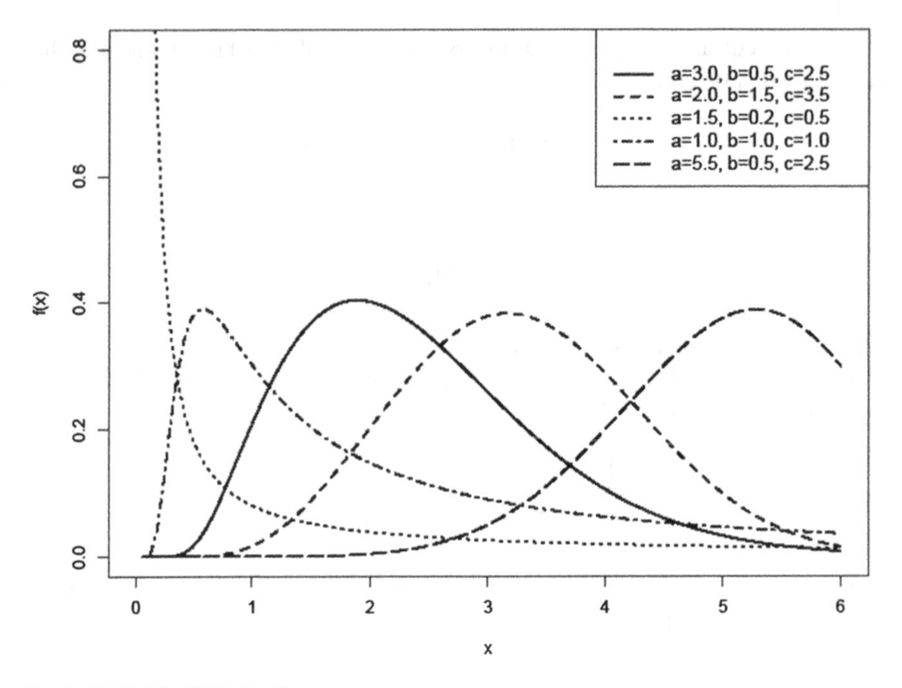

Fig. 1 PDF of the WIE distribution

3.1 Survival Function

The mathematical expression for survival (or reliability) function is:

$$\bar{F}(x) = 1 - F(x) \tag{8}$$

So, the survival function of the WIE is derived as:

$$\bar{F}(x) = \exp\left\{-a\left[\frac{\exp(-\frac{c}{x})}{1-\exp(-\frac{c}{x})}\right]^{b}\right\} \quad ; \; x > 0, a, b, c > 0 \tag{9}$$

3.2 Hazard Function

The mathematical expression for hazard function (or failure rate) is:

$$h(x) = \frac{f(x)}{\bar{F}(x)} \tag{10}$$

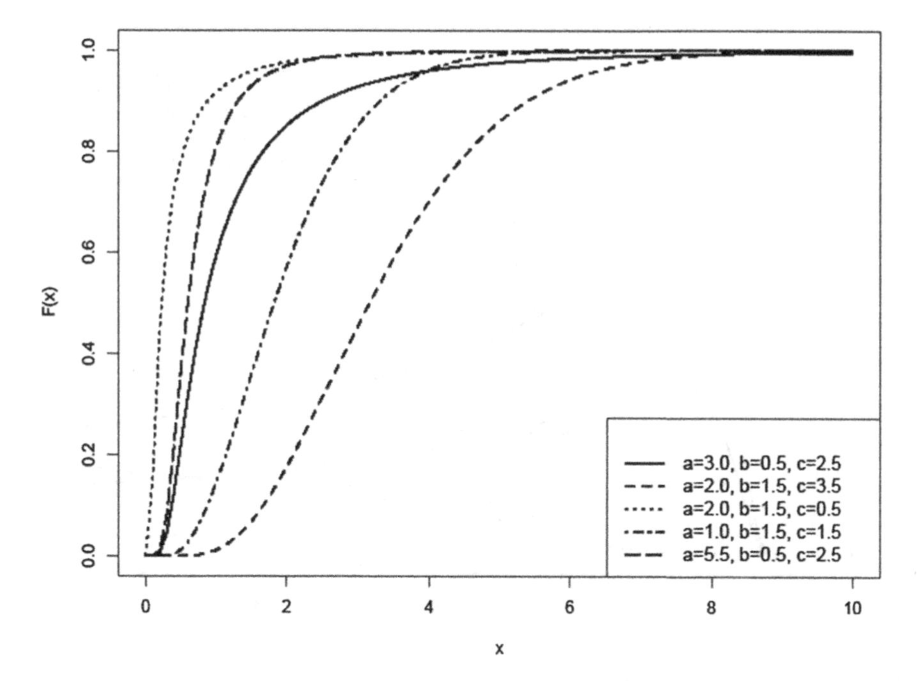

Fig. 2 CDF of the WIE distribution

So, the hazard function of the WIE is derived as:

$$h(x) = ab\frac{c}{x^2}\frac{\left\{\exp\left(-\frac{c}{x}\right)\right\}^b}{\left[1-\exp\left(-\frac{c}{x}\right)\right]^{b+1}} \quad ; x > 0, a, b, c > 0 \tag{11}$$

Figure 3 shows the graphical representation of the hazard function for the WIE distribution at different parameter values.

As shown in Fig. 3, the failure rate of the WIE has the following shapes; increasing, decreasing and unimodal (or inverted bathtub).

3.3 *Reversed Hazard Function*

The mathematical expression for reversed hazard function is:

$$r(x) = \frac{f(x)}{F(x)} \tag{12}$$

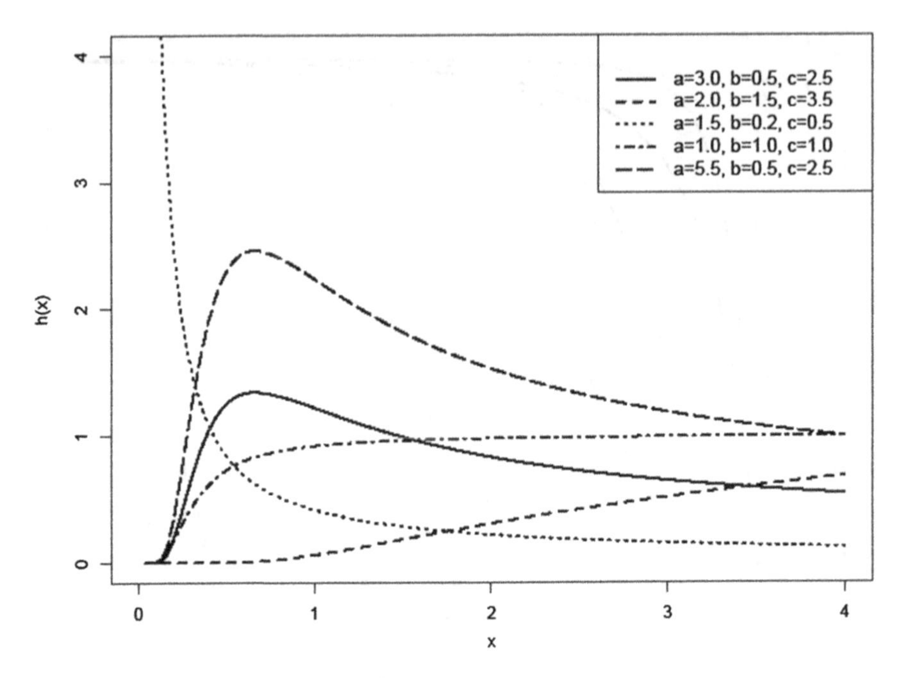

Fig. 3 Failure Rate of the WIE distribution

So, the reversed hazard function of the WIE is derived as:

$$r(x)=\frac{ab\frac{c}{x^2}\frac{\{\exp(-\frac{c}{x})\}^b}{[1-\exp(-\frac{c}{x})]^{b+1}}\exp\left\{-a\left[\frac{\exp(-\frac{c}{x})}{1-\exp(-\frac{c}{x})}\right]^b\right\}}{1-\exp\left\{-a\left[\frac{\exp(-\frac{c}{x})}{1-\exp(-\frac{c}{x})}\right]^b\right\}} \tag{13}$$

For $x > 0, a, b, c > 0$

3.4 Odds Function

The mathematical expression for odds hazard function is:

$$O(x)=\frac{F(x)}{\bar{F}(x)} \tag{14}$$

So, the odds function of the WIE is derived as:

$$O(x) = \frac{1 - \exp\left\{-a\left[\frac{\exp(-\frac{c}{x})}{1-\exp(-\frac{c}{x})}\right]^b\right\}}{\exp\left\{-a\left[\frac{\exp(-\frac{c}{x})}{1-\exp(-\frac{c}{x})}\right]^b\right\}}\ ; x > 0, a, b, c > 0 \tag{15}$$

3.5 Quantile Function

This is usually derived from:

$$Q(u) = F^{-1}(u) \tag{16}$$

So, the quantile function of the WIE is derived as:

$$Q(u) = \frac{c}{\log\left\{\left[-a^{-1}\log(1-u)\right]^{-\frac{1}{b}} + 1\right\}} \tag{17}$$

where; $u \sim Uniform(0, 1)$.

3.6 Median

The median is obtained directly form Eq. (17) by substituting u = 0.5 as follows:

$$Median = \frac{c}{\log\left\{\left[-a^{-1}\log(0.5)\right]^{-\frac{1}{b}} + 1\right\}} \tag{18}$$

It is good to note that the median is the second quartile, other quartiles can as well be derived from Eq. (17) by substituting appropriate values of u.

3.7 Estimation of Parameters

Supposing that $x_1, x_2, \ldots, x_n$ denote random samples drawn from the WIE distribution, then the likelihood function is of the form:

$$f(x_1, x_2, \ldots x_n; a, b, c) = \prod_{i=1}^{n}\left[ab\frac{c}{x_i^2}\frac{\left[\exp\left(-\frac{c}{x_i}\right)\right]^b}{\left[1-\exp\left(-\frac{c}{x_i}\right)\right]^{b+1}}\exp\left\{-a\left[\frac{\exp\left(-\frac{c}{x_i}\right)}{1-\exp\left(-\frac{c}{x_i}\right)}\right]^b\right\}\right] \tag{19}$$

The log-likelihood function which is denoted by L is of the form:

$$l = n\log(a) + n\log(b) + n\log(c) - 2\sum_{i=1}^{n}\log(x_i) - b\sum_{i=1}^{n}\left(\frac{c}{x_i}\right)$$
$$-(b+1)\sum_{i=1}^{n}\log\left[1-\exp\left(-\frac{c}{x_i}\right)\right] - a\sum_{i=1}^{n}\left[\frac{\exp\left(-\frac{c}{x_i}\right)}{1-\exp\left(-\frac{c}{x_i}\right)}\right]^{b} \quad (20)$$

Differentiating Eq. (20) with respect to parameters a, b and c gives the following:

$$\frac{dL}{da} = \frac{n}{a} - \sum_{i=1}^{n}\left[\frac{\exp\left(-\frac{c}{x_i}\right)}{1-\exp\left(-\frac{c}{x_i}\right)}\right]^{b} \quad (21)$$

$$\frac{dL}{db} = \frac{n}{b} - \sum_{i=1}^{n}\left(\frac{c}{x_i}\right) - \sum_{i=1}^{n}\log\left[1-\exp\left(-\frac{c}{x_i}\right)\right]$$
$$-a\sum_{i=1}^{n}\left[\frac{\exp\left(-\frac{c}{x_i}\right)}{1-\exp\left(-\frac{c}{x_i}\right)}\right]^{b}\log\left[\frac{\exp\left(-\frac{c}{x_i}\right)}{1-\exp\left(-\frac{c}{x_i}\right)}\right] \quad (22)$$

$$\frac{dL}{dc} = \frac{n}{c} - b\sum_{i=1}^{n}\left(\frac{1}{x_i}\right) - (b+1)\sum_{i=1}^{n}\frac{\frac{1}{x_i}\exp\left(-\frac{c}{x_i}\right)}{\left[1-\exp\left(-\frac{c}{x_i}\right)\right]}$$
$$-ab\sum_{i=1}^{n}\left[\frac{\exp\left(-\frac{c}{x_i}\right)}{1-\exp\left(-\frac{c}{x_i}\right)}\right]^{b-1} \times W(x;c) \quad (23)$$

Meanwhile, $W(x;c) = \frac{d}{dc}\left[\frac{\exp\left(-\frac{c}{x_i}\right)}{1-\exp\left(-\frac{c}{x_i}\right)}\right]$

Equating Eqs. (21), (22) and (23) to zero and solving the nonlinear system of equations gives the maximum likelihood estimates of parameters a, b and c. The solution cannot be obtained in closed form; therefore statistical software can be adopted to estimate the parameters numerically. In particular, R software was used in this research for this purpose.

3.8 Order Statistics

Supposing that $x_1, x_2, \ldots, x_n$ denote random samples drawn from the WIE, the pdf of the jth order statistics for the WIE distribution is derived as follows:

$$f_{j:n}(x) = \frac{n!}{(n-j)!\,(j-1)!} f(x)[F(x)]^{j-1}\left[\bar{F}(x)\right]^{n-j} \tag{24}$$

Therefore;

$$f_{j:n}(x) = abn\frac{c}{x^2}\exp\left(-\frac{c}{x}\right)\frac{\left[\exp\left(-\frac{c}{x}\right)\right]^{b-1}}{\left[1-\exp\left(-\frac{c}{x}\right)\right]^{b+1}}$$
$$\times \exp\left\{-a\left[\frac{\exp\left(-\frac{c}{x}\right)}{1-\exp\left(-\frac{c}{x}\right)}\right]^b\right\} \times \left\langle 1-\exp\left\{-a\left[\frac{\exp\left(-\frac{c}{x}\right)}{1-\exp\left(-\frac{c}{x}\right)}\right]^b\right\}\right\rangle^{j-1}$$
$$\times \left\langle \exp\left\{-a\left[\frac{\exp\left(-\frac{c}{x}\right)}{1-\exp\left(-\frac{c}{x}\right)}\right]^b\right\}\right\rangle^{n-j}$$

The distributions of minimum and maximum order statistics for the WIE are respectively given as:

$$f_{1:n}(x) = abn\frac{c}{x^2}\exp\left(-\frac{c}{x}\right)\frac{\left[\exp\left(-\frac{c}{x}\right)\right]^{b-1}}{\left[1-\exp\left(-\frac{c}{x}\right)\right]^{b+1}}\left\langle \exp\left\{-a\left[\frac{\exp\left(-\frac{c}{x}\right)}{1-\exp\left(-\frac{c}{x}\right)}\right]^b\right\}\right\rangle^{n} \tag{25}$$

and

$$f_{n:n}(x) = abn\frac{c}{x^2}\exp\left(-\frac{c}{x}\right)\frac{\left[\exp\left(-\frac{c}{x}\right)\right]^{b-1}}{\left[1-\exp\left(-\frac{c}{x}\right)\right]^{b+1}}\exp\left\{-a\left[\frac{\exp\left(-\frac{c}{x}\right)}{1-\exp\left(-\frac{c}{x}\right)}\right]^b\right\}$$
$$\times \left\langle 1-\exp\left\{-a\left[\frac{\exp\left(-\frac{c}{x}\right)}{1-\exp\left(-\frac{c}{x}\right)}\right]^b\right\}\right\rangle^{n-1} \tag{26}$$

4 Real Life Applications

In a bid to demonstrate the potentials of the WIE distribution, it was applied to two real life datasets. The Negative Log-likelihood (NLL) and the Akaike Information Criteria (AIC) are used to select the best distribution. The lower the value of these criteria, the better the model.

First Data: This has been used previously by [14, 15] and it relates to the survival times of a group of patients suffering from Head and Neck cancer diseases and treated using a combination of radiotherapy and chemotherapy (RT + CT). The observations are:

Table 1 Table showing the results from the first data

Distributions	$\hat{a}$	$\hat{b}$	$\hat{c}$	NLL	AIC
Weibull inverse exponential	0.7609	0.2488	0.0018	200.40	406.81
Inverse exponential	–	–	76.7000	279.58	561.15

Table 2 Table showing the results from the second data

Distributions	$\hat{a}$	$\hat{b}$	$\hat{c}$	NLL	AIC
Weibull inverse exponential	0.3989	0.3105	0.0047	227.98	461.95
Inverse exponential	–	–	2.4847	460.38	922.76

12.20, 23.56, 23.74, 25.87, 31.98, 37, 41.35, 47.38, 55.46, 58.36, 63.47, 68.46, 78.26, 74.47, 81.43, 84, 92, 94, 110, 112, 119, 127, 130, 133, 140, 146, 155, 159, 173, 179, 194, 195, 209, 249, 281, 319, 339, 432, 469, 519, 633, 725, 817, 1776.

The result of the analysis with respect to the first data is as given in Table 1.

Second Data: This has been used previously by [16, 17] and it relates to the remission times of a random sample of 128 bladder cancer patients. The observations are:

0.08, 2.09, 3.48, 4.87, 6.94, 8.66, 13.11, 23.63, 0.20, 2.23, 3.52, 4.98, 6.97, 9.02, 13.29, 0.40, 2.26, 3.57, 5.06, 7.09, 9.22, 13.80, 25.74, 0.50, 2.46, 3.64, 5.09, 7.26, 9.47, 14.24, 25.82, 0.51, 2.54, 3.70, 5.17, 7.28, 9.74, 14.76, 26.31, 0.81, 2.62, 3.82, 5.32, 7.32, 10.06, 14.77, 32.15, 2.64, 3.88, 5.32, 7.39, 10.34, 14.83, 34.26, 0.90, 2.69, 4.18, 5.34, 7.59, 10.66, 15.96, 36.66, 1.05, 2.69, 4.23, 5.41, 7.62, 10.75, 16.62, 43.01, 1.19, 2.75, 4.26, 5.41, 7.63, 17.12, 46.12, 1.26, 2.83, 4.33, 5.49, 7.66, 11.25, 17.14, 79.05, 1.35, 2.87, 5.62, 7.87, 11.64, 17.36, 1.40, 3.02, 4.34, 5.71, 7.93, 11.79, 18.10, 1.46, 4.40, 5.85, 8.26, 11.98, 19.13, 1.76, 3.25, 4.50, 6.25, 8.37, 12.02, 2.02, 3.31, 4.51, 6.54, 8.53, 12.03, 20.28, 2.02, 3.36, 6.76, 12.07, 21.73, 2.07, 3.36, 6.93, 8.65, 12.63, 22.69.

The result of the analysis with respect to the second data is as given in Table 2.

From Tables 1 and 2, the distribution that corresponds to the lowest NLL and AIC values is the WIE distribution. Therefore, the WIE distribution can be selected as the best distribution among the distributions considered.

5 Conclusion and Future Work

The Inverse Exponential distribution has been successfully extended using the Weibull generalized family of distributions. Explicit expressions for the densities and structural properties of the Weibull Inverse Exponential (WIE) distribution have been successfully derived. Real life applications to demonstrate the potentials of the WIE distribution reveal that the WIE distribution is a viable alternative to the Inverse

Exponential distribution. Further works would involve investigating the behaviors of the WIE parameters using a simulation study.

Acknowledgements This work was supported by Covenant University, Nigeria.

References

1. M. Bourguignon, R.B. Silva, G.M. Cordeiro, The Weibull-G family of probability distributions. J. Data Sci. **12**, 53–68 (2014)
2. A.Z. Keller, A.R. Kamath, Reliability analysis of CNC machine tools. Reliab. Eng. **3**, 449–473 (1982)
3. N. Eugene, C. Lee, F. Famoye, Beta-normal distribution and its applications. Commun. Stat.: Theory Methods **31**, 497–512 (2002)
4. G.M. Cordeiro, M. de Castro, A new family of generalized distributions. J. Stat. Comuput. Simul. **81**, 883–898 (2011)
5. E.A. Owoloko, P.E. Oguntunde, A.O. Adejumo, Performance rating of the transmuted exponential distribution: an analytical approach. SpringerPlus **4**, 818 pp. (2015)
6. P.E. Oguntunde, A.O. Adejumo, H.I. Okagbue, M.K. Rastogi, Statistical properties and applications of a new Lindley exponential distribution. Gazi Univ. J. Sci. **29**(4), 831–838 (2016)
7. P.E. Oguntunde, M.A. Khaleel, M.T. Ahmed, A.O. Adejumo, O A. Odetunmibi, A new generalization of the Lomax distribution with increasing, decreasing and constant failure rate. Modell. Simul. Eng. **2017**, Article ID 6043169, 6 pp. (2017)
8. P.E. Oguntunde, A.O. Adejumo, E.A. Owoloko, The Weibull-inverted exponential distribution: a generalization of the inverse exponential distribution. In: Lecture Notes in Engineering and Computer Science: Proceedings of the World Congress on Engineering 2017, WCE 2010, 5–7 July 2017, London, U.K., pp. 16–19
9. F. Merovci, I. Elbatal, Weibull rayleigh distribution: theory and applications. Appl. Math. Inf. Sci. **9**(4), 2127–2137 (2015)
10. P.E. Oguntunde, O.S. Balogun, H.I. Okagbue, S.A. Bishop, The Weibull-exponential distribution: its properties and applications. J. Appl. Sci. **15**(11), 1305–1311
11. M.H. Tahir, G.M. Cordeiro, M. Mansoor, Z. Zubair, The Weibull-Lomax distribution: properties and applications. Hacettepe J. Math. Stat. **44**(2), 461–480 (2015)
12. N.A. Ibrahim, M.A. Khaleel, F. Merovci, A. Kilicman, M. Shitan, Weibull Burr X distribution: properties and application. Pak. J. Stat. **33**(5), 315–336 (2017)
13. P.E. Oguntunde, A.O. Adejumo, E.A. Owoloko, Exponential inverse exponential (EIE) distribution with applications to lifetime data. Asian J. Sci. Res. **10**, 169–177 (2017)
14. B. Efron, Logistic regression, survival analysis and the Kaplan-Meier curve. J. Am. Stat. Assoc. **83**(402), 414–425 (1988)
15. R. Shanker, H. Fasshaye, S. Selvaraj, On modeling lifetimes data using exponential and Lindley distributions. Biometr. Biostat. Int. J. **2**(5), 00042 (2015)
16. E.T. Lee, J.W. Wang, *Statistical Methods for Survival Data Analysis*, 3rd edn. (Wiley, New York, USA, 2003)
17. P.E. Oguntunde, A.O. Adejumo, K.A. Adepoju, Assessing the flexibility of the exponentiated generalized exponential distribution. Pac. J. Sci. Technol. **17**(1), 49–57 (2016)

On Computing the Inverse of Vandermonde Matrix via Synthetic Divisions

Yiu-Kwong Man

Abstract A simple method for computing the inverse of Vandermonde matrix via synthetic divisions is introduced. It can be applied to compute each row of the inverse of Vandermonde matrix systematically and effectively. Some illustrative examples are provided.

Keywords Linear algebra · Mathematical computation · Matrix inverse
Polynomial interpolation · Synthetic divisions · Vandermonde matrix

1 Introduction

Vandermonde matrix (VDM) has important applications in various areas such as polynomial interpolation, signal processing, curve fitting, coding theory and control theory [1–8], etc. The study of efficient methods for computing the inverse of VDM or its generalized version (i.e. confluent VDM) is still an important research topic. In this chapter, we introduce a novel and simple method for computing the inverse of VDM via synthetic divisions, based on the works reported in [9–13].

This chapter is organized like this. The basic mathematical background is introduced in Sect. 2. The method of synthetic divisions for computing the inverse of VDM is described in Sect. 3. Some illustrative examples are provided in Sect. 4. Then, some final remarks are given in Sect. 5.

Y.-K. Man (✉)
Department of Mathematics and Information Technology, The Education University of Hong Kong, 10 Lo Ping Road, Tai Po, New Territories, Hong Kong
e-mail: ykman@eduhk.hk

S.-I. Ao et al. (eds.), *Transactions on Engineering Technologies*,
https://doi.org/10.1007/978-981-13-0746-1_9

2 Basic Mathematical Background

Consider the following polynomial

$$\begin{aligned} f(x) &= (x-\lambda_1)(x-\lambda_2)\cdots(x-\lambda_n) \\ &= x^n + a_1x^{n-1} + a_2x^{n-2} + \cdots + a_n \end{aligned}$$

where a_i, λ_i are constants and λ_i are distinct. We are interested to find the inverse of the Vandermonde matrix below:

$$V = \begin{pmatrix} 1 & 1 & \cdots & 1 \\ \lambda_1 & \lambda_2 & \cdots & \lambda_n \\ \lambda_1^2 & \lambda_2^2 & \cdots & \lambda_n^2 \\ \vdots & \vdots & \ddots & \vdots \\ \lambda_1^{n-1} & \lambda_2^{n-1} & \cdots & \lambda_n^{n-1} \end{pmatrix}$$

According to [12], we can apply the formula $V^{-1} = W \times A$ to compute the inverse of V, where the matrices W and A are defined by:

$$W = \begin{pmatrix} \frac{\lambda_1^{n-1}}{\prod\limits_{j\neq 1}(\lambda_1-\lambda_j)} & \frac{\lambda_1^{n-2}}{\prod\limits_{j\neq 1}(\lambda_1-\lambda_j)} & \cdots & \frac{1}{\prod\limits_{j\neq 1}(\lambda_1-\lambda_j)} \\ \frac{\lambda_2^{n-1}}{\prod\limits_{j\neq 2}(\lambda_2-\lambda_j)} & \frac{\lambda_2^{n-2}}{\prod\limits_{j\neq 2}(\lambda_2-\lambda_j)} & \cdots & \frac{1}{\prod\limits_{j\neq 2}(\lambda_2-\lambda_j)} \\ \vdots & \vdots & \ddots & \vdots \\ \frac{\lambda_n^{n-1}}{\prod\limits_{j\neq n}(\lambda_n-\lambda_j)} & \frac{\lambda_n^{n-2}}{\prod\limits_{j\neq n}(\lambda_n-\lambda_j)} & \cdots & \frac{1}{\prod\limits_{j\neq n}(\lambda_n-\lambda_j)} \end{pmatrix}$$

$$A = \begin{pmatrix} 1 & 0 & 0 & \cdots 0 \\ a_1 & 1 & 0 & \cdots 0 \\ a_2 & a_1 & 1 & \cdots 0 \\ \vdots & \vdots & \vdots & \ddots \vdots \\ a_{n-1} & a_{n-2} & a_{n-3} & \cdots 1 \end{pmatrix}$$

with $a_1 = -\sum \lambda_j$, $a_2 = \sum_{j\neq m} \lambda_j\lambda_m$, $a_3 = -\sum_{j\neq m\neq s} \lambda_j\lambda_m\lambda_s$, ..., and $a_n = (-1)^n \prod \lambda_j$. However, the computational cost could be high if direct matrix multiplications are used to compute V^{-1}. In the next section, we will introduce a novel and simple method to compute V^{-1} instead, based on synthetic divisions only.

3 The Method of Synthetic Divisions

Let us explain how the method of synthetic divisions work, by considering the following two cases.

(i) For $n=2$, if we apply the formula $V^{-1} = W \times A$ directly, the result will be

$$V^{-1} = \begin{pmatrix} \frac{\lambda_1+a_1}{\lambda_1-\lambda_2} & \frac{1}{\lambda_1-\lambda_2} \\ \frac{\lambda_2+a_1}{\lambda_2-\lambda_1} & \frac{1}{\lambda_2-\lambda_1} \end{pmatrix}$$

However, if we apply synthetic division to $[f(x) - a_2] \div (x - \lambda_1)$, where $f(x) = (x - \lambda_1)(x - \lambda_2) = x^2 + a_1 x + a_2$, then we have

$$\begin{array}{c|cc} \lambda_1 & 1 & a_1 \\ & & \lambda_1 \\ \hline & 1 & \lambda_1 + a_1 \\ & & \lambda_1 \\ \hline & 1 & \boxed{\lambda_1 - \lambda_2} \end{array}$$

since $2\lambda_1 + a_1 = \lambda_1 - \lambda_2$. We can see that the elements obtained by the first synthetic division is equal to the numerators of the elements in the first row of V^{-1}, except in the reverse order. Also, the last element obtained by the second synthetic division, namely $\lambda_1 - \lambda_2$, is equal to the denominators of the elements in the first row of V^{-1}. Similarly, the elements of the second row of V^{-1} can be determined by applying synthetic divisions to $[f(x) - a_2] \div (x - \lambda_2)$ as shown below:

$$\begin{array}{c|cc} \lambda_2 & 1 & a_1 \\ & & \lambda_2 \\ \hline & 1 & \lambda_2 + a_1 \\ & & \lambda_2 \\ \hline & 1 & \boxed{\lambda_2 - \lambda_1} \end{array}$$

since $2\lambda_2 + a_1 = \lambda_2 - \lambda_1$. In other words, each row of V^{-1} can be computed easily and effectively by synthetic divisions, without having to use matrix multiplication or other methods for doing matrix inversion.

(ii) For $n=3$, if we apply the formula $V^{-1} = W \times A$ directly to compute the inverse of V, the result will be

$$V^{-1} = \begin{pmatrix} \frac{\lambda_1^2+a_1\lambda_1+a_2}{(\lambda_1-\lambda_2)(\lambda_1-\lambda_3)} & \frac{\lambda_1+a_1}{(\lambda_1-\lambda_2)(\lambda_1-\lambda_3)} & \frac{1}{(\lambda_1-\lambda_2)(\lambda_1-\lambda_3)} \\ \frac{\lambda_2^2+a_1\lambda_2+a_2}{(\lambda_2-\lambda_1)(\lambda_2-\lambda_3)} & \frac{\lambda_2+a_1}{(\lambda_2-\lambda_1)(\lambda_2-\lambda_3)} & \frac{1}{(\lambda_2-\lambda_1)(\lambda_2-\lambda_3)} \\ \frac{\lambda_3^2+a_1\lambda_3+a_2}{(\lambda_3-\lambda_1)(\lambda_3-\lambda_2)} & \frac{\lambda_3+a_1}{(\lambda_3-\lambda_1)(\lambda_3-\lambda_2)} & \frac{1}{(\lambda_3-\lambda_1)(\lambda_3-\lambda_2)} \end{pmatrix}$$

Similar to the above case, if we apply synthetic divisions to $[f(x) - a_3] \div (x - \lambda_i)$, where $f(x) = (x - \lambda_1)(x - \lambda_2)(x - \lambda_3) = x^3 + a_1x^2 + a_2x + a_3$ and $i = 1, 2$ and 3, we can determine the elements of V^{-1}, as shown below.

$$\begin{array}{c|ccc} \lambda_1 & 1 & a_1 & a_2 \\ & & \lambda_1 & \lambda_1^2 + a_1\lambda_1 \\ \hline & 1 & \lambda_1 + a_1 & \lambda_1^2 + a_1\lambda_1 + a_2 \\ & & \lambda_1 & \lambda_1^2 - \lambda_1\lambda_2 - \lambda_1\lambda_3 \\ \hline & 1 & \lambda_1 - \lambda_2 - \lambda_3 & \boxed{(\lambda_1 - \lambda_2)(\lambda_1 - \lambda_3)} \end{array}$$

since $2\lambda_1^2 + a_1\lambda_1 + a_2 - \lambda_1\lambda_2 - \lambda_1\lambda_3 = (\lambda_1 - \lambda_2)(\lambda_1 - \lambda_3)$.

Next,

$$\begin{array}{c|ccc} \lambda_2 & 1 & a_1 & a_2 \\ & & \lambda_2 & \lambda_2^2 + a_1\lambda_2 \\ \hline & 1 & \lambda_2 + a_1 & \lambda_2^2 + a_1\lambda_2 + a_2 \\ & & \lambda_2 & \lambda_2^2 - \lambda_1\lambda_2 - \lambda_2\lambda_3 \\ \hline & 1 & \lambda_2 - \lambda_1 - \lambda_3 & \boxed{(\lambda_2 - \lambda_1)(\lambda_2 - \lambda_3)} \end{array}$$

since $2\lambda_2^2 + a_1\lambda_2 + a_2 - \lambda_1\lambda_2 - \lambda_1\lambda_3 = (\lambda_2 - \lambda_1)(\lambda_2 - \lambda_3)$.

Also,

$$\begin{array}{c|ccc} \lambda_3 & 1 & a_1 & a_2 \\ & & \lambda_3 & \lambda_3^2 + a_1\lambda_3 \\ \hline & 1 & \lambda_3 + a_1 & \lambda_3^2 + a_1\lambda_3 + a_2 \\ & & \lambda_3 & \lambda_3^2 - \lambda_1\lambda_3 - \lambda_2\lambda_3 \\ \hline & 1 & \lambda_3 - \lambda_1 - \lambda_2 & \boxed{(\lambda_3 - \lambda_1)(\lambda_3 - \lambda_2)} \end{array}$$

since $2\lambda_3^2 + a_1\lambda_3 + a_2 - \lambda_1\lambda_3 - \lambda_2\lambda_3 = (\lambda_3 - \lambda_1)(\lambda_3 - \lambda_2)$. Therefore, the method of synthetic divisions works equally well for the case $n = 3$. In fact, it can be proved that this method is applicable to find the inverse of Vandermonde matrix of order n (≥ 2). The proof is left as an exercise.

4 Some Examples

Example 1 Find the inverse of the following Vandermonde matrix by synthetic divisions.

$$V = \begin{pmatrix} 1 & 1 & 1 \\ 1 & 2 & 3 \\ 1 & 4 & 9 \end{pmatrix}.$$

Solution. Consider the polynomial

$$f(x) = (x-1)(x-2)(x-3) = x^3 - 6x^2 + 11x - 6.$$

By synthetic divisions, we have:

$$\begin{array}{c|ccc} 1 & 1 & -6 & 11 \\ & & 1 & -5 \\ \hline & 1 & -5 & 6 \\ & & 1 & -4 \\ \hline & 1 & -4 & \boxed{2} \end{array}$$

So, the first row of V^{-1} is $(3\ {-5/2}\ 1/2)$.
Similarly,

$$\begin{array}{c|ccc} 2 & 1 & -6 & 11 \\ & & 2 & -8 \\ \hline & 1 & -4 & 3 \\ & & 2 & -4 \\ \hline & 1 & -2 & \boxed{-1} \end{array}$$

So, the second row of V^{-1} is $(-3\ 4\ {-1})$.
Finally,

$$\begin{array}{c|ccc} 3 & 1 & -6 & 11 \\ & & 3 & -9 \\ \hline & 1 & -3 & 2 \\ & & 3 & 0 \\ \hline & 1 & 0 & \boxed{2} \end{array}$$

So, the last row of V^{-1} is $(1\ {-3/2}\ 1/2)$.

Hence,

$$V^{-1} = \begin{pmatrix} 3 & -5/2 & 1/2 \\ -3 & 4 & -1 \\ 1 & -3/2 & 1/2 \end{pmatrix}.$$

Example 2 Determine a cubic polynomial that can pass through the points $(-2, -3)$, $(-1, 3)$, $(1, 3)$ and $(2, 9)$.

Solution. Suppose the cubic polynomial is $p(x) = a_0 + a_1x + a_2x^2 + a_3x^3$. With the given points, we can formulate an equation as follows:

$$\begin{pmatrix} -3 \\ 3 \\ 3 \\ 9 \end{pmatrix} = \begin{pmatrix} 1 & -2 & 4 & -8 \\ 1 & -1 & 1 & -1 \\ 1 & 1 & 1 & 1 \\ 1 & 2 & 4 & 8 \end{pmatrix} \begin{pmatrix} a_0 \\ a_1 \\ a_2 \\ a_3 \end{pmatrix}.$$

The associated Vandermonde matrix is

$$V = \begin{pmatrix} 1 & 1 & 1 & 1 \\ -2 & -1 & 1 & 2 \\ 4 & 1 & 1 & 4 \\ -8 & -1 & 1 & 8 \end{pmatrix}.$$

Consider a polynomial as follows:

$$f(x) = (x + 2)(x + 1)(x - 1)(x - 2) = x^4 - 5x^2 + 4.$$

By synthetic divisions, we have:

-2	1	0	-5	0
		-2	4	2
	1	-2	-1	2
		-2	8	-14
	1	-4	7	-12

So, the first row of V^{-1} is $(-1/6\ 1/12\ 1/6\ -1/12)$.
Next,

-1	1	0	-5	0
		-1	1	4
	1	-1	-4	4
		-1	2	2
	1	-2	-2	6

So, the second row of V^{-1} is (2/3 −2/3 −1/6 1/6).
Similarly,

1	1	0	-5	0
		1	1	-4
	1	1	-4	-4
		1	2	-2
	1	2	-2	-6

So, the third row of V^{-1} is (2/3 2/3 −1/6 −1/6).
Finally,

2	1	0	-5	0
		2	4	-2
	1	2	-1	-2
		2	8	14
	1	4	7	12

So, the last row of V^{-1} is (−1/6 −1/12 1/6 1/12).

Hence, the inverse of the associated Vandermonde matrix is given by

$$V^{-1} = \begin{pmatrix} -1/6 & 1/12 & 1/6 & -1/12 \\ 2/3 & -2/3 & -1/6 & 1/6 \\ 2/3 & 2/3 & -1/6 & -1/6 \\ -1/6 & -1/12 & 1/6 & 1/12 \end{pmatrix}$$

Hence,

$$\begin{pmatrix} a_0 \\ a_1 \\ a_2 \\ a_3 \end{pmatrix} = \begin{pmatrix} -1/6 & 2/3 & 2/3 & -1/6 \\ 1/12 & -2/3 & 2/3 & -1/12 \\ 1/6 & -1/6 & -1/6 & 1/6 \\ -1/12 & 1/6 & -1/6 & 1/12 \end{pmatrix} \begin{pmatrix} -3 \\ 3 \\ 3 \\ 9 \end{pmatrix} = \begin{pmatrix} 3 \\ -1 \\ 0 \\ 1 \end{pmatrix}.$$

Therefore, $p(x) = 3 - x + x^3$ is the required cubic polynomial.

5 Final Remarks

We have introduced a novel and simple method for computing the inverse of Vandermonde matrix (VDM) of order $n \geq 2$ via synthetic divisions. The proposed method is able to compute the inverse of VDM row by row. Due to its simplicity and algorithmic nature, it is suitable for either hand or machine calculation, as well as applications to areas such as polynomial interpolations, signal processing, coding theory or control theory, etc.

Acknowledgements The author would like to acknowledge that this chapter is a revised version of the paper presented orally at the International Conference of Applied and Engineering Mathematics (ICAEM 2017) held on 5–7 July 2017 at Imperial College, London, United Kingdom [13].

References

1. T. Kailath, *Linear Systems* (Prentice Hall, Englewood Cliffs, NJ, 1980)
2. J.J. Rushanan, On the Vandermonde matrix. Am. Math. Monthly **96**, 921–924 (1989)
3. V.E. Neagoe, Inversion of the Vandermonde matrix. IEEE Signal Process. Lett. **3**, 119–120 (1996)
4. C. Pozrikidis, *Numerical Computation in Science and Engineering* (Oxford University Press, New York, 1998)
5. H.K. Garg, *Digital Signal Processing Algorithms: Number Theory, Convolution, Fast Fourier Transform and Applications* (CRC Press, Boca Raton, 1998)
6. M. Goldwurm, V. Lonati, Pattern statistics and Vandermonde matrices. Theor. Comput. Sci. **356**, 153–169 (2006)
7. H. Hakopian, K. Jetter, G. Zimmermann, Vandermonde matrices for intersection points of curves. J. Approx. **1**, 67–81 (2009)
8. A.M. Elhosary, N. Hamdy, I.A. Farag, A.E. Rohiem, Optimum dynamic diffusion of block cipher based on maximum distance separable matrices. Int. J. Inf. Netw. Secur. **2**, 327–332 (2013)
9. Y.K. Man, A simple algorithm for computing partial fraction expansions with multiple poles. Int. J. Math. Educ. Sci. Technol. **38**, 247–251 (2007)
10. Y.K. Man, Partial fraction decomposition by synthetic division and its applications, in *Current Themes in Engineering Science 2008* (AIP, New York, 2009)
11. Y.K. Man, A cover-up approach to partial fractions with linear or irreducible quadratic factors in the denominators. Appl. Math. and Comput. **219**, 3855–3862 (2012)
12. Y.K. Man, On the inversion of Vandermonde matrix via partial fraction decomposition, in *IAENG Transaction on Engineering Sciences: Special Issue for the International Association of Engineers Conferences 2014* (World Scientific, Singapore, 2015)
13. Y.K. Man, On computing the Vandermonde Matrix inverse, in *Proceedings of the World Congress on Engineering 2017, WCE 2017, Lecture Notes in Engineering and Computer Science*, 5–7 July 2017, London, U.K., pp. 127–129
14. A. Eisinberg, G. Fedele, On the inversion of the Vandermonde matrix. Appl. Math. Comput. **174**, 1384–1397 (2006)

Application and Generation of the Univariate Skew Normal Random Variable

Dariush Ghorbanzadeh, Philippe Durand and Luan Jaupi

Abstract In this paper, for the generating the Skew normal random variables, we propose a new method based on the combination of minimum and maximum of two independent normal random variables. The estimation of parameters using the maximum likelihood estimation and the methods of moments estimation method. A real data set has been considered to illustrate the practical utility of the paper.

Keywords Centered moment · Maximum likelihood estimation · Methods of moments estimation · Min–Max method · Simulation method · Skew normal distribution

1 Introduction

In this paper presented at the World Congress on Engineering in London [1], we are interested by the univariate Skew Normal distribution has been studied by [2–5], has been studied and generalized extensively. The skew normal distribution family is well known for modeling and analyzing skewed data. It is the distribution family that extends the normal distribution family by adding a shape parameter to regulate the skewness, which has the higher exibility in setting a real data where some skewness is present.

D. Ghorbanzadeh · P. Durand (✉) · L. Jaupi
Département Mathématiques-Statistiques, Conservatoire National des Arts et Métiers, 292 rue Saint martin, 75141 Paris, France
e-mail: philippe.durand@cnam.fr

D. Ghorbanzadeh
e-mail: dariush.ghorbanzadeh@cnam.fr

L. Jaupi
e-mail: leon.jaupi@cnam.fr

S.-I. Ao et al. (eds.), *Transactions on Engineering Technologies*,
https://doi.org/10.1007/978-981-13-0746-1_10

The density function of the Skew Normal distribution with parameters (ξ, τ, θ) is given by

$$\frac{2}{\tau}\varphi\Big(\frac{x-\xi}{\tau}\Big)\Phi\Big(\theta\Big(\frac{x-\xi}{\tau}\Big)\Big) \tag{1}$$

where ξ is location parameters, $\tau > 0$ is scale parameters, θ is asymmetry parameter and φ and Φ are, respectively, the density and cumulative distribution functions of the standard normal distribution $\mathcal{N}(0, 1)$. In this paper we use the notation $\mathcal{SN}\big(\xi, \tau, \theta\big)$ to denote this distribution.

For the simulation of the $\mathcal{SN}\big(0, 1, \theta\big)$ distribution, [4], in his paper showed that, if U_1 and U_2 are identically and independently distributed $\mathcal{N}(0, 1)$ random variables, then $\frac{\theta|U_1|+U_2}{\sqrt{1+\theta^2}}$ has the $\mathcal{SN}\big(0, 1, \theta\big)$ distribution.

For the simulation of $\mathcal{SN}\big(0, 1, \theta\big)$ distribution, recently [6], have developed a method, called the Min–Max method, which consists in taking the combination of minimum and maximum of two independent random variables distributed $\mathcal{N}(0, 1)$. In this parpier, for the simulation of $\mathcal{SN}\big(\xi, \tau, \theta\big)$ distribution, we will use the Min–Max method.

2 Generating Random Variables by Min–Max Method

Let $m = \sqrt{\frac{1+\theta^2}{2}}\,\xi$, U_1 and U_2 two independent and identically distributed $\mathcal{N}(m, \tau^2)$ random variables and $U = \max(U_1, U_2)$ and $V = \min(U_1, U_2)$. For simulation of the random variable $X \sim \mathcal{SN}\big(\xi, \tau, \theta\big)$, we take the combination of U and V. First note that:

- if $\theta = 0$, the density (1) becomes: $\frac{1}{\tau}\varphi\Big(\frac{x-\xi}{\tau}\Big)$, simply simulate $X \sim \mathcal{N}(\xi, \tau^2)$.
- if $\theta = -1$, the density (1) becomes:

$$\frac{2}{\tau}\varphi\Big(\frac{x-\xi}{\tau}\Big)\Big(1-\Phi\Big(\frac{x-\xi}{\tau}\Big)\Big)$$

  we take $X = \min(X_1, X_2)$.
- if $\theta = 1$, the density (1) becomes:

$$\frac{2}{\tau}\varphi\Big(\frac{x-\xi}{\tau}\Big)\Phi\Big(\frac{x-\xi}{\tau}\Big)$$

  we take $X = \max(X_1, X_2)$.

where X_1 and X_2 are identically and independently distributed $\mathcal{N}(\xi, \tau^2)$ random variables.

For $\theta \notin \{-1, 0, 1\}$, note :

$$\lambda_1 = \frac{1+\theta}{\sqrt{2(1+\theta^2)}} \quad , \quad \lambda_2 = \frac{1-\theta}{\sqrt{2(1+\theta^2)}} \tag{2}$$

We note that: $\lambda_1^2 + \lambda_2^2 = 1$. For simulation of the random variable $X \sim \mathcal{SN}(\xi, \tau, \theta)$, we take the combination of U and V in the form:

$$X = \lambda_1 U + \lambda_2 V \tag{3}$$

Proposition 1 *The random variable X defined in the Eq. (3) has the Skew Normal distribution* $\mathcal{SN}(\xi, \tau, \theta)$.

Proof The pair (U, V) has density:

$$f_{U,V}(u, v) = \frac{2}{\tau^2}\varphi\Big(\frac{u-m}{\tau}\Big)\,\varphi\Big(\frac{v-m}{\tau}\Big)\,\mathbf{1}_{\{v \le u\}}(u, v) \tag{4}$$

where $\mathbf{1}$ is the indicator function.

Consider the transformation: $x = \lambda_1 u + \lambda_2 v$, $y = \lambda_1 u$. The inverse transform is defined by: $u = \frac{y}{\lambda_1}$, $v = \frac{x-y}{\lambda_2}$ and the corresponding Jacobian is: $J = \frac{1}{|\lambda_1\lambda_2|}$. X density is defined by:

$$\begin{aligned} f(x) &= \frac{1}{\mid \lambda_1\lambda_2 \mid}\int_{\Delta} f_{U,V}\Big(\frac{y}{\lambda_1}, \frac{x-y}{\lambda_2}\Big)\,dy \\ &= \frac{2}{\tau^2 \mid \lambda_1\lambda_2 \mid}\int_{\Delta}\varphi\Big(\frac{y-m\lambda_1}{\lambda_1\tau}\Big)\,\varphi\Big(\frac{y-x+m\lambda_2}{\lambda_2\tau}\Big)\,dy \end{aligned} \tag{5}$$

where $\Delta = \{\frac{x-y}{\lambda_2} \le \frac{y}{\lambda_1}\}$.

For the density of the standard normal distribution $\mathcal{N}(0, 1)$, we have the following classical property,

$$\begin{aligned} &\varphi\Big(\frac{x-\mu_1}{\sigma_1}\Big)\,\varphi\Big(\frac{x-\mu_2}{\sigma_2}\Big) = \\ &\varphi\Bigg(\frac{\sqrt{\sigma_1^2+\sigma_2^2}}{\sigma_1\sigma_2}\Big(x - \frac{\mu_1\sigma_2^2+\mu_2\sigma_1^2}{\sigma_1^2+\sigma_2^2}\Big)\Bigg)\,\varphi\Bigg(\frac{\mu_2-\mu_1}{\sqrt{\sigma_1^2+\sigma_2^2}}\Bigg) \end{aligned} \tag{6}$$

By using relation (6), the relation (5) becomes

$$\begin{aligned} f(x) &= \frac{2}{\tau^2 \mid \lambda_1\lambda_2 \mid}\,\varphi\Big(\frac{x-m(\lambda_1+\lambda_2)}{\tau}\Big) \\ &\times \int_{\Delta}\varphi\Big(\frac{1}{\lambda_1\lambda_2\tau}\big(y - x\lambda_1^2 - m\lambda_1\lambda_2(\lambda_2-\lambda_1)\big)\Big)\,dy \end{aligned} \tag{7}$$

Taking account of $m(\lambda_1 + \lambda_2) = \xi$, we obtain,

$$f(x) = \frac{2}{\tau^2 \mid \lambda_1\lambda_2 \mid} \varphi\Big(\frac{x-\xi}{\tau}\Big) \times \int_{\Delta} \varphi\Big(\frac{1}{\lambda_1\lambda_2\tau}\Big(y - x\lambda_1^2 - m\lambda_1\lambda_2(\lambda_2 - \lambda_1)\Big)\Big)\, dy \tag{8}$$

For the domain Δ, we have the following three cases:

case 1 $\theta \in (-1, 0) \cup (0, 1)$, we have:
$\mid \lambda_1\lambda_2 \mid = \frac{1-\theta^2}{2(1+\theta^2)}$ and $\Delta = \{y \geq \frac{\lambda_1 x}{\lambda_1+\lambda_2}\}$.

case 2 $\theta < -1$, we have: $\mid \lambda_1\lambda_2 \mid = \frac{\theta^2-1}{2(1+\theta^2)}$ and

$$\Delta = \{y \geq \frac{\lambda_1 x}{\lambda_1 + \lambda_2}\}$$

case 3 $\theta > 1$, we have: $\mid \lambda_1\lambda_2 \mid = \frac{\theta^2-1}{2(1+\theta^2)}$ and

$$\Delta = \{y \leq \frac{\lambda_1 x}{\lambda_1 + \lambda_2}\}$$

Using Eq. (8) and the three cases above, we get the result. □

3 Inference

Let $X_1, \ldots, X_n$ be a sample of size n from a $\mathscr{SN}(\xi, \tau, \theta)$ distribution.

3.1 Maximum Likelihood Estimation

The log-likelihood is given by

$$\ell(\xi, \tau, \theta) = n \log 2 - n \log\tau + \sum_{i=1}^{n} \log\Big(\varphi\Big(\frac{x_i - \xi}{\tau}\Big)\Big) + \sum_{i=1}^{n} \log\Big(\Phi\Big(\theta\Big(\frac{x_i - \xi}{\tau}\Big)\Big)\Big)$$

The maximum likelihood estimators (MLE) of (ξ, τ, θ), denoted $(\hat{\xi}, \hat{\tau}, \hat{\theta})$, are the numerical solution of the system of equations:

$$\begin{cases} \tau^2 = \frac{1}{n}\sum\limits_{i=1}^{n}\left(x_i - \xi\right)^2 \\ \tau\theta \sum\limits_{i=1}^{n} \dfrac{\varphi\left(\theta\left(\frac{x_i-\xi}{\tau}\right)\right)}{\Phi\left(\theta\left(\frac{x_i-\xi}{\tau}\right)\right)} = \sum\limits_{i=1}^{n}\left(x_i - \xi\right) \\ \sum\limits_{i=1}^{n} \dfrac{\left(x_i - \xi\right)\varphi\left(\theta\left(\frac{x_i-\xi}{\tau}\right)\right)}{\Phi\left(\theta\left(\frac{x_i-\xi}{\tau}\right)\right)} = 0 \end{cases} \tag{9}$$

3.2 Methods of Moments Estimation

Let m_k be the centered moment of order k of data defined by

$$m_k = \frac{1}{n}\sum_{i=1}^{n}\left(x_i - \overline{x}_n\right)^k$$

where $\overline{x}_n = \frac{1}{n}\sum\limits_{i=1}^{n} x_i$, and denote by γ_1 the skewness coefficient defined by $\gamma_1 = \frac{m_3}{m_2^{3/2}}$.

The method of moment estimators (MME) of (ξ, τ, θ), denoted $(\tilde{\xi}, \tilde{\tau}, \tilde{\theta})$, are obtained by solving the set of three equations:

$$\begin{cases} \xi + a\tau\delta = \overline{x}_n & \text{(10a)} \\ \tau^2\left(1 - a^2\delta^2\right) = m_2 & \text{(10b)} \\ \dfrac{ba^3\delta^3}{\left(1 - a^2\delta^2\right)^{3/2}} = \gamma_1 & \text{(10c)} \end{cases}$$

where $a = \sqrt{\frac{2}{\pi}}$, $b = \frac{4-\pi}{2}$ and $\delta = \frac{\theta}{\sqrt{1+\theta^2}}$.

We note that the sign of δ is the same as the sign of γ_1. The Eq. (10c) admits the solution

$$\tilde{\delta} = sign(\gamma_1)\sqrt{\frac{|\gamma_1|^{2/3}}{a^2\left(b^{2/3} + |\gamma_1|^{2/3}\right)}} \tag{11}$$

where $sign(\gamma_1)$ is the sign of γ_1. Consequently,

Table 1 Statistics of estimators of (ξ, τ, θ) for size $n = 5000$

$\xi = -6, \tau = 8, \theta = -7$						
MLE	mean of $\hat{\xi}$	−6.0017	mean of $\hat{\tau}$	8.0045	mean of $\hat{\theta}$	−7.2612
	std of $\hat{\xi}$	0.2109	std of $\hat{\tau}$	0.2930	std of $\hat{\theta}$	1.2125
MME	mean of $\tilde{\xi}$	−6.1489	mean of $\tilde{\tau}$	7.9178	mean of $\tilde{\theta}$	−7.5284
	std of $\tilde{\xi}$	0.8805	std of $\tilde{\tau}$	0.4288	std of $\tilde{\theta}$	2.8694
$\xi = 2, \tau = 5, \theta = 3$						
MLE	mean of $\hat{\xi}$	2.0025	mean of $\hat{\tau}$	4.9891	mean of $\hat{\theta}$	3.0745
	std of $\hat{\xi}$	0.2396	std of $\hat{\tau}$	0.2382	std of $\hat{\theta}$	0.5178
MME	mean of $\tilde{\xi}$	2.0054	mean of $\tilde{\tau}$	5.0546	mean of $\tilde{\theta}$	3.2742
	std of $\tilde{\xi}$	0.0788	std of $\tilde{\tau}$	0.0795	std of $\tilde{\theta}$	0.1930

$$\begin{cases} \tilde{\theta} = \dfrac{\tilde{\delta}}{\sqrt{1-\tilde{\delta}^2}} \\ \tilde{\tau} = \sqrt{\dfrac{m_2}{1-a^2\tilde{\delta}^2}} \\ \tilde{\xi} = \overline{x}_n - a\tilde{\tau}\tilde{\delta} \end{cases} \tag{12}$$

4 Simulation Results

In order to study the performance of the method, we simulated 100-samples of sizes $n = 5000$. For each sample we calculated the parameter estimators, MLE and MME, the Table 1 summarizes the results obtained for different values of (ξ, τ, θ).

5 Applications

5.1 Application at Hepatobiliary System and Pancreas

In this section, we illustrate the use of the estimation procedures described in the previous section. The variable to be considered is the average length of stay for

Table 2 Histogram corresponds to the data divided into 11 classes. The dashed line represent fitted distributions using the maximum likelihood estimators and the dotted line represent fitted distributions using the method of moment estimators

Parameter estimates	MLE	MME
ξ	4.1192	3.8856
τ	2.3006	2.4731
θ	2.0608	2.6286

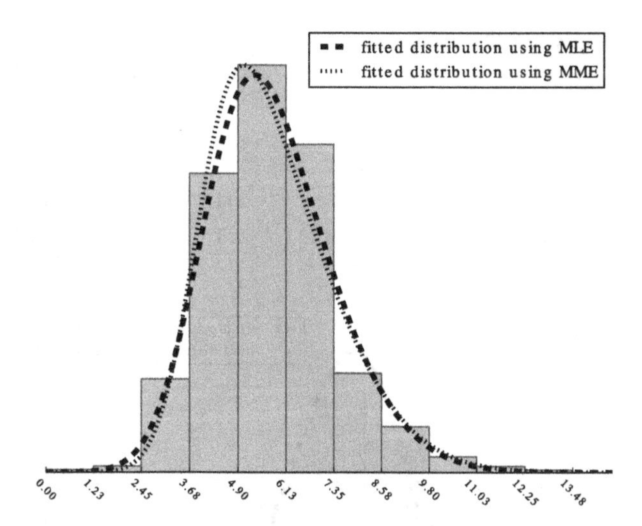

Fig. 1 Histogram corresponds to the data divided into 11 classes. The dashed line represent fitted distributions using the maximum likelihood estimators and the dotted line represent fitted distributions using the method of moment estimators

Table 3 The Otis IQ scores data from [9]

Otis IQ scores for whites						Otis IQ scores for non-whites					
124	106	108	112	113	122	91	102	100	117	122	115
100	108	108	94	102	120	97	109	108	104	108	118
101	118	113	117	100	106	103	123	123	103	106	102
111	107	112	120	102	135	118	100	103	107	108	107
125	98	121	117	124	114	97	95	119	102	108	103
103	122	122	113	113	104	102	112	99	116	114	102
103	113	120	106	132	106	111	104	122	103	111	101
112	118	113	112	112	121	91	99	121	97	109	106
112	85	117	109	104	129	102	104	107	95		
140	106	115	109	122	108						
119	121	113	107	122	103						
97	116	114	131	94	112						
108	118	112	116	113	111						
122	112	136	116	108	112						
108	116	103									

Table 4 Please write your table caption here

	Otis IQ scores for whites		Otis IQ scores for non-whites	
Parameter estimates	MLE	MME	MLE	MME
ξ	105.78	101.81	97.59	100.69
τ	11.94	14.59	11.27	12.23
θ	1.14	3.03	1.11	2.47

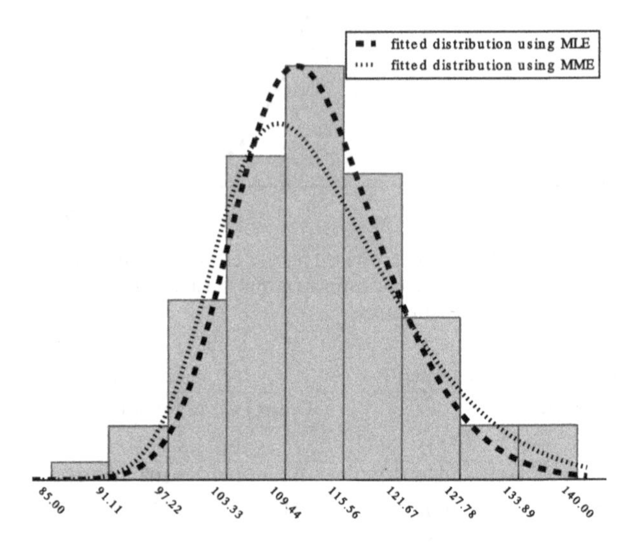

Fig. 2 Histogram corresponds to the data of Otis IQ scores for whites, divided into 9 classes. The dashed line represent fitted distributions using the maximum likelihood estimators and the dotted line represent fitted distributions using the method of moment estimators

patients who are in hospital for acute care because of problems, hepatobiliary system and pancreas, and die for this cause. The sample, under study, corresponds to 1082 hospitals in 10 states of the United States. For more information see columns 4 in http://lib.stat.cmu.edu/data-expo/1997/ascii/p07.dat. The data were used by [7] by modeling with a model "alpha-Skew-Normal distribution".

Table 2 summarizes the results obtained by the two methods of estimation. By the MLE method, we get the same results as [7]. The results obtained by the MME method are similar as the results obtained by the MLE method (Fig. 1).

5.2 Otis IQ scores data

Arnold et al. [8] applied the skew normal distribution to a portion of an IQ score data set from [9]. In this section we expand the application to the full data set. The Roberts IQ data gives the Otis IQ scores for 87 white males and 52 non-white males

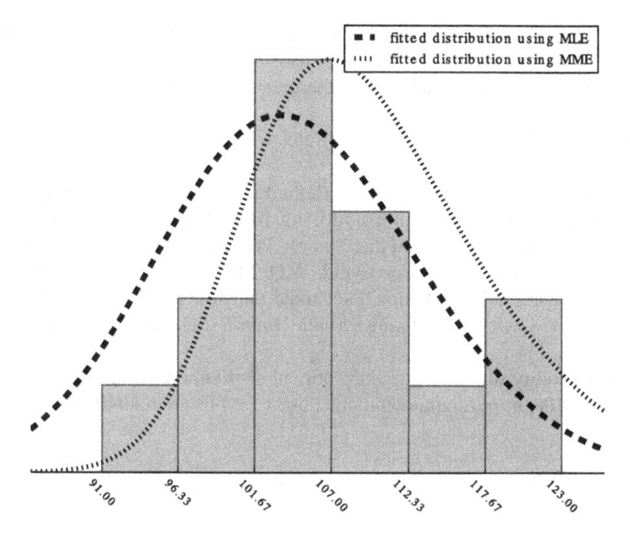

Fig. 3 Histogram corresponds to the data of Otis IQ scores for non-whites, divided into 6 classes. The dashed line represent fitted distributions using the maximum likelihood estimators and the dotted line represent fitted distributions using the method of moment estimators

hired by a large insurance company in 1971. We fit a Skew Normal distribution to both data sets. The datasets are used by [10] by modeling with normal and skew normal distribution. The data is given in the following (Table 3).
Table 4 summarizes the results obtained by the two methods of estimation (Figs. 2 and 3).

6 Conclusion

The simulation method proposed in this paper is simpler to use than the three methods, the inverse transform method, the composition method and the acceptance-rejection method. The results of estimation of the parameters, by the two estimation methods, from the simulations are very satisfactory. For the application, we obtain the same estimation values as other author using the same data.

References

1. D. Ghorbanzadeh, P. Durand, L. Jaupi, Generating the skew normal random variable, in *Lecture Notes in Engineering and Computer Science: Proceedings of The World Congress on Engineering 2017*, 5–7 July 2017, London, UK, pp. 113–116 (2017)
2. A. Azzalini, A class of distributions which includes the normal ones. Scandinavian J. Stat. **12**, 171–178 (1985)

3. A. Azzalini, Further results on the class of distributions which includes the normal ones. Statistica **46**, 199–208 (1986)
4. N. Henze, A probabilistic representation of the skew-normal distribution. Scandinavian J. Stat. **13**, 271–275 (1986)
5. A. Pewsey, Problems of inference for Azzalini's skew-normal distribution. J. Appl. Stat. **27**, 859–870 (2000)
6. D. Ghorbanzadeh, L. Jaupi, P. Durand, A Method to simulate the skew normal distribution. Appl. Math. **5**, 2073–2076 (2014). https://doi.org/10.4236/am.2014.513201
7. D. Elal-Olivero, Alpha-skew-normal distribution. Proyecc. J. Math. **29**(3), 224–240 (2010)
8. B.C. Arnold, R.J. Beaver, R.A. Groeneveld, W.Q. Meeker, The nontruncated marginal of a truncated bivariate normal distribution. Psychometrika **58**(3), 471–488 (1993)
9. H.V. Roberts, *Data Analysis for Managers with Minitab* (Scientific Press, Redwood City, CA, 1988)
10. N.D. Brown, Eliability studies of the skew normal distribution. Electronic Theses and Dissertations, 2001, p. 408. http://digitalcommons.library.umaine.edu/etd/408

Semi-analytical Methods for Higher Order Boundary Value Problems

A. A. Opanuga, H. I. Okagbue, O. O. Agboola, S. A. Bishop and P. E. Oguntunde

Abstract This work considers the solution of higher order boundary value problems using Homotopy perturbation method (HPM) and modified Adomian decomposition method (MADM). HPM is applied without any transformation or calculation of Adomian polynomials. The differential equations are transformed into an infinite number of simple problems without necessarily using the perturbation techniques. Two numerical examples are solved to illustrate the method and the results are compared with the exact and MADM solutions. The accuracy, simplicity and rapid convergence of HPM in handling the boundary value problems reveals its advantage over MADM.

Keywords Boundary value problems · Exact solution · Numerical solution MADM · HPM · Recursive relations

1 Introduction

Many phenomena in sciences and engineering are modeled by differential equations and are expressed in terms of boundary value problems. Incidentally, most of the boundary value problems do not have closed form solution, this has led to the development of various semi-analytical techniques such as Adomian decomposition

---

A. A. Opanuga (✉) · H. I. Okagbue · O. O. Agboola · S. A. Bishop · P. E. Oguntunde
Department of Mathematics, Covenant University, Ota, Nigeria
e-mail: abiodun.opanuga@covenantuniversity.edu.ng

H. I. Okagbue
e-mail: hilary.okagbue@covenantuniversity.edu.ng

O. O. Agboola
e-mail: ola.agboola@covenantuniversity.edu.ng

S. A. Bishop
e-mail: bishop.sheila@covenantuniversity.edu.ng

P. E. Oguntunde
e-mail: pelumi.oguntunde@covenantuniversity.edu.ng

S.-I. Ao et al. (eds.), *Transactions on Engineering Technologies*,
https://doi.org/10.1007/978-981-13-0746-1_11

method: Adesanya et al. [1], Opanuga et al. [2–5]. Differential transform method: Agboola et al. [6], Opanuga et al. [7]. Spline method: Lamnii et al. [8], Akram and Siddiqi [9]. Exp-function method: Mohyud-Din [10]. Generalized differential quadrature rule (GDQR): Liua and Wub [11]. Others include variational iteration technique: Siddiqi et al. [12]. Finite-difference method: Opanuga et al. [13]. Runge-kutta method: Mohamad-Jawad [14] and Block method: Anake et al. [15]. All these methods have some difficulties in their application. For instance, differential transform method requires transformation of the equations while Adomian decomposition method involves calculations of Adomian polynomials.

The Homotopy Perturbation Method (HPM) applied in this work was proposed by the Chinese researcher He [16, 17] by coupling the perturbation method and homotopy in topology. The method was developed to eliminate the limitations placed by the traditional perturbation technique viz:

- Presence of small parameters in the equations whereas most non-linear problems do not contain the so-called small parameters.
- Identification of small parameters in equations requires the mastery of some special techniques because wrong choice will affect the results.
- The approximate solution by perturbation method is valid only for the small values of the parameters.

The presence of the so-called small parameters places restrictions on the application of the perturbation method since most of the linear and nonlinear problems have no small parameters. To overcome the drawbacks homotopy perturbation method was developed. According to the method, a nonlinear problem is transformed into an infinite number of simple problems without necessarily using the perturbation techniques. This is done by letting the small parameter float and converge to unity, the problem will be converted into a special perturbation problem.

Homotopy perturbation technique has since then been developed and applied to numerous models. For instance Edeki et al. [18] used it to obtain the solution Navier-Stokes equation, El-Shahed [19] applied it to solve volterra's integro-differential equation, Biazar et al. [20] in Zakharov–Kuznetsov equations, Biazar et al. [21] applied it to hyperbolic partial differential equations, Darvishi et al. [22] and Aminikhaha et al. [23] used it to solve stiff systems of ordinary differential equations. Furthermore, Chun and Sakthivel [24] used it to obtain the solution of two-point boundary value problems in comparison with other methods, Mohyud-Din and Noor [25] used it to solve Flierl–Petviashivili equation and Ganji and Sadighi [26] in nonlinear heat transfer and porous media equations.

The objective of this work is to compare the HPM and MADM for the solution of higher order boundary value problems. Wazwaz [27] presented the modification to Adomian decomposition method which reduces the size of computations involved in the method and thereby enhances the rapidity of its convergence.

2 Analysis of Methods

2.1 Analysis of Homotopy Perturbation

Considering the following function

$$\psi(u) - g(r) = 0,\ r \in \Omega \tag{1}$$

$$\gamma\left(u, \frac{\partial u}{\partial t}\right) = 0,\ r \in \Gamma \tag{2}$$

Note ψ is a general differential operator, $g(r)$ is a known analytic function, γ is a boundary operator and Γ is the boundary of the domain Ω. Splitting the operator ψ into linear $L(u)$ and nonlinear part $N(u)$, written as

$$L(u) + N(u) - g(r) = 0 \tag{3}$$

Applying the technique of homotopy, then homotopy can be constructed as $v(r, p) : \Omega \times [0, 1] \to R$ which satisfies

$$\left.\begin{array}{l} H(v, p) = (1 - p)[L(v) - L(u_0)] + p[\Psi(v) - g(r)] = 0, \\ p \in [0, 1], r \in \Omega \quad \text{or} \\ H(v, p) = L(v) - L(u_0) + pL(u_0) + p[N(v) - g(r)] = 0 \end{array}\right\} \tag{4}$$

where $p \in [0, 1]$ is called homotopy parameter, and u_0 is an initial approximation for the solution of Eq. (1), which satisfies the boundary conditions. It is obvious from Eq. (4) that

$$H(v, 0) = L(v) - L(u_0) = 0 \tag{5}$$

$$H(v, 1) = A(v) - g(r) = 0 \tag{6}$$

It can be assumed that the solution of (4) can be written as a series in p as shown below

$$p : v = v_0 + pv_1 + p^2 v_2 + \cdots \tag{7}$$

At $p = 1$, Eq. (7) yields the approximate solution of Eq. (1) of the form

$$u = \lim_{p \to 1} v = v_0 + v_1 + v_2 + \ldots \tag{8}$$

2.2 Analysis of Adomian Decomposition Method

Considering the differential equation below in an operator form as

$$Ly + Ry + Ny = g \tag{9}$$

L is the highest order derivative which is assumed to be invertible, R is a linear differential operator, N is the non-linear term and g is the source term. Applying L^{-1} to both sides of (9) and imposing the given condition yields

$$y = h - L^{-1}(Ry) - L^{-1}(Ny) \tag{10}$$

where h represents given conditions and the source term. Adomian decomposition method gives the solution of $y(v)$ by an infinite series of components written as

$$y = \sum_{m=0}^{\infty} y_n \tag{11}$$

Here, components $y_0, y_1, y_2, \ldots$ are determined recursively. Using (11) in (10) yields

$$\sum_{m=0}^{\infty} y_n = h - L^{-1}\left(R\sum_{m=0}^{\infty} y_n\right) - L^{-1}\left(N\sum_{m=0}^{\infty} y_n\right) \tag{12}$$

Equation (12) can also be expressed as

$$y_0 + y_1 + y_2 + y_3 + \cdots = h - L^{-1}(N(y_0 + y_1 + y_2 + y_3 + \cdots)) \tag{13}$$

The zeroth component is stated as

$$y_0 = h \tag{14}$$

and the remaining components are written as the recursive relation

$$y_{n+1} = -L^{-1}\left(R\sum_{m=0}^{\infty} y_n\right) - L^{-1}\left(N\sum_{m=0}^{\infty} y_n\right), n \geq 0 \tag{15}$$

so that

$$y_1 = -L^{-1}\left(R\sum_{m=0}^{\infty} y_0\right) - L^{-1}\left(N\sum_{m=0}^{\infty} y_0\right), for\ n = 0 \tag{16}$$

$$y_2 = -L^{-1}\left(R\sum_{m=0}^{\infty} y_1\right) - L^{-1}\left(N\sum_{m=0}^{\infty} y_1\right), \; for \; n = 1 \tag{17}$$

$$\vdots$$

The modification by Wazwaz [27] splits the function h into two parts say h_0 and h_1,

$$h = h_0 + h_1 \tag{18}$$

We will then have the zeroth component as

$$y_0 = h_0 \tag{19}$$

and the rest terms written as

$$y_1 = h_1 - L^{-1}(Ry_0) - L^{-1}(Ny_0),\; y_2 = -L^{-1}(Ry_1) - L^{-1}(Ny_1), \ldots \tag{20}$$

The above modification reduces the computational involvement in the application of the method thereby enhancing the rapidity of its convergence. The nonlinear term Ny can be determined by an infinite series of Adomian polynomials.

$$Ny = \sum An \tag{21}$$

3 Test Examples

3.1 Example 1

Consider the third order three-point boundary value problem

$$y''' - 25y' + 1 = 0 \tag{22}$$

with the following boundary conditions

$$y'(0) = 0,\; y'(1) = 0,\; y(0,5) = 0 \tag{23}$$

The exact solution of the above problem is

$$y(t) = \frac{1}{25^3}\left(\sinh\frac{25}{2} - \sinh 25t\right) + \frac{1}{25^2}\left(t - \frac{1}{2}\right) + \frac{1}{25^3}\tanh\frac{25}{2}\left(\cosh 25h - \cosh\frac{25}{2}\right) \tag{24}$$

3.1.1 Solution by Homotopy Perturbation Method

Transforming Eq. (22) and the boundary conditions (23) as system of integral equations gives

$$y_1 = A + \int_0^x y_2(t)dt;\ y_2 = 0 + \int_0^x y_3(t)dt;\ y_3 = B + \int_0^x [25y_1 - 1]dt \tag{25}$$

The equations above can be expressed as

$$y_{10} + py_{11} + p^2y_{12} + \cdots = A + p\int_0^x (y_{20} + py_{21} + p^2y_{22} + \cdots)dt$$

$$y_{20} + py_{21} + p^2y_{22} + \cdots = 0 + p\int_0^x (y_{30} + py_{31} + p^2y_{32} + \cdots)dt$$

$$y_{30} + py_{31} + p^2y_{32} + \cdots = B + p\int_0^x \left[25\left(y_{10} + py_{11} + p^2y_{12} + \cdots\right) - 1\right]dt \tag{26}$$

Equating the coefficients of equal powers of p, we have the following

$$p^0 : \begin{cases} y_{10} = A \\ y_{20} = 0 \\ y_{30} = B \end{cases};\ p^1 : \begin{cases} y_{11} = 0 \\ y_{21} = Bx \\ y_{31} = -x \end{cases};\ p^2 : \begin{cases} y_{12} = \frac{Bx^2}{2} \\ y_{22} = -\frac{x^2}{2} \\ y_{32} = 0 \end{cases};\ p^3 : \begin{cases} y_{13} = -\frac{x^3}{6} \\ y_{23} = 0 \\ y_{33} = \frac{25x^2}{2} \end{cases}, \tag{27}$$

and so on

Combining all the first terms yields

$$y(x) = A + \frac{Bx^2}{2} - \frac{x^3}{6} + \frac{25Bx^4}{24} - \frac{5x^5}{24} + \frac{125Bx^6}{144} - \frac{125x^7}{1008} + \cdots \tag{28}$$

Applying the boundary conditions at $x = 1$ for $y'(1)$ and $x = 0.5$ for $y(0.5)$ gives the system of equations below

$$\begin{aligned} &0.2052915792B + A = -0.02840163584, \\ &14.84064212B = 2.928397942 \end{aligned} \tag{29}$$

Solving the equations above, yields the following

$$A = -0.01210708565, \quad B = 0.1973228596 \tag{30}$$

Using (30) in (28) gives the series solution,

$$\begin{aligned} y(x) = &-0.01210708565 + 0.09866142980x^2 - \frac{x^3}{6} \\ &+ 0.2055446454x^4 - \frac{5x^5}{24} + 0.1712872045x^6 + \cdots \end{aligned} \tag{31}$$

3.1.2 Solution by Modified Adomian Decomposition Method

Writing Eq. (22) in operator form, yields

$$Ly = 25y' - 1 \tag{32}$$

Applying L^{-1} to Eq. (32) gives

$$y = 25L^{-1}(y') - L^{-1}(1) \tag{33}$$

Using the boundary conditions we obtain

$$y(x) = Ax + \frac{x^2 B}{2} - \frac{x^3}{6} + 25L^{-1}(y') \tag{34}$$

Following Wazwaz [27], the zeroth component is identified as

$$y_0 = Ax \tag{35}$$

Table 1 Numerical result for Example 1

S/N	Exact	HPM error	MADM error
0	−0.012107086	3.5211E−11	4.7886E−12
0.1	−0.011268507	3.5368E−11	4.6323E−12
0.2	−0.009222206	3.5897E−11	4.1031E−12
0.3	−0.006466868	3.6995E−11	3.0054E−12
0.4	−0.003320195	3.9038E−11	9.6175E−13
0.5	0	4.267E−11	2.6704E−12
0.6	0.003320195	4.8934E−11	8.934E−12
0.7	0.006466868	5.9484E−11	1.9484E−11
0.8	0.009222206	7.693E−11	3.6929E−11
0.9	0.011268507	1.0542E−10	6.5374E−11
1	0.012107086	1.5215E−10	1.1133E−10

While the remaining recursive relation is written as;

$$y_{n+1} = \frac{x^2 B}{2} - \frac{x^3}{6} + 25L^{-1}(y_n') \tag{36}$$

$$y_1 = \frac{x^2 B}{2} - \frac{x^3}{6} + 25L^{-1}(y_0') = \frac{x^2 B}{2} - \frac{x^3}{6} \tag{37}$$

$$y_2(x) = 25L^{-1}(y_1') = \frac{25Bx^4}{24} - \frac{5x^5}{24} \tag{38}$$

$$y_3(x) = 25L^{-1}(y_2') = \frac{125Bx^6}{144} - \frac{125x^7}{1008} \tag{39}$$

$$\begin{aligned} y(x) = A + \frac{Bx^2}{2} - \frac{x^3}{6} + \frac{25Bx^4}{24} - \frac{5x^5}{24} + \frac{125Bx^6}{144} \\ - \frac{125x^7}{1008} + \frac{3125Bx^8}{8064} + \cdots \end{aligned} \tag{40}$$

Applying the boundary conditions at $x = 1$ for $y'(1)$ and $x = 0.5$ for $y(0.5)$ to obtain A and B, which gives the system of equations below.

$$\left. \begin{aligned} A + 0.205291579B = 0.02840163584 \\ 14.84064211B = 2.928397941 \end{aligned} \right\} \tag{41}$$

Solving the system of equations gives the values of constants A and B as

$$A = -0.01210708561, \quad B = 0.1973228596 \tag{42}$$

Using (42) in (40) yields the series solution as (see Table 1 for comparism with HPM solution).

$$y(x) = -0.01210708561 + 0.09866142980x^2 - \frac{x^3}{6} + 0.2055446454x^4 - \frac{5x^5}{24} + 0.1712872045x^6 - \frac{125x^7}{1008} + \cdots \quad (43)$$

3.2 *Example 2*

Consider a fifth-order, two-point boundary value problems

$$y^{v}(x) = y - 15e^x - 10xe^x \quad (44)$$

The boundary conditions are

$$y(0) = 0,\ y'(0) = 1,\ y''(0) = 0,\ y(1) = 0,\ y'(1) = -e \quad (45)$$

The theoretical solution is given as

$$y(x) = x(1-x)e^x \quad (46)$$

3.2.1 Solution by Homotopy Perturbation Method

Transforming Eq. (44) together with the boundary conditions (45) to a system of integral equations

$$y_1 = 0 + \int_0^x y_2(t)dt;\ y_2 = 1 + \int_0^x y_3(t)dt$$

$$y_3 = 0 + \int_0^x y_4(t)dt;\ y_4 = A + \int_0^x y_5(t)dt$$

$$y_5 = B + \int_0^x \left(y_1 - 15e^x - 10xe^x\right)dt \quad (47)$$

Equation (47) can then be written in the form

$$y_{10} + py_{11} + p^2y_{12} + \cdots = 0 + p\int_0^x (y_{20} + py_{21} + p^2y_{22} + \cdots)dt;$$

$$y_{20} + py_{21} + p^2y_{22} + \cdots = 1 + p\int_0^x (y_{30} + py_{31} + p^2y_{32} + \cdots)dt;$$

$$y_{30} + py_{31} + p^2y_{32} + \cdots = 0 + p\int_0^x \left(y_{40} + py_{41} + p^2y_{42} + \cdots\right)dt;$$

$$y_{40} + py_{41} + p^2y_{42} + \cdots = A + p\int_0^x \left(y_{50} + py_{51} + p^2y_{52} + \cdots\right)dt;$$

$$y_{50} + py_{51} + p^2y_{52} + \cdots = B + p\int_0^x \begin{bmatrix} \left(y_{10} + py_{11} + p^2y_{12} + \cdots\right) - \\ 15e^x - 10xe^x \end{bmatrix} dt \tag{48}$$

Equating the coefficients of like powers of P, yields the following

$$p^0 : \begin{cases} y_{10} = 0 \\ y_{20} = 1 \\ y_{30} = 0 \\ y_{40} = A \\ y_{50} = B \end{cases} ; p^1 : \begin{cases} y_{11} = x \\ y_{21} = 0 \\ y_{31} = Ax \\ y_{41} = Bx \\ y_{51} = -15x - \\ \frac{25x^2}{2} - \frac{35x^3}{6} - \\ \frac{15x^4}{8} - \frac{11x^5}{24} - \frac{13x^6}{144} + \cdots \end{cases} \quad p^2 : \begin{cases} y_{12} = 0 \\ y_{22} = \frac{Ax^2}{2} \\ y_{32} = \frac{Bx^2}{2} \\ y_{42} = -\frac{15x^2}{2} - \frac{25x^3}{6} - \frac{35x^4}{24} - \\ \frac{3x^5}{8} - \frac{11x^6}{144} + \cdots \\ y_{52} = \frac{x^2}{2}, \text{ etc.} \end{cases} \tag{49}$$

Combining all the first terms, we get

$$y(x) = x + \frac{Ax^3}{6} + \frac{Bx^4}{24} - \frac{x^5}{8} - \frac{x^6}{30} - \frac{x^7}{144} - \frac{x^8}{896} - \frac{11x^9}{72576} - \frac{x^{10}}{45360} - \cdots \tag{50}$$

We proceed as shown above to evaluate the values of A and B by imposing the boundary conditions (56) at $x = 1$, giving rise to system of equations below

$$\left.\begin{aligned} 0.1666914685A + 0.04166942240B = 0.8334297846 \\ 0.5001984148A + 0.1666914683B = 0.1158451633 \end{aligned}\right\} \tag{51}$$

The equations yield

$$A = -2.999999995, \quad B = -8.000000019 \tag{52}$$

Applying (52) in (50) gives

$$\begin{aligned} y(x) = x &- 0.499999999x^3 - 0.33333333x^4 - \frac{x^5}{8} - \frac{x^6}{30} - \frac{x^7}{144} \\ &- 0.001190476191x^8 - 0.00011736111111x^9 - \frac{x^{10}}{45360} - \cdots \end{aligned} \tag{53}$$

3.2.2 Solution by Modified Adomian Decomposition Method

The first stage is to express Eq. (44) in operator as

$$Ly = y - 15e^x - 10xe^x \tag{54}$$

Applying L^{-1} on Eq. (54) to give

$$y = L^{-1}y - 15L^{-1}(e^x) - 10L^{-1}(xe^x) \tag{55}$$

Using the boundary conditions to obtain

$$y(x) = x + \frac{Ax^3}{6} + \frac{Bx^4}{24} + L^{-1}y - 15L^{-1}(e^x) - 10L^{-1}(xe^x) \tag{56}$$

The zeroth component is identified according the modification made by Wazwaz [27] as

$$y_0 = x \tag{57}$$

While the remaining recursive relations are written as

$$y_{n+1} = \frac{Ax^3}{6} + \frac{Bx^4}{24} + L^{-1}(y_n) - 15L^{-1}(e^x) - 10L^{-1}(xe^x), \tag{58}$$

$$y_1 = \frac{Ax^3}{6} + \frac{Bx^4}{24} + L^{-1}(y_0) - 15L^{-1}(e^x) - 10L^{-1}(xe^x), \; y_2 = L^{-1}(y_1), \ldots \tag{59}$$

Then

$$y(x) = A + \frac{Bx^2}{2} - \frac{x^3}{6} + \frac{25Bx^4}{24} - \frac{5x^5}{24} + \frac{125Bx^6}{144} - \frac{125x^7}{1008} + \frac{3125Bx^8}{8064} + \cdots \tag{60}$$

Table 2 Numerical result for Example 2

S/N	Exact	HPM error	MADM error
0	0	0	0
0.1	0.099465383	7.2331E−13	2.8332E−13
0.2	0.195424441	5.17328E−12	2.1333E−12
0.3	0.28347035	1.539E−11	6.7499E−12
0.4	0.358037927	3.15732E−11	1.4933E−11
0.5	0.412180318	5.20873E−11	2.7081E−11
0.6	0.437308512	7.3509E−11	4.3191E−11
0.7	0.422888069	9.09748E−11	6.2854E−11
0.8	0.356086549	9.99623E−11	8.5248E−11
0.9	0.22136428	1.03564E−10	1.0914E−10
1	0	1.37377E−10	1.3289E−10

We now impose the boundary conditions at $x = 1$, which gives the system of equations below

$$\left.\begin{array}{l} 0.04166942241B + 0.1666914684B = -0.8334297846 \\ 0.1666914684B + 0.5001984148A = -0.1158451632 \end{array}\right\} \tag{61}$$

Solving the equations yields

$$A = -2.999999995, \quad B = -8.000000003 \tag{62}$$

Using (62) in Eq. (60) yields (Table 2)

$$\begin{aligned} y(x) = x &- 0.4999999997x^3 - 0.3333333335x^4 - \frac{x^5}{8} - \frac{x^6}{30} - \frac{x^7}{144} \\ &- 0.001190476191x^8 - 0.0001736111111x^9 - \frac{x^{10}}{45360} - \cdots \end{aligned} \tag{63}$$

4 Conclusion

In this work, homotopy perturbation method and modified Adomian decomposition method are applied to obtain the solution of higher order boundary value problems. The homotopy perturbation method is implemented without linearization, transformation or discretization. Its rapid convergence to the exact solution with few terms is its main advantage over modified Adomian decomposition method.

Acknowledgements Authors are grateful to Covenant University for the financial support and the anonymous reviewers for their constructive comments.

References

1. S.O. Adesanya, E.S. Babadipe, S.A. Arekete, A new result on Adomian decomposition method for solving Bratu's problem. Math. Theory Model. **3**(2), 116–120 (2013)
2. A.A. Opanuga, O.O. Agboola, H.I. Okagbue, Approximate solution of multipoint boundary value problems. J. Eng. Appl. Sci. **10**(4), 85–89 (2015)
3. A.A. Opanuga, J.A. Gbadeyan, S.A. Iyase, H.I. Okagbue, Effect of thermal radiation on the entropy generation of hydromagnetic flow through porous channel. Pac J. Sci. Technol. **17**(2), 59–68 (2016)
4. A.A. Opanuga, E.A. Owoloko, O.O. Agboola, H.I. Okagbue, Application of homotopy perturbation and modified Adomian decomposition methods for higher order boundary value problems, in *Proceedings of the World Congress on Engineering 2017 (WCE 2017)*. Lecture Notes in Engineering and Computer Science, 5–7 July 2017, London, U.K., pp. 130–134
5. A.A. Opanuga, H.I. Okagbue, E.A. Owoloko, O.O. Agboola, Modified Adomian decomposition method for thirteenth order boundary value problems. Gazi Univ. J. Sci. **30**(4), 454–461 (2017)
6. O.O. Agboola, A.A. Opanuga, J.A. Gbadeyan, Solution of third order ordinary differential equations using differential transform method. Glob. J. Pure Appl. Math. **11**(4), 2511–2517 (2015)
7. A.A. Opanuga, O.O. Agboola, H.I. Okagbue, G.J. Oghonyon, Solution of differential equations by three semi-analytical techniques. Int. J. Appl. Eng. Res. **10**(18), 39168–39174 (2015)
8. A. Lamnii, A.H. Mraoui, D. Sbibih, A. Tijini, A. Zidna, Spline solution of some linear boundary value problems. Appl. Math. E-Notes **8**, 171–178 (2008)
9. G. Akram, S.S. Siddiqi, Nonic spline solutions of eighth order boundary value problems. Appl. Math. Comput. **182**, 829–845 (2006)
10. S.T. Mohyud-Din, M.A. Noor, K.I. Noor, Exp-Function method for solving higher-order boundary value problems. Bull. Inst. Math. Academia Sinica (New Series) **4**(2), 219–234 (2009)
11. G.R. Liua, T.Y. Wub, Differential quadrature solutions of eighth-order boundary-value differential equations. J. Comput. Appl. Math. **145**, 223–235 (2002)
12. S.S. Siddiqi, G. Akram, M. Iftikhar, Solution of seventh order boundary value problems by variational iteration technique. Appl. Math. Sci. **6**(94), 4663–4672 (2012)
13. A.A. Opanuga, E.A. Owoloko, H.I. Okagbue, O.O. Agboola, Finite difference method and Laplace transform for boundary value problems, in *Proceedings of The World Congress on Engineering 2017 (WCE 2017)*. Lecture Notes in Engineering and Computer Science, 5–7 July, 2017, London, U.K., pp. 65–69
14. A.J. Mohamad-Jawad, Solving second order non-linear boundary value problems by four numerical methods. Eng. Tech. J. **28**(2), 12 (2010)
15. T.A. Anake, D.O. Awoyemi, A.O. Adesanya, One-step implicit hybrid block method for the direct solution of general second order ordinary differential equations. IAENG Int. J. Appl. Math. **42**(4), 224–228 (2012)
16. J.H. He, A coupling method of Homotopy technique and perturbation technique for nonlinear problems. Int. J. Non-Linear Mech. **35**(1), 37–43 (2000)
17. J.H. He, Application of homotopy perturbation method to nonlinear wave equations. Chaos, Solitons Fractal **26**, 695–700 (2005)
18. S.O. Edeki, G.O. Akinlabi, M.E. Adeosun, Analytical solutions of the Navier-Stokes model by He's polynomials, in *Proceedings of the World Congress on Engineering 2016*. Lecture Notes in Engineering and Computer Science, 29 June–1 July, 2016, London, U.K., pp. 16–19
19. M. El-Shahed, Application of He's homotopy perturbation method to Volterra's integro-differential equation. Int. J. Nonlinear Sci. Numer. Simul. **6**(2), 163–168 (2005)
20. J. Biazar, F. Badpeima, F. Azimi, Application of the Homotopy perturbation method to ZakharovKuznetsov equations. Comput. Math Appl. **58**, 2391–2394 (2009)
21. J. Biazar, H. Ghazvini, Homotopy perturbation method for solving hyperbolic partial differential equations. Comput. Math. Appl. **56**, 453–458 (2008)
22. M.T. Darvishi, F. Khani, Application of He's homotopy perturbation method to stiff systems of ordinary differential equations. Zeitschrift fuer Naturforschung **63**(1–2), 19–23 (2008)

23. H. Aminikhaha, M. Hemmatnezhad, An effective modification of the Homotopy perturbation method for stiff systems of ordinary differential equations. Appl. Math. Lett. **24**, 1502–1508 (2011)
24. C. Chun, R. Sakthivel, Homotopy perturbation technique for solving two-point boundary value problems-comparison with other methods. Comput. Phys. Commun. **181**, 1021–1024 (2010)
25. S.T. Mohyud-Din, M.A. Noor, Homotopy perturbation method and Padé approximants for solving Flierl-Petviashivili equation. Appl. Math. **3**(2), 224–234 (2008)
26. D.D. Ganji, A. Sadighi, Application of Homotopy-perturbation and variational iteration methods to nonlinear heat transfer and porous media equations. J. Comput. Appl. Math. **207**(1), 24–34 (2007)
27. A.M. Wazwaz, Approximate solutions to boundary value problems of higher order by the modified Adomian decomposition method. Comput. Math. Appl. **40**, 679–691 (2000)

Developing Indonesian Highway Capacity Manual Based on Microsimulation Model (A Case of Urban Roads)

Ahmad Munawar, Muhammad Zudhy Irawan and Andrean Gita Fitrada

Abstract Recently, the 1997 Indonesian Highway Capacity Manual is commonly used to analyze the performance of road and intersection. However, since the motorized vehicle grows rapidly and significant change of driver behavior, several parameters must be validated so that the calculation result could represent the actual condition. This paper aims to improve the Indonesian HCM formula especially in regards to the value of free flow speed and initial capacity. To support our research, traffic surveys have been carried out on various urban roads in some Indonesian cities for four lane roads with divider. The flows then have been simulated by using VISSIM software (VerkehrInStädten—Simulations Modell). The measured speeds and headways have been compared to the simulation results. The Wiedemann parameters in VISSIM have been calibrated such that there is no significant difference between the measured speeds and headways and the simulation results. After calibration, the flow-speed relationships have been compared to Indonesian HCM formula, by simulating various traffic flows. Our results show that the values of free flow speed need to be verified become 37 and 40 kph for car and motorcycle respectively, while the basic capacity becomes 1750 pcu/h/lane.

Keywords Flow · Headway · Simulation · Speed · Traffic · Vissim

1 Introduction

The significant changes of traffic flow in the last few decades bring the Indonesia Highway Capacity Manual, which was prepared in 1997, less suitable for use at this

A. Munawar (✉) · M. Z. Irawan · A. G. Fitrada
Gadjah Mada Univesity, Jalan Grafika 2, Sleman 55283, Indonesia
e-mail: munawar@ugm.ac.id

M. Z. Irawan
e-mail: zudhyirawan@ugm.ac.id

A. G. Fitrada
e-mail: andreanfitrada@gmail.com

S.-I. Ao et al. (eds.), *Transactions on Engineering Technologies*,
https://doi.org/10.1007/978-981-13-0746-1_12

time. For example, looking into the factor of vehicle composition, out of 132 million motorized vehicles in 1994, 39.57% of them are motorcycle. While, in 2011, 70% of total vehicles on the road are motorcycle. Due to this, many studies evaluate the performance of IHCM 1997 both on road [1–5] and intersection [6] by comparing between IHCM calculation results and field condition, such as: speed, queue, and delay.

This chapter will examine the use of Indonesian HCM for predicting free flow speed and initial capacity at urban roads for four lane-divided road (two lanes for each direction). A VISSIM microsimulation software was used with a consideration that VISSIM can control the unique characteristic of traffic flow in developing counties, especially in Indonesia (such as: mixed traffic and driver behaviour). By this, our simulation model can be flexible cover a wide range of highway and traffic conditions.

IHCM 1997 itself was prepared by a comprehensive study, which was carried out as an Indonesian Highway Capacity Manual Project and reported by Bang et al. [3] under the consultancy of Swedish National Road Consulting AB, SweRoad. It identified significant effects of geometric factors (i.e. carriageway width, shoulder width, median), traffic and environmental factors (directional split, city size) and side friction factors (i.e. pedestrians, non-motorized vehicles, public transport vehicles) on speed-flow relationships on Indonesian urban/suburban road links [3].

2 IHCM 1997 Evaluation

The evaluation of IHCM have been extensively studied in the last few decades. Irawan et al. [2] evaluate the volume delay function of IHCM and proposed a IHCM based new formula with a least square method. The proposed formula of volume delay function is then applied using a field data on a simple network and simulated with EMME/2 [2].

Kusnandar [7] identifies several parameters that have to be revised on IHCM. He concluded that conversion variables of passenger car have to be changed related to the current mixed traffic [7]. In the similar point of view, Iskandar [8] found that the value of initial capacity in IHCM must be raised up to 51% and 25% for the divided road type of two ways two lanes and four lanes respectively. The equivalent value of passenger car also must be higher than 1.2 for heavy vehicle and less than 0.25 for motorcycle [8].

Winnetou and Munawar [9] measure the performance of an urban road in Yogyakarta with a microsimulation model. From this study, they found that there were significant differences in the value of speed of the car and motorcycle resulted from IHCM 1997 with the value of speed in the field. To make a comparison, they simulate the field traffic flow on VISSIM software. The simulation result shows that refer to the t-test statistical analysis, the value of the speed of cars and motorcycles on VISSIM software were no significant differences with the real traffic flow [9]. However, a comprehensive study method was carried by Munawar et al. [1] in evaluating the performance of two ways four lanes urban road by comparing the model and field travel speed and road capacity. This study results some important findings and will be detailed described in the next section [1].

3 The Proposed Model to Develop IHCM 1997 for Two-Way Four Lanes Urban Road

As previously explained, Munawar et al. [1] conducted a study to evaluate the performance of IHCM 1997 and proposed some changed parameters.

A. Data Collection

The data is collected in five different cities in Indonesia, there are Banda Aceh, Pekanbaru, Yogyakarta, Surabaya, and Banjarmasin. It is expected that by taking many samples from those five different cities, this research can represent the real traffic condition in Indonesia. Geographically, the locations of those five cities is shown in Fig. 1.

The secondary data in this study are obtained from data on the number of population in each city and aerial photographs of the road network which is used as a reference in designing of road section models. Meanwhile, the primary data are obtained through direct survey in the form of traffic volume, speed of each vehicle type, and roads geometrics. The survey is conducted on weekdays from morning to evening by using cctv cameras, in order that the survey can produce the data that represent the daily circumstances on the road network. The next step is modeling data to VISSIM software. The modeling of road network is based on the result of

Fig. 1 Traffic survey locations

Table 1 Calibrated parameters

Parameters	Calibration value	
	Before	After
Desired position at free flow	Middle of lane	Any
Overtake on same lane: on left and on right	off	on
Distance standing (at 0 km h) (m)	1	0.3
Distance standing (at 50 kmh) (m)	1	0.5
Average standstill distance	2	0.45
Additive part of safety' distance	2	0.45
Multiplicative part of safety distance	3	4
Waiting time before diffusion (s)	60	40
Min. headway (front rear) (m)	0.5	0.4
Safety distance reduction factor	0.6	0.4
Maximum deceleration for cooperative braking (m s:)	−3.00	−3.00

road geometric survey and the data input is resulted from the survey of the traffic volume during peak hours.

B. Simulating and Modeling Using VISSIM

VISSIM is a microscopic multi-modal traffic flow simulation software package developed by PTV (Planung Transports Verkehr) AG in Karlsruhe, Germany. The name is derived from "Verkehr In Städten—SIMulations modell" (German for "Traffic in cities—simulation model"). VISSIM was first developed in 1992. VISSIM has used the Wiedemann approach, psycho physical model for driving behavior [10]. These following parameters have been calibrated to meet the Indonesian behavior (Table 1).

The next step is making the model based on road geometric and traffic volume in the five different cities resulted from the traffic survey. Next, the speed result data of VISSIM software are compared with the Indonesian Highway Capacity Manual (IHCM) calculation result.

According to the Indonesian Highway Capacity Manual, the free flow speed can be calculated as shown below [11]:

$$FV = (FV_0 + FV_W) \times FFV_{SF} \times FFV_{CS} \tag{1}$$

With

FV free flow speed (km/hour)
FV_0 basic free flow speed (km/hour)
FV_W road width factor (km/hour)
FFV_{SF} side friction factor
FFV_{CS} city size factor

Table 2 Calibrated parameter

City	Capacity (pcu)	
	IHCM	VISSIM
Yogyakarta 1	2951.52	2784.78
Yogyakarta 2	2894.76	2823.87
Banda Aceh 1	3175.50	3394.63
Banda Aceh 2	3175.50	3373.33
Banjarmasin 1	3383.66	3415.08
Banjarmasin 2	3383.66	3679.03
Pekanbaru 1	3039.96	3439.28
Pekanbaru 2	3039.96	3253.97
Surabaya 1	3333.00	3647.04
Surabaya 2	3333.00	3574.10

The analysis is then, proceeded further by the following steps:

(1) The traffic volume of each road in each different city is gradually increased from the existing volume, and then modelled back using VISSIM software.
(2) The data of speed estimation from VISSIM software for each traffic volume is then noted and analysed.
(3) The estimation of speed data resulted from VISSIM software is then compared with the speed data resulted from the calculation using IHCM.
(4) Some changes are made on IHCM parameters such as basic capacity and basic speed, so that data resulted from IHCM calculation is suitable with the data from VISSIM software result.

C. Data Analysis

The speed estimation data from the VISSIM software is then used in capacity calculation analysis based on Greenshield method [12] by using a linear relationship between density and speed. The capacity calculation result is then compared with the capacity calculation using IHCM. The comparison from the IHCM calculation result and VISSIM software result can be seen in Table 2 and Fig. 2.

From the Table 2 and Fig. 2, it can be concluded that the IHCM based capacities lower compared to the result from VISSIM software. So that, the constants of the IHCM needs to be calibrated. In this research, the trial and error to change the constants in IHCM were conducted. At the end of the trial, the basic capacity for the divided road increased from 1650 passenger car unit (pcu) to 1750 pcu. The result of the trial can be seen in Table 3 and Fig. 3.

Table 3 and Fig. 3 show that by changing the basic capacity of the IHCM resulting better relation between IHCM and VISSIM software result.

The next analysis is comparing the resulted data from VISSIM software and IHCM in estimating the speed of car and motorcycle in different number of traffic volume. The speed of heavy vehicle has not been analyzed, because there is only a few heavy

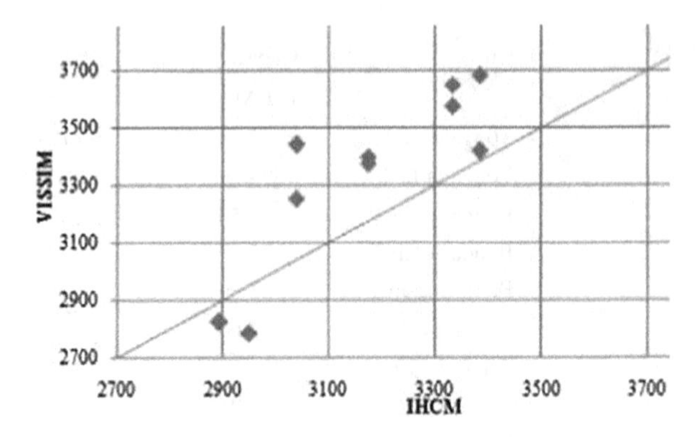

Fig. 2 The relationship between IHCM and VISSIM in basic capacity calculation

Table 3 Comparison between calibrated IHCM and VISSIM results in capacity calculation

City	Capacity (pcu)	
	IHCM	VISSIM
Yogyakarta 1	3130.40	2784.78
Yogyakarta 2	3070.20	2823.87
Banda Aceh 1	3367.98	3394.63
Banda Aceh 2	3367.98	3373.33
Banjarmasin 1	3588.73	3415.08
Banjarmasin 2	3588.73	3679.03
Pekanbaru 1	3224.20	3439.28
Pekanbaru 2	3224.20	3253.97
Surabaya 1	3535.00	3647.04
Surabaya 2	3535.00	3574.10

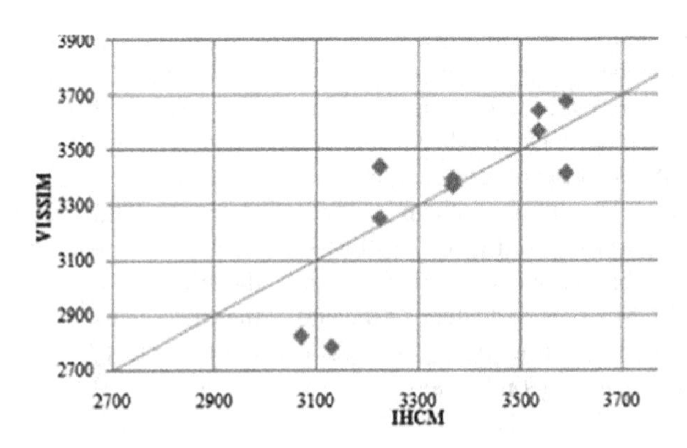

Fig. 3 Relationship between IHCM and VISSIM results in capacity calculation

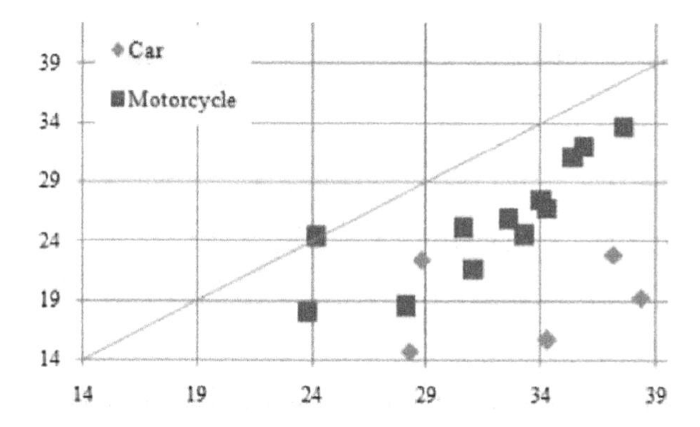

Fig. 4 Relationship between Speed prediction at Yogyakarta from IHCM calculation and VISSIM results

vehicles on the road. The increase of traffic volume will in line with increase of the degree of saturation. In IHCM we can predict the speed of car and motorcycle by calculating the free flow speed and then with the graphic relation between the degree of saturation and the free flow speed, we can predict the actual speed on a road section. These calculations result is then compared with the estimation of speed from VISSIM software that has been simulated before. The result of the comparison between IHCM calculation and data from VISSIM software is shown in Fig. 4.

As shown from Fig. 4, speed prediction from IHCM calculation are mostly too high compared with the data result from VISSIM software. Furthermore, the speed prediction from IHCM calculation showed that motorcycle has lower speed than car, whereas the data from the VISSIM software in line with data from the field observation show that in urban area motorcycle has higher speed than car. These cases were happened in all five cities, so it is necessary to change the constants on IHCM formula to predict the actual speed on the field.

Once again, trial and error to change the constants on IHCM formula were conducted. At the end of the trial, IHCM give the best result when the change was made in decreasing basic free flow speed. Based on the survey analysis, the basic free flow speed for car in divided road was changed from 57 to 37 kph, while the basic free flow speed for motorcycle in divided road was changed from 47 to 40 kph. And the result from the last trial is shown at the Fig. 5.

As shown in Fig. 5, the trial by changing the basic free flow speed in urban area of the IHCM resulting better relation between IHCM and VISSIM software result in speed estimation. It can be proved by the relation between these two has been approaching the 45° reference line (blue line). Comparison between calibrated IHCM result and VISSIM software is shown in Figs. 6 and 7.

As shown in Figs. 6 and 7, the graphics show that there is no significant difference between IHCM result after correction and VISSIM software result. It can be proved by the relation between these two has been approaching the reference line. On the

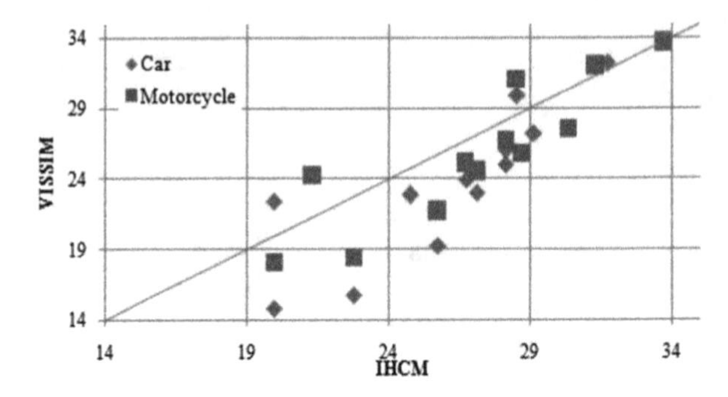

Fig. 5 Relationship between Speed prediction at Yogyakarta from calibrated IHCM and VISSIM software results

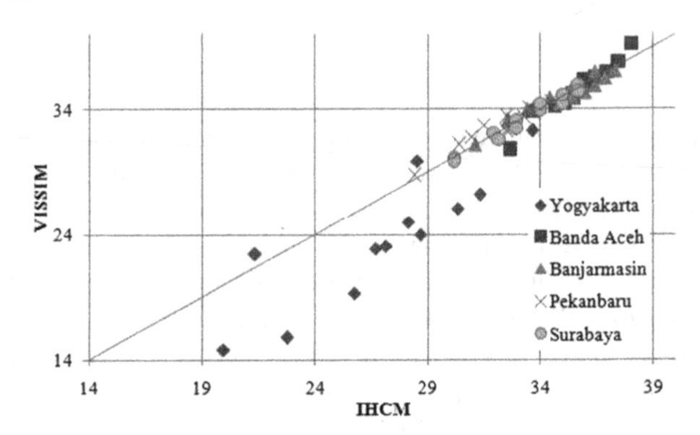

Fig. 6 Relationship between calibrated IHCM result and VISSIM software results for car speed estimation

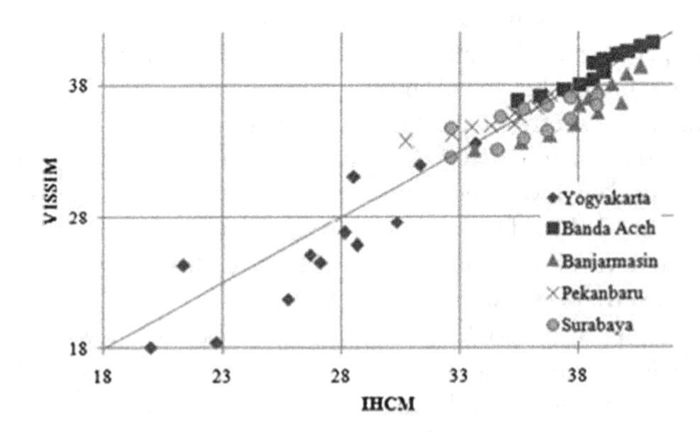

Fig. 7 Relationship between calibrated IHCM and VISSIM software result for motorcycle speed estimation

Table 4 Correlation between predicted and actual speed at Kaliurang Street (evening peak, side friction high)

Predicted speed (km/h)	Actual speed (km/h)
40.5	32.2
40.5	32.0
40.5	31.5
40.5	30.7
40.5	29.7
40.5	28.7
40.5	29.3
40.5	28.1

other words, the IHCM result already resemble with the real condition, since the model in VISSIM software has been validated before.

Because of the high traffic flow, the speed of the car is lower than the speed of motorcycle. The motorcycle can move easily among the cars with less distance than in the past. It is quite different from the traffic condition in the past, when the traffic flow in most urban areas is low and more distance between motorcycle while standing and overtaking.

4 New Formula for Speed Prediction

It is also observed that although there are so many side frictions in site, but by using the Indonesian Highway Capacity Manual formula, it is found that the side friction is still medium or high. It is rarely found that the side friction is very high.

The predicted speed according to the Indonesian Highway Capacity Manual [11] is significantly different to the actual speed, when the side friction value is high or medium. An example of that is shown Table 4 for Kaliurang Street in Yogyakarta during evening peak hour, when the side friction is high.

The friction class category, which has been employed by Indonesian HCM, does not take into account the speed sensitivity reduction because of the side friction.

It has been analysed, therefore, to use the multi regression formula to predict the speed. The first proposed formula is as follows:

$$Y = a_1x_1 + a_2x_2 + a_3x_3 + a_4x_4 + a_5x_5 + a_6x_6 + a_7x_7 + a_8x_8 + a_9x_9 + a_{10}x_{10} + k \ldots \quad (2)$$

with

Y	speed (km/hour)
x_1	number of unmotorized vehicles (veh/hour)
x_2	number of stopping city buses (veh/200 m/h)
x_3	pedestrian movement (pedestrian/200 m/h)

x_4	number of parking/stopping passenger car (veh/200 m/h)
x_5	number of entry vehicles into the street (veh/200 m/h)
x_6	number of exit vehicles from the street (veh/200 m/h)
x_7	number of passenger cars per hour (veh/hour)
x_8	number of heavy vehicles per hour (veh/hour)
x_9	number of motorcycles per hour (veh/hour)
x_{10}	number of total vehicles per hour (veh/hour)
$a_1, a_2, a_3, \ldots a_{10}$	coefficient factor
k	constant factor

By using trial and error method, it is found that the best formula is as follows:

$$Y = -0.132\,x_2 - 0.126\,x_3 - 0.280\,x_4 - 0.126\,x_5 - 0.153\,x_8 + 39.458\ldots \quad (3)$$

It is also found that parking/stopping vehicles is the most influence factor in reducing the speed. Parking/stopping vehicles could be taken into account as a factor in reducing capacity (lane width reduction) not as a side friction factor. The impact of parking vehicles is very high in reducing speed and capacity. The second factor is stopping city buses. The city buses can stop anywhere and their stopping behavior was considered to be of special importance to the operation of urban road. However, there is a new policy of the government to enforce city buses to stop only at the bus stop. If it is succeeded, this factor can be omitted as a friction factor.

5 Conclusion

(1) According to the Indonesian Highway Capacity Manual, the speed of car is higher than the speed of motorcycle. But according to the real condition and simulation, the speed of car is lower than the speed of motorcycle.
(2) The capacity according to the simulation, which is based on the reality, is higher than the capacity according to the IHCM, because of the change of the driver behaviour. The driver can take more risks by taking less distance when standing and overtaking.
(3) It is, therefore, recommended to change:
    (a) the basic free flow speed for urban road for car: from 57 (IHCM 1997) to 37 kph.
    (b) the basic free flow speed for urban road for motorcycle: from 47 (IHCM 1997) to 40 kph.
    (c) the basic capacity for urban road from 1650 (IHCM 1997) to 1750 pcu/h/lane.
(4) It is recommended to use new formula for speed prediction by using multi regression analysis.

References

1. A. Munawar, M.Z. Irawan, A.G. Fitrada, Development of urban road capacity and speed estimation methods in Indonesia, in *Proceedings of The World Congress on Engineering 2017, WCE 2017*. Lecture Notes in Engineering and Computer Science, 30 June–2 July, 2010 (London, U.K.) pp 564–567
2. M.Z. Irawan, T. Sumi, A. Munawar, Implementation of the 1997 Indonesian highway capacity manual (MKJI) volume delay function. J. East. Asia Soc. Transp. Stud. **8**, 350–360 (2010)
3. K.L. Bang, A. Carlsson, Palgunadi, Development of speed-flow relationships for Indonesian rural roads using empirical data and simulation, in *Transportation Research Record 1484* (Transportation Research Board, National Academy Press, Washington, D.C, 1995)
4. A. Munawar, Speed and capacity for urban roads, Indonesian experience. Proc.—Soc. Behav. Sci. **16**, 362–387 (2011), http://www.sciencedirect.com/science/article/pii/S1877042811010056
5. N. Marler, G Harahap, E. Novara, Speed-flow relationship and side friction on Indonesian urban highways, in *The Second International Symposium on Highway Capacity*, vol. 2, ed. by R. Akcelik (Transportation Research Board and Australian Road Research Board, Sydney, Australia, 1994)
6. M.Z. Irawan, N.H. Putri, Kalibrasi Vissim Untuk Mikrosimulasi Arus Lalu Lintas Tercampur Pada Simpang Bersinyal (Studi Kasus: Simpang Tugu, Yogyakarta. Jurnal Transportasi Multimoda) **13**(3), 97–106 (2017)
7. E. Kusnandar, Pengkinian Manual Kapasitas Jalan Indonesia 1997. Jurnal Jalan dan Jembatan **26** (2) (2009)
8. H. Iskandar, *Ekuivalen Kendaraan Ringan dan Kapasitas Dasar Jalan Perkotaan* (Pusta Litbang Jalan dan Jembatan, Bandung, 2011)
9. I.A. Winnetou, A. Munawar, Penggunaan Software Vissim Untuk Evaluasi Hitungan MKJI 1997 Kinerja Ruas Jalan Perkotaan (Studi Kasus: Jalan Affandi, Yogyakarta), in *The 18th FSTPT International Symposium*, Bandar Lampung, Indonesia (2015)
10. R. Wiedemann, *Simulations des Strassenverkehrsflusses* (Universitataet (TH) Karlslruhe, Heft, Schriftenreihe des Institutsfuer Verkeherswesen, 1974), p. 8
11. Directorate General of Highways, Republic Indonesia, Indonesian Highway Capacity Manual, Jakarta (1997)
12. B.D. Greenshields, A study of traffic capacity. Proc. Highw. Res. Board **14**(1), 448–474 (1934)

[illegible] Indonesian Highway Capacity Manual [illegible]

References

[illegible]

Natural Frequencies of an Euler-Bernoulli Beam with Special Attention to the Higher Modes via Variational Iteration Method

Olasunmbo O. Agboola, Jacob A. Gbadeyan, Abiodun A. Opanuga, Michael C. Agarana, Sheila A. Bishop and Jimevwo G. Oghonyon

Abstract In this chapter, natural frequencies of an Euler-Bernoulli prismatic beam on different supports are analyzed. The variational iteration method (VIM) is employed to compute the said natural frequencies especially for the higher modes of vibration. Some numerical examples are presented with the view to demonstrating excellent agreement between the results obtained using VIM and other methods.

Keywords Cantilever beam · Clamped-clamped beam · Euler-Bernoulli beam Natural frequency · Prismatic beam · Variational iteration method

1 Introduction

Several techniques have been used to carry out the vibration analysis of beams with a view to determining their vibration characteristics. Lai et al. [1] analysed the free vibration of uniform Euler-Bernoulli beam with different elastically supported con-

O. O. Agboola (✉) · A. A. Opanuga · M. C. Agarana · S. A. Bishop · J. G. Oghonyon
Department of Mathematics, Covenant University, Ota, Nigeria
e-mail: ola.agboola@covenantuniversity.edu.ng

A. A. Opanuga
e-mail: abiodun.opanuga@covenantuniversity.edu.ng

S. A. Bishop
e-mail: sheila.bishop@covenantuniversity.edu.ng

J. G. Oghonyon
e-mail: godwin.oghonyon@covenantuniversity.edu.ng

J. A. Gbadeyan
Department of Mathematics, University of Ilorin, Ilorin, Nigeria
e-mail: j.agbadeyan@yahoo.com

M. C. Agarana
Department of Mechanical Engineering Science, University of Johannesburg, Johannesburg, South Africa
e-mail: michael.agarana@covenantuniversity.edu.ng

S.-I. Ao et al. (eds.), *Transactions on Engineering Technologies*,
https://doi.org/10.1007/978-981-13-0746-1_13

ditions using Adomian decomposition method (ADM). Li [2] had earlier studied the vibration characteristics of a beam having general boundary conditions. The displacement of the beam was sought in form of a linear combination of a Fourier series and an auxiliary polynomial function. Kim and Kim [3] also applied Fourier series to determine the natural frequencies of beams having generally restrained boundary conditions. Later, Liu and Gurram [4] adopted the He's variational iteration method (VIM) to estimate the vibration frequencies of a uniform Euler-Bernoulli beam for various supporting end conditions. Natural frequencies for the first six modes of vibration were presented in their work.

Malik and Dang [5] employed the differential transform method (DTM) to obtain the natural frequencies and mode shapes of a uniform Euler-Bernoulli beam. The natural frequencies and the mode shape functions for the first three modes of vibration were derived for all the possible combinations of the classical end conditions of a beam. Opanuga et al. [6] employed DTM, ADM and Homotopy Perturbation Method so solve initial value and boundary value problems. Agboola et al. [7] also used DTM to solve third-order differential equations. He's polynomials approach is used by Edeki et al. [8] to obtain approximate and exact solutions of the Navier-Stokes model.

The primary aim of the work in this chapter is to, therefore, explore further the versatility of the VIM in determining the vibration characteristics of a uniform Euler-Bernoulli beam under different boundary conditions especially for higher modes. The VIM is a modification of a general Lagrange multiplier method which was proposed by He [9]. The method had earlier been used in [10] to determine the natural frequencies of a cantilever Euler-Bernoulli beam.

2 Problem Formulation

The governing equation of motion for free vibration of a uniform Euler-Bernoulli beam is given as:

$$EI\frac{\partial^4 y(x,t)}{\partial x^4} + \rho A\frac{\partial^2 y(x,t)}{\partial t^2} = 0 \tag{1}$$

where $y(x, t)$ is the transverse displacement of the beam at any distance x along the length of the beam at time t. In Eq. (1), E, and ρ are Young's modulus of elasticity and mass density of the beam material respectively while A and I are the cross-sectional area and the area moment of inertia of the cross section of the beam, respectively.

The boundary conditions considered in this chapter are:

(a) Pinned-pinned beam

$$\begin{aligned} y(0,t) = 0,\ y''(0,t) = 0, \\ y(l,t) = 0,\ y''(l,t) = 0, \end{aligned} \tag{2}$$

(b) Clamped-pinned beam

$$\begin{aligned} y(0,t) = 0,\ y'(0,t) = 0, \\ y(l,t) = 0,\ y''(l,t) = 0, \end{aligned} \tag{3}$$

(c) Clamped-clamped beam

$$\begin{aligned} y(0,t) = 0,\ y'(0,t) = 0, \\ y(l,t) = 0,\ y'(l,t) = 0, \end{aligned} \tag{4}$$

(d) Clamped-free (cantilevered) beam

$$\begin{aligned} y(0,t) = 0,\ y'(0,t) = 0, \\ y''(l,t) = 0,\ y'''(l,t) = 0, \end{aligned} \tag{5}$$

where l is the length of the beam.

For Eq. (1) and any of Eqs. (2)–(5), we assume a solution of the form:

$$y(x,t) = Y(x)(\cos \omega t + \theta) \tag{6}$$

where $Y(x)$ is the mode shape of the beam and ω is the angular frequency of the system.

By using Eq. (6) in Eq. (1) and any of Eqs. (2)–(5) and after carrying out non-dimensionalization, the equation of motion reduces to

$$Y^{iv}(\xi) - \Omega^2 Y(\xi) = 0,\ 0 < \xi < 1 \tag{7}$$

such that

$$\xi = \frac{x}{L},\ Y(\xi) = \frac{Y(x)}{L},\ \Omega^2 = \frac{\rho A \omega^2 L^4}{EI}, \tag{8}$$

where Ω is the nondimensional natural frequency of the cantilever beam.

The boundary conditions in Eqs. (2)–(5) can also be written in non-dimensional form as:

(a) Pinned-pinned beam

$$\begin{aligned} &Y(0) = 0,\ Y''(0) = 0, \\ &Y(1) = 0,\ Y''(1) = 0, \end{aligned} \tag{9}$$

(b) Clamped-pinned beam

$$\begin{aligned} &Y(0) = 0,\ Y'(0) = 0, \\ &Y(1) = 0,\ Y''(1) = 0, \end{aligned} \tag{10}$$

(c) Clamped-clamped beam

$$\begin{aligned} &Y(0) = 0,\ Y'(0) = 0, \\ &Y(1) = 0,\ Y'(1) = 0, \end{aligned} \tag{11}$$

(d) Clamped-free (cantilevered) beam

$$\begin{aligned} &Y(0) = 0,\quad Y'(0) = 0, \\ &Y''(1) = 0,\ Y'''(1) = 0, \end{aligned} \tag{12}$$

3 Basic Review of the Variational Iteration Method

The basic principle of variational iteration method (VIM) is presented in this section. Consider the differential equation of the form:

$$Lw(t) + Nw(t) = g(t), \tag{13}$$

where L and N are respectively linear and nonlinear operators, $g(t)$ is a known continuous function referred to as the nonhomogeneous term. A correction functional presented by the variational iteration method for Eq. (13) is given by

$$w_{n+1}(\xi) = w_n(\xi) + \int_0^{\xi} \lambda(\tau)(Lw_n(\tau) + N\tilde{w}_n(\tau))d\tau \tag{14}$$

where λ is a general Lagrange multiplier, which can be optimally determined by variational theory, $\tilde{w}_n$ is a restricted variation which implies that $\delta\tilde{w}_n = 0$, where δ is the variational derivative.

4 Solution by VIM

The correction functional for the governing equation of the vibration problem can be written as

$$Y_{n+1}(\xi) = Y_n(\xi) + \int_0^{\xi} \lambda(\tau)\left(\frac{d^4Y_n(\tau)}{d\tau^4} - \Omega^2 Y_n(\tau)\right)d\tau \tag{15}$$

which is obtained by comparing Eq. (7) with Eq. (13).

Using the method of integration by parts, Eq. (12) can be written as

$$\begin{aligned} Y_{n+1}(\xi) = Y_n(\xi) &+ \lambda(\xi)\frac{d^3Y_n(\xi)}{d\xi^3} - \frac{d\lambda(\xi)}{d\xi}\frac{d^2Y_n(\xi)}{d\xi^2} + \frac{d^2\lambda(\xi)}{d\xi^2}\frac{dY_n(\xi)}{d\xi} \\ &- \frac{d^3\lambda(\xi)}{d\xi^3}Y_n(\xi) + \int_0^{\xi}\left[\frac{d^4\lambda(\tau)}{d\tau^4} - \Omega^2\lambda(\tau)\right]Y_n(\tau)d\tau \end{aligned} \tag{15}$$

Taking the variation of Eq. (13) with respect to Y_n gives

$$\begin{aligned} \delta Y_{n+1}(\xi) = \delta Y_n(\xi) &+ \lambda(\xi)\delta\frac{d^3Y_n(\xi)}{d\xi^3} - \frac{d\lambda(\xi)}{d\xi}\delta\frac{d^2Y_n(\xi)}{d\xi^2} \\ &+ \frac{d^2\lambda(\xi)}{d\xi^2}\delta\frac{dY_n(\xi)}{d\xi} - \frac{d^3\lambda(\xi)}{d\xi^3}\delta Y_n(\xi) \\ &+ \int_0^{\xi}\left[\frac{d^4\lambda(\tau)}{d\tau^4} - \Omega^2\lambda(\tau)\right]\delta Y_n(\tau)\,d\tau \end{aligned} \tag{17}$$

It can be readily shown that the Lagrange multiplier is

$$\lambda(\tau) = \frac{(\tau-\xi)^3}{3!} \tag{18}$$

(Wazwaz [11]).
Substituting this value of the Lagrange multiplier in Eq. (18) into the correction functional stated in Eq. (15), the following iteration formula can be obtained:

$$Y_{n+1}(\xi) = Y_n(\xi) + \int_0^{\xi} \frac{(\tau-\xi)^3}{3!}\left(\frac{d^4Y_n(\tau)}{d\tau^4} - \Omega^2 Y_n(\tau)\right)d\tau \tag{19}$$

For fast convergence, the function $Y_0(\xi)$ is required to be selected by using the initial conditions as follows:

$$Y_0(\xi) = Y(0) + \frac{dY(0)}{d\xi}\xi + \frac{1}{2!}\frac{d^2Y(0)}{d\xi^2}\xi^2 + \frac{1}{3!}\frac{d^3Y(0)}{d\xi^3}\xi^3 \tag{20}$$

for fourth order $\frac{d^4Y_n}{d\xi^4}$.

Several approximations can be obtained from Eq. (18) as follows:

$$\begin{aligned} Y_1(\xi) &= Y_0(\xi) + \int_{\sim 0}^{\xi} \frac{(\tau-\xi)^3}{3!}\left(\frac{d^4Y_0(\tau)}{d\tau^4} - \Omega^2 Y_0(\tau)\right)d\tau \\ Y_2(\xi) &= Y_1(\xi) + \int_0^{\xi} \frac{(\tau-\xi)^3}{3!}\left(\frac{d^4Y_1(\tau)}{d\tau^4} - \Omega^2 Y_1(\tau)\right)d\tau \\ &\vdots \\ Y_k(\xi) &= Y_{k-1}(\xi) + \int_0^{\xi} \frac{(\tau-\xi)^3}{3!}\left(\frac{d^4Y_{k-1}(\tau)}{d\tau^4} - \Omega^2 Y_{k-1}(\tau)\right)d\tau \end{aligned} \tag{21}$$

Thus, the solution to Eq. (7) is obtained as the limit of the above resulting successive approximations expressed by

$$Y(\xi) = \lim_{k\to\infty} Y_k(\xi) \tag{22}$$

In practice, a large number, say s, which is to be determined based on the required accuracy, is chosen to replace infinity in Eq. (22). So, one has

$$Y(\xi) = Y_s(\xi) \tag{23}$$

In order to demonstrate the procedure for a clamped-free (cantilever) beam, we proceed as follows:

First part of Eq. (12) is substituted into Eq. (20) to give

$$Y_0(\xi) = \frac{1}{2!}\frac{d^2Y(0)}{d\xi^2}\xi^2 + \frac{1}{3!}\frac{d^3Y(0)}{d\xi^3}\xi^3 \tag{24}$$

or

$$Y_0(\xi) = a_1\xi^2 + a_2\xi^3 \tag{25}$$

where

$$a_1 = \frac{1}{2!}\frac{d^2Y(0)}{d\xi^2},\ a_2 = \frac{1}{3!}\frac{d^3Y(0)}{d\xi^3} \tag{26}$$

Substituting Eq. (25) into Eq. (21) leads to the following successive approximations

$$Y_1(\xi) = \left(\xi^2 + \frac{1}{360}\Omega^2\xi^6\right)a_1 + \left(\xi^3 + \frac{1}{840}\Omega^2\xi^7\right)a_2 \tag{27}$$

$$Y_2(\xi) = \left(\xi^2 + \frac{1}{360}\Omega^2\xi^6 + \frac{1}{1814400}\Omega^4\xi^{10}\right)a_1 + \left(\xi^3 + \frac{1}{840}\Omega^2\xi^7 + \frac{1}{6652800}\Omega^4\xi^{11}\right)a_2 \tag{28}$$

$$\begin{aligned} Y_3(\xi) = &\left(\xi^2 + \frac{1}{360}\Omega^2\xi^6 + \frac{1}{1814400}\Omega^4\xi^{10} + \frac{1}{43589145600}\Omega^6\xi^{14}\right)a_1 \\ &+\left(\xi^3 + \frac{1}{840}\Omega^2\xi^7 + \frac{1}{6652800}\Omega^4\xi^{11} + \frac{1}{217945728000}\Omega^6\xi^{15}\right)a_2 \end{aligned} \tag{29}$$

$$\begin{aligned} Y_4(\xi) &= \left(\xi^2 + \frac{1}{360}\Omega^2\xi^6 + \frac{1}{1814400}\Omega^4\xi^{10} + \frac{1}{43589145600}\Omega^6\xi^{14} + \frac{1}{3201186852864000}\Omega^8\xi^{18}\right)a_1 \\ &+ \left(\xi^3 + \frac{1}{840}\Omega^2\xi^7 + \frac{1}{6652800}\Omega^4\xi^{11} + \frac{1}{217945728000}\Omega^6\xi^{15} + \frac{1}{20274183401472000}\Omega^8\xi^{19}\right)a_2 \end{aligned} \tag{30}$$

$$\begin{aligned} Y_5(\xi) = &\left(\xi^2 + \frac{1}{360}\Omega^2\xi^6 + \frac{1}{1814400}\Omega^4\xi^{10} + \frac{1}{43589145600}\Omega^6\xi^{14} \right. \\ &\left. + \frac{1}{3201186852864000}\Omega^8\xi^{18} + \frac{1}{5620003638888038400000}\Omega^{10}\xi^{22}\right)a_1 \\ &+ \left(\xi^3 + \frac{1}{840}\Omega^2\xi^7 + \frac{1}{6652800}\Omega^4\xi^{11} + \frac{1}{217945728000}\Omega^6\xi^{15} \right. \\ &\left. + \frac{1}{20274183401472000}\Omega^8\xi^{19} + \frac{1}{43086694564808294400000}\Omega^{10}\xi^{23}\right) + a_2 \end{aligned} \tag{31}$$

and so on.

Thus, it is seen that $Y_n(\xi)$ takes the form

$$Y_n(\xi) = a_1 f^{[n]}(\xi, \Omega) + a_2 g^{[n]}(\xi, \Omega) \tag{32}$$

where $f^{[n]}(\xi, \Omega)$ and $g^{[n]}(\xi, \Omega)$ are functions of ξ and Ω associated with n.

Let $Y_n(\xi) = Y(\xi)$, then the second part of Eq. (12) becomes

$$a_1 \frac{d^2 f^{[n]}(1, \Omega)}{d\xi^2} + a_2 \frac{d^2 g^{[n]}(1, \Omega)}{d\xi^2} = 0 \tag{33}$$

and

$$a_1 \frac{d^3 f^{[n]}(1, \Omega)}{d\xi^3} + a_2 \frac{d^3 g^{[n]}(1, \Omega)}{d\xi^3} = 0 \tag{34}$$

Equations (33) and (34) can be put in the vector-matrix form

$$\begin{bmatrix} \frac{d^2 f^{[n]}(1,\Omega)}{d\xi^2} & \frac{d^2 g^{[n]}(1,\Omega)}{d\xi^2} \\ \frac{d^3 f^{[n]}(1,\Omega)}{d\xi^3} & \frac{d^3 g^{[n]}(1,\Omega)}{d\xi^3} \end{bmatrix} \begin{bmatrix} a_1 \\ a_2 \end{bmatrix} = \begin{bmatrix} 0 \\ 0 \end{bmatrix} \tag{35}$$

The system of equations in (35) possesses nontrivial solutions provided the determinant of the coefficient matrix is equal to zero.

That is,

$$\begin{vmatrix} \frac{d^2 f^{[n]}(1,\Omega)}{d\xi^2} & \frac{d^2 g^{[n]}(1,\Omega)}{d\xi^2} \\ \frac{d^3 f^{[n]}(1,\Omega)}{d\xi^3} & \frac{d^3 g^{[n]}(1,\Omega)}{d\xi^3} \end{vmatrix} = 0 \tag{36}$$

The values of Ω yielded by solving Eq. (36) are the nondimensional natural frequencies of the beam under consideration. In essence, one can obtain the jth estimated nondimensional frequency, $\Omega_j^{[n]}$ corresponding to n. The value of n would be decided by this inequality:

$$\left|\Omega_j^{[n]} - \Omega_j^{[n-1]}\right| \leq \varepsilon \tag{37}$$

where ε is the allowable error tolerance dependent on the required accuracy.

The vibration analysis for pinned-pinned, clamped-pinned and clamped-clamped Euler-Bernoulli beam can be carried out following the same procedure outlined above by using the corresponding boundary conditions. For instance, the successive approximations in Eq. (21) that correspond to a pinned-pinned beam can be obtained as:

$$Y_1(\xi) = \left(\xi + \frac{1}{120}\Omega^2\xi^5\right)a_3 + \left(\xi^3 + \frac{1}{840}\Omega^2\xi^7\right)a_1 \tag{38}$$

$$Y_2(\xi) = \left(\xi + \frac{1}{120}\Omega^2\xi^5 + \frac{1}{362880}\Omega^4\xi^9\right)a_3 + \left(\xi^3 + \frac{1}{840}\Omega^2\xi^7 + \frac{1}{6652800}\Omega^4\xi^{11}\right)a_1 \tag{39}$$

$$\begin{aligned} Y_3(\xi) = &\left(\xi + \frac{1}{120}\Omega^2\xi^5 + \frac{1}{362880}\Omega^4\xi^9 + \frac{1}{6227020800}\Omega^6\xi^{13}\right)a_3 \\ &+\left(\xi^3 + \frac{1}{840}\Omega^2\xi^7 + \frac{1}{6652800}\Omega^4\xi^{11} + \frac{1}{217945728000}\Omega^6\xi^{15}\right)a_1 \end{aligned} \tag{40}$$

$$\begin{aligned} Y_4(\xi) = &\left(\xi + \frac{1}{120}\Omega^2\xi^5 + \frac{1}{362880}\Omega^4\xi^9 + \frac{1}{6227020800}\Omega^6\xi^{13} + \frac{1}{355687428096000}\Omega^8\xi^{17}\right)a_3 \\ &+\left(\xi^3 + \frac{1}{840}\Omega^2\xi^7 + \frac{1}{6652800}\Omega^4\xi^{11} + \frac{1}{217945728000}\Omega^6\xi^{15} + \frac{1}{20274183401472000}\Omega^8\xi^{19}\right)a_1 \end{aligned} \tag{41}$$

$$\begin{aligned} Y_5(\xi) = &\left(\xi + \frac{1}{120}\Omega^2\xi^5 + \frac{1}{362880}\Omega^4\xi^9 + \frac{1}{6227020800}\Omega^6\xi^{13} + \frac{1}{355687428096000}\Omega^8\xi^{17} + \frac{1}{51090942171709440000}\Omega^{10}\xi^{21}\right)a_3 \\ &+\left(\xi^3 + \frac{1}{840}\Omega^2\xi^7 + \frac{1}{6652800}\Omega^4\xi^{11} + \frac{1}{217945728000}\Omega^6\xi^{15}\right. \\ &\left.+\frac{1}{20274183401472000}\Omega^8\xi^{19} + \frac{1}{4308669456480829440000}\Omega^{10}\xi^{23}\right)a_1 \end{aligned} \tag{42}$$

and so on, such that a_1 is as defined in Eq. (25), and also

$$a_3 = \frac{dY(0)}{d\xi} \tag{43}$$

5 Numerical Analysis

In this section, the procedures of variational iteration method for solving the vibration problem in this study earlier discussed are implemented numerically with a view to computing the nondimensional frequencies of an Euler-Bernoulli beam with the clamped-free supports and three other conditions. The results obtained for the first three vibration modes are compared with the results obtained analytically and by differential transform method. These are displayed in Table 1. It can be observed from Table 1 that there is an excellent agreement between the results obtained by DTM and VIM. The nondimensional frequencies for higher modes of vibration for a cantilever beam which are obtained using VIM are reported in Table 2.

Table 1 Comparison of the first three nondimensional natural frequencies by methods for clamped-free (cantilever) Euler-Bernoulli beam

Nondimensional frequencies	Analytical method [12]	DTM [5]	VIM (Present)
Ω_1	3.51563	3.516015	3.516015
Ω_2	22.03364	22.034492	22.034492
Ω_3	61.7010	61.697214	61.697214

Table 2 Nondimensional natural frequencies by methods for clamped-free (cantilever) Euler-Bernoulli beam for higher modes of vibration

Nondimensional frequencies	VIM (Present)
Ω_4	120.901916
Ω_5	199.859530
Ω_6	298.555531
Ω_7	416.990786
Ω_8	555.165247
Ω_9	713.0789222
Ω_{10}	890.731667
Ω_{11}	1088.127450
Ω_{12}	1305.168455
Ω_{13}	1544.058323
Ω_{14}	1771.488309
Ω_{15}	2401.026455

Table 3 Comparison of the first three nondimensional natural frequencies by methods for pinned-pinned Euler-Bernoulli beam

Nondimensional frequencies	DTM [5]	VIM (Present)
Ω_1	9.869604	9.869604
Ω_2	39.478418	39.478418
Ω_3	88.826440	88.826440

The set of Tables 3, 4, 5, 6, 7 and 8 shows the results for the natural frequencies of an Euler-Bernoulli beam with pinned-pinned, clamped-pinned and clamped-clamped boundary conditions.

Table 4 Nondimensional natural frequencies by methods for pinned-pinned Euler-Bernoulli beam for higher modes of vibration

Nondimensional frequencies	VIM (Present)
Ω_4	157.913670
Ω_5	246.740110
Ω_6	355.305758
Ω_7	483.610616
Ω_8	631.654681
Ω_9	799.437959
Ω_{10}	986.960503
Ω_{11}	1194.220034
Ω_{12}	1421.232865
Ω_{13}	1668.598482
Ω_{14}	1937.205909
Ω_{15}	

Table 5 Comparison of the first three nondimensional natural frequencies by methods for clamped-pinned Euler-Bernoulli beam

Nondimensional frequencies	DTM [5]	VIM (Present)
Ω_1	15.418206	15.418206
Ω_2	49.964862	49.964862
Ω_3	178.269729	178.269730

Table 6 Nondimensional natural frequencies by methods for clamped-pinned Euler-Bernoulli beam for higher modes of vibration

Nondimensional frequencies	VIM (Present)
Ω_4	272.030971
Ω_5	385.531422
Ω_6	518.771081
Ω_7	671.749950
Ω_8	844.468027
Ω_9	1036.925312
Ω_{10}	1249.121807
Ω_{11}	1481.057510
Ω_{12}	1732.732423
Ω_{13}	2004.146544
Ω_{14}	2295.299928
Ω_{15}	2605.797645

Table 7 Comparison of the first three nondimensional natural frequencies by methods for clamped-clamped Euler-Bernoulli beam

Nondimensional frequencies	DTM [5]	VIM (Present)
Ω_1	22.373285	22.373285
Ω_2	61.672823	61.672823
Ω_3	199.859448	199.859448

Table 8 Nondimensional natural frequencies by methods for clamped-clamped Euler-Bernoulli beam for higher modes of vibration

Nondimensional frequencies	VIM (Present)
Ω_4	298.555535
Ω_5	416.990786
Ω_6	555.165248
Ω_7	713.078918
Ω_8	890.731776
Ω_9	1088.124298
Ω_{10}	1305.267266
Ω_{11}	1541.144296

6 Conclusion

The variational iteration method (VIM) has been used to estimate the natural frequencies of an Euler-Bernoulli beam with different boundary conditions. The emphasis has been placed on the natural frequencies for higher modes of vibration. The efficiency and the reliability of the method was also established by showing that there is an excellent agreement between VIM results and the existing results in the literature.

Acknowledgements This work was financially supported by Covenant University, Ota, Nigeria under her Center for Research, Innovation and Discovery.

References

1. H.-Y. Lai, J.-C. Hsu, C.-K. Chen, An innovative eigenvalue problem solver for free vibration of Euler-Bernoulli beam by using the Adomian decomposition method. Comput. Math Appl. **56**, 3204–3220 (2008)
2. W.L. Li, Free vibrations of beams with general boundary conditions. J. Sound Vib. **237**(4), 709–725 (2000). https://doi.org/10.1006/jsvi.2000.3150
3. H.K. Kim, M.S. Kim, Vibration of beams with generally restrained boundary conditions using Fourier series. J. Sound Vib. **245**(5), 771–784 (2001)
4. Y. Liu, C.S. Gurram, The use of He's variational iteration method for obtaining the free vibration of an Euler-Bernoulli beam. Math. Comput. Model. **50**, 1545–1552 (2009)
5. M. Malik, H.Y. Dang, Vibration analysis of continuous systems by differential transformation. Appl. Math. Comput. **96**, 17–26 (1998)
6. A.A. Opanuga, O.O. Agboola, H.I. Okagbue, J.G. Oghonyon, Solution of differential equations by three semi-analytical techniques. Int. J. Appl. Eng. Res. **10**(18), 39168–39174 (2015)
7. O.O. Agboola, A.A. Opanuga, J.A. Gbadeyan, Solution of third order ordinary differential equations using differential transform method. Global J. Pure Appl. Math. **11**(4), 2511–2516 (2015)
8. S.O. Edeki, G.O. Akinlabi, M.E. Adeosun, Analytical solutions of the Navier-Stokes model by He's polynomials, in *Proceedings of The World Congress on Engineering 2016*. Lecture Notes in Engineering and Computer Science, 29 June–1 July, 2016, London, U.K., pp. 16–19
9. J.H. He, Variational iteration method—a kind of non-linear analytical technique: some examples. Int. J. Non-Linear Mech. **34**(4), 699–708 (1999)

10. O.O. Agboola, J.A. Gbadeyan, A.A. Opanuga, M.C. Agarana, S.A. Bishop, J.G. Oghonyon, Variational iteration method for natural frequencies of a cantilever beam with special attention to the higher modes, in *Proceedings of The World Congress on Engineering 2017*. Lecture Notes in Engineering and Computer Science, 5–7 July, 2017, London, U.K., pp. 148–151
11. A.M. Wazwaz, in *A First Course in Integral Equations*, 2nd edn. (World Scientific Publishing Co. Plc. Ltd., Singapore, 2015). ISBN 978-9814675116
12. A.A. Shabana, *Theory of Vibration: An Introduction*, 2nd edn. (Springer, New York, 1995). ISBN 978-1-14612-8456-7

Learning Noise in Web Data Prior to Elimination

Julius Onyancha, Valentina Plekhanova and David Nelson

Abstract This research work explores how noise in web data is currently addressed. We establish that current research works eliminate noise in web data mainly based on the structure and layout of web pages i.e. they consider noise as any data that does not form part of the main web page. However, not all data that form part of the main web page is of a user interest and not every data considered noise is actually noise to a given user. The ability to determine what is useful from noise data taking into account dynamic change of user interests has not been fully addressed by current research works. We aim to justify a claim that it is important to learn noise prior to elimination, not only to decrease levels of noise but also reduce the loss of useful information otherwise eliminated as noise. This is because if the process of eliminating noise in web data is not user-driven, the interestingness of web data available to a user will not reflect their interests given the time of the request.

Keywords Dynamic session identification · Noise web data learning
User interest · User profile · Web log data · Web usage mining

1 Introduction

The amount of information available on the web is increasing rapidly in line with the explosive growth of the World Wide Web [1, 2]. While users are provided with more information, due to the scale and diversity of the data, it is becoming difficult to extract useful information that actually meets a given user's interests [3]. The

J. Onyancha (✉) · V. Plekhanova · D. Nelson
Faculty of Computer Science, University of Sunderland, St Peter's Campus,
DG Building, St Peter's Way, Sunderland SR6 0DD, UK
e-mail: julius.onyancha@research.sunderland.ac.uk

V. Plekhanova
e-mail: valentina.plekhanova@sunderland.ac.uk

D. Nelson
e-mail: david.nelson@sunderland.ac.uk

S.-I. Ao et al. (eds.), *Transactions on Engineering Technologies*,
https://doi.org/10.1007/978-981-13-0746-1_14

extraction of information from the web that meets the needs of a web user otherwise referred to as Web Usage Mining (WUM) [4, 5] has become a popular research area. WUM attempts to discover useful information from web logs which record the activities of users on the web [5]. In the real world, it is practically impossible to extract web log data and thus create a user profile which is free from noise. A user profile is defined as a description of user interests, characteristics, and preferences in relation to a given website [6–8]; user interest is measured by looking at user web log data to determine what time is spent on what web pages and the frequency of visits to particular web pages [9, 10]. Web data is noisy, inconsistent and often irrelevant by nature [11]. The presence of noise in web data hinders the extraction of useful information in relation to users' interests [5, 9].

The existing research on data mining defines noise data in terms of the application domain; for example, [12] defines noise data as irrelevant or meaningless data that typically does not reflect the main trends but instead makes the identification of these trends more difficult. Sunithaa et al. [13] argues that noise data is meaningless or corrupted data: i.e., any data that cannot be understood and interpreted correctly by machine or data that is incorrectly typed or is dissimilar to other entries. In web usage mining, noise is defined as any irrelevant data that is not part of the main content of a web page [1, 14, 15]. The content pages are web pages wherein a user can find relevant information [16]. Examples of noise in web data include advertisements, banners, graphics, web page links from external websites etc.

Existing research studies have contributed towards addressing problems associated with noise in web data: for example, use machine learning tools which can identify and eliminate noise in web data based on the structure and layout of web pages [1, 14, 17, 18]. The main web page content is considered as useful information and any data that does not form part of the main web page content is noise. Htwe and Kham [19], Velloso and Dorneles [20] proposed a mechanism whereby noise associated with web pages is matched to stored noise data for classification and subsequent elimination. This, therefore, suggests that the elimination of noise in web data can be based on pre-existing noise data patterns. However, the view taken by this proposed research is that noise is not necessarily data that does not form part of the main content of a web page but can also be otherwise useful information that does not reflect the user interest.

1.1 Problem Statement

In the previous section, the contribution made by existing research studies is acknowledged; however, there are still critical issues that have not so far been fully addressed. For example, as user interests change over time, the interestingness of web data is also affected; this makes it difficult to identify and eliminate noise in web data. In the evolving web, the current noise data can be seen as useful if the user's interests prior to the point of elimination are considered. From existing research, there is no evidence that noise in web data has been addressed, taking into account changes in user interest. The influence of web users and their evolving interests in the available data also opens up challenges in determining if the main web content itself is useful or noise. The dynamic aspects of user interest are critical to the web data reduction process because the quality of web data needs to be determined in relation to what web users are interested in. Therefore, based on the existing research's contribution vis-à-vis the current state of the art, the elimination of noise in web data prior to learning its interestingness to a user can lead to the loss of useful information [21].

1.2 Research Contribution

In this work, we establish, based on the existing literature, how existing research studies address the problems associated with noise in web data. The proposed research argues that not all data that form part of the main web page is relevant to a user's interest and not every data which can be considered noise is actually noise to a given user. Therefore, without learning the interestingness of web data based on a user's interest, the process of eliminating noise web data is limited to simply recognizing how the data on web pages is presented and not what users are interested in at any given time. The position established in this work contributes towards proposing machine learning algorithms capable of learning noise in web data for elimination. Noise web data reduction depends not only on pre-existing noise data patterns or relationships in data on a web page but also must be determined by the dynamic changes in user interest as well as the evolving web data.

2 Current Research Work

This section explores how existing research addresses the problem of noise in web data. Existing machine learning tools as applied by current research studies are critically evaluated. In order to justify the position taken by the proposed research, this section will undertake the following critical investigations:

(1) Establish from current research work the different types of web data noise eliminated by existing tools.
(2) Identify the measures used by currently available tools to eliminate noise from web log data.
(3) Find out how currently available tools take into consideration the interests of a user prior to eliminating web data noise.

Current tools developed to identify and eliminate noise from web pages are mainly based on two different approaches: the underlying structure of the document as appraised using a document object model (DOM) tree [1, 22], and the approach which is solely dependent on the visual layout of web pages [23]. The DOM tree is a data structure used to represent the structure of a web page; it is built using a web page's html parse from which a web content structure is created to distinguish areas of a website with relevant and with noise data [17, 24]. For example, [1] proposed a tool known as a Site Style Tree (SST) to detect and eliminate noise data from web pages based on the observation that the main web page contents usually shares the same presentation style and that any other page with different presentation styles may be considered as noise. To eliminate noise from web pages, SST simply maps the page to the main web page to determine if the page is useful or noise based on its presentation style. Htwe and Kham [19] developed a tool using case-based reasoning (CBR) and neural networks to eliminate noise data from web pages. CBR is a machine learning approach which makes use of past experience to solve future problems: i.e., in this case it detects noise from web pages using existing stored web data noise for reference. Different noise patterns presented by websites are stored in the form of DOM trees; this 'case base' can then searched for similar existing noise patterns. An Artificial Neural Network is used to match noise patterns with those stored in the Case-Base. This approach is based on the idea of using case-based reasoning to identify noise data by matching existing noise patterns stored in the case-base; however, it is difficult to determine if such content is relevant or noise to a particular user at a particular time regardless of the fact that it matches existing patterns of noise data. This is because web data is dynamic and so is the user's interest. If the quality of data is determined using such a case-based approach, then the output can be misleading. Narwal [25] proposed a pattern tree algorithm to eliminate noise data from web pages; this pattern tree algorithm is based on the DOM tree concept and uses the assumption that data present on the web can be considered noise if its pattern is dissimilar from the main content of a web page. Garg and Kaur [24] applied the Least Recently Used paging algorithm (LRU) to detect and remove noise from web pages. LRU takes into account both visual and non-visual characteristics of a web page and is able to remove web data noise such as news, blogs and discussions. The LRU algorithm determines frequently visited pages and those that have not been visited over a certain period.

The proposed research work does not rely on the structure and layout of web data in order to identify and eliminate noise but instead, it focuses on the interests of a user in relation to the available web data. However, the above tools play a significant role in defining user interest level from web log data. The reduction of noise from web pages based on the structure of data on a given website will subsequently affect the quality of extracted web log data. The proposed research work argues that noise in web data should be determined based on the web user's level of interest in the available data. In order to understand user interest from the extracted web log data, pre-processing is essential for the purposes of improving the quality of the output from the web usage mining processes [26].

A number of machine learning tools have been proposed for extracting useful information, given a user interest. For example, [27] applied a Naïve Bayesian Classification algorithm to identify the interests of users based on web log data extracted from a website. Those authors' main objective was to classify extracted web data logs and study how useful the extracted information is, based on a user's interest. Their initial processing phase involved removing noise data such as advertisement banners, images and screensavers from the extracted web data logs. They used a Naïve Bayesian classification model to classify useful as opposed to noise data based on the number of pages viewed and the time taken on a specific page. Neural networks (NN) are another tool widely applied in web usage mining to reduce noise from web logs. [28] states that neural networks can use the frequency of occurrence of a web page in web log data to determine its weight. The authors define weight as a statistical measure used to evaluate how important information is to a given user. Azad et al. [29] applied kNN on web data logs to find information concerning a user's interest by removing noise data; their main focus was on local noise: for example, advertisements, banners, navigational links etc. Web log data was extracted and surveyed in regard to which web server they belonged. If the address belonged to a list of already defined advertisement servers, then the link was removed. Malarvizhi and Sathiyabhama [30] proposed a 'Weighted Association Rule Mining' method for extracting useful information from web log data. Their objective was to find web pages visited by a user and assign weights based on interest level. The user-interest based page weight is used to eliminate noise web pages from useful information. In their research work, the weight of a web page in relation to a user interest is estimated from the frequency of page visits and the number of different pages visited. Where pages are visited only once by only one user, for instance, they will be assigned low weights and subsequently considered noise.

2.1 Influence of User Interest on Noise Web Data Reduction Process

User interest can be determined in two ways, i.e. explicitly or implicitly [31–33]. Explicit interest is where users provide feedback concerning what they think about the information that they have received. However, many users may not be willing to state what their true intentions are in terms of the visited websites. For this reason, the implicit learning of user interest is also considered by this proposed research. The implicit method uses logs created by a user's visit to a website in order to learn their (the user's) interest instead of requesting user feedback. According to [34, 35], there are two ways to capture implicit user interest: i.e., from browsing behaviour and from browser history. Browsing histories capture the relationship between a user's interests and their click history; this is necessary for the identification of useful information from the extracted web log data. Wu [36] acknowledge that the evaluation of user interest relies on the use of the visiting time of a page as an indicator of the level of user interest concerning visited web pages. The amount of time spent with a set of pages requested by the user within a single session forms the aggregate interest of that user in that session.

Learning user interests plays a fundamental role in understanding the usefulness of available web data to a user, thereby improving the process of noise web data reduction. It is therefore important to understand how the dynamic nature of the web and varying user interests influence the identification of noise in web data. In the proposed research work, the focus is on learning the interest of a user in relation to available web data with the aim of reducing the amount of useful information eliminated as noise.

2.2 Contribution—Existing Research

Current research acknowledges that the noise web data reduction process is aimed at improving the process of mining useful information from the web [9, 12, 37]. The proposed research outlines some of the major contributions made by the currently available noise web data reduction tools. For example: (1) the existing tools applied in the noise web data reduction processes focus mainly on improving the quality of data concerning web pages by removing noise at the pre-processing stage; (2) automatically detecting and removing noise data by matching noise data in currently logged web data with previously stored web data noise patterns. The contribution of this present research is not only to remove multiple noise patterns from logs relating to a web page but also to enable classification of the noise encountered, based on defined patterns. (3) Protecting useful data regions by identifying boundaries between noise and useful data [20, 38–40]. This is tied to the assumption that only the main web data contains useful information.

Despite efforts made in recent research work to address problems with noise in extracted web data logs, this work identifies a research gap that has not been fully addressed in relation to web data noise reduction; this is to do with creating a dynamic and interest-specific web-user profile. The proposed work argues that eliminating noise from extracted web data logs should be user-driven rather than determined by the relationship of data to the web page in which it resides. In the next section, we establish some critical perspectives that justify our position, based on current research contributions.

2.3 Discussion of Critical Aspects

Based on the discussion in the previous section, various critical perspectives related to problems with noise in web data are identified. Firstly, the existing research work mainly focuses on eliminating noise in web data based on the layout and structure of websites. This is to ensure that the core information which forms the main content of the web is isolated from noise data [14]. Moreover, the machine learning algorithms which have been proposed by the existing research are based on the observation that web pages usually share some common layouts and presentation styles [26, 37, 41]. However, the dynamic nature of the web makes it difficult to rely on presentation attributes for the identification and elimination of noise data.

Secondly, the common objective identified from existing research in web-data noise reduction is to improve the performance of the web data mining process [2, 42]. However, there are no clear discussions on how the performance of their proposed tools is evaluated and if user interests are considered. According to [25, 29, 43], the performance of a web usage mining process is evaluated based on the discovery of useful information that characterises the interests of end users. Therefore, eliminating noise in web data should consider the interests of web users. Finally, there is a great deal of focus on identifying and eliminating noise such as advertisement banners, failed https links, mirror sites, duplicated web pages, copyrights, external links etc. Since the process of identifying and eliminating these type of noise is mainly based on its relationship with the main web page content, the process is not user-driven and hence results in loss of useful information.

In summary, the proposed research's main focus is to learn to recognize noise in web data so as to reduce the loss of useful information otherwise considered as noise—as well as to decrease noise levels. Rather than isolating the main web page content and relying on its layout and content, the proposed research aims to concentrate mainly on how user interest can influence the type of noise present in web logs.

3 Noise Web Data Learning

A key focus in the proposed research is to learn, identify and eliminate noise in web data, taking into account dynamic changes in user interests. The proposed approach is user driven: i.e. dynamic changes in user interest determine the 'interestingness' of web data. The proposed research considers various indicators which can be applied to learning user interests and their influence in the web data noise reduction process. A research methodology framework which describes various development phases to be used in the proposed research is presented in Fig. 1.

A user profile is a collection of information that describes various characteristics of a user while visiting a website [44]. For example, time of entry to a website, duration of web page visit, exit time from the website, etc. This type of information is recorded in form of user session. Creating a web user profile aid in learning interests of a user on visited web pages. Currently, user sessions are mainly determined based on a fixed time-out value [45]. Sessions derived from fixed time-out value will impact noise web data reduction process because, (1) they tends to ignore user's varying time spent on a web page as well as their familiarity with the website. (2) The frequency of a user visit may be higher but the time spent on the web page is lower. The proposed research considers dynamic time-out threshold value so as to ensure user sessions reflect varying time of visits to a web page. In addition, user interest level determined by weighted web pages involves aspects such as depth, frequency and duration of visit. Therefore, page visit duration based on dynamic session time-out value learns how user interest vary over the time, as well as impact on information considered noise or useful.

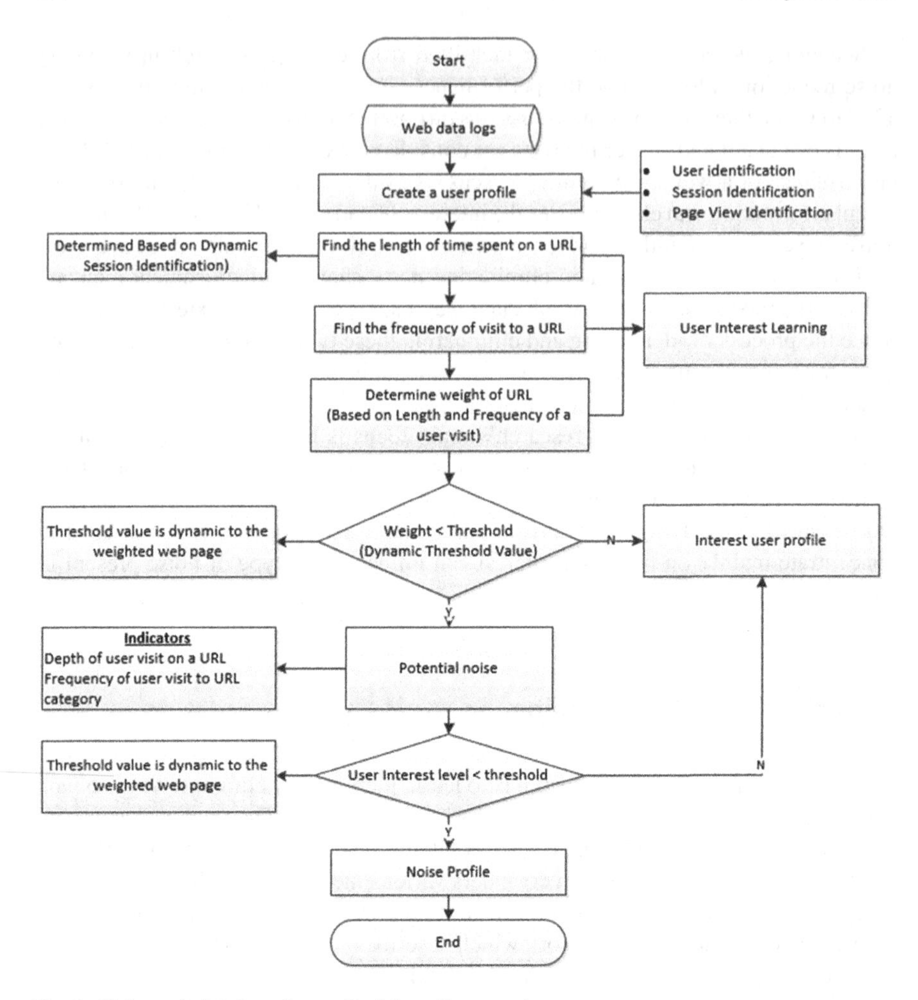

Fig. 1 Noise web data learning methodology framework

Identifying new or potentially noise web pages is a problem encountered when classifying web data, this is due to lack of any previous interest from a user. The proposed algorithms consider interest indicators such as frequency of visit to a 'web page category'. Web page category in this proposed work include, *sports, fashion, music, health,* etc. For instance, a user can only be interested in football during World Cup, therefore the frequency and duration of visit to pages associated with this event will demonstrate interest only for a specific time period.

Existing research works mainly consider uniform support threshold [46] when classifying web data. But where a low threshold value is assigned, it may lead to high level of noise web pages identified as useful and vice versa. The results obtained will not necessarily conform to a user change of interests. For this reason, dynamic threshold values in web data classification play a critical role in ensuring data available on the web is user-driven.

4 Conclusion and Future Work

This research work aims to justify why learning to recognize noise in web data prior to its elimination is critical. It is important to take into account both the available web data and the interests of web users in order to determine which data is noise and which is useful—given the varying interest levels of a user. Our position is based on the fact that the interests of a web user are dynamic and so is web data. In our ongoing research work, we propose a noise web data learning tool capable of learning to recognize noise via a web user profile so that it can be eliminated. In addition to measures that existing research have applied to determine the interestingness of a web page prior to classification, we derive a number of algorithms based on measures such as depth of user visit, and frequency of visit to a web page category in order to demonstrate how dynamic changes in user interest influence the noise web data reduction process. Subsequently, a number of experimental directions will be considered to validate the performance of the proposed algorithms. Measurement metrics such as Confusion Matrix, Precise, Recall and F-Measure will be considered for evaluating the proposed algorithm's performance.

Acknowledgements Special thanks to the University of Sunderland, Computing and Engineering Library Services for their financial support towards publication of this research work.

References

1. L. Yi, B. Liu, X. Li, Eliminating noisy information in web pages for data mining, in *Proceedings of the Ninth ACM SIGKDD International Conference on Knowledge Discovery and Data Mining* (ACM, New York, NY, USA, 2003), pp. 296–305. (KDD '03). http://doi.acm.org/10.1145/956750.956785. Accessed 4 Apr 2017
2. C. Ramya, G. Kavitha, D.K. Shreedhara, Preprocessing: A Prerequisite for Discovering Patterns in Web Usage Mining Process (2011). ArXiv:11050350. Accessed 4 Apr 2017
3. S. Dias, J. Gadge, Identifying informative web content blocks using web page segmentation, in *Entropy*, vol. 1 (2014), p. 2
4. J. Srivastava, R. Cooley, M. Deshpande, P.-N. Tan, Web usage mining: discovery and applications of usage patterns from web data. SIGKDD Explor Newsl. **1**(2), 12–23 (2000)
5. M. Jafari, F. SoleymaniSabzchi, S. Jamali, Extracting users' navigational behavior from web log data: a survey. J. Comput. Sci. Appl. **1**(3), 39–45 (2013)
6. S. Gauch, M. Speretta, A. Chandramouli, A. Micarelli, User profiles for personalized information access, in *The adaptive web* (Springer, 2007), pp. 54–89, http://link.springer.com/chapter/10.1007/978-3-540-72079-9_2. Accessed 4 Apr 2017
7. P. Peñas, R. del Hoyo, J. Vea-Murguía, C. González, S. Mayo, Collective knowledge ontology user profiling for twitter—automatic user profiling, in *2013 IEEE/WIC/ACM International Joint Conferences on Web Intelligence (WI) and Intelligent Agent Technologies (IAT)* (2013), pp. 439–444

8. S. Kanoje, S. Girase, D. Mukhopadhyay, User profiling trends, techniques and applications (2015). ArXiv:150307474. Accessed 4 Apr 2017
9. H. Xiong, G. Pandey, M. Steinbach, V. Kumar, Enhancing data analysis with noise removal. IEEE Trans. Knowl. Data Eng. **18**(3), 304–319 (2006)
10. H. Liu, V. Kešelj, Combined mining of web server logs and web contents for classifying user navigation patterns and predicting users' future requests. Data Knowl. Eng. **61**(2), 304–330 (2007)
11. S.K. Dwivedi, B. Rawat, A review paper on data preprocessing: a critical phase in web usage mining process, in *2015 International Conference on Green Computing and Internet of Things (ICGCIoT)* (2015), pp. 506–510
12. H. Yang, S. Fong, Moderated VFDT in stream mining using adaptive tie threshold and incremental pruning, in *Data Warehousing and Knowledge Discovery*. Lecture Notes in Computer Science (Springer, Berlin, Heidelberg, 2011), pp. 471–483, https://link.springer.com/chapter/10.1007/978-3-642-23544-3_36. Accessed 18 July 2017
13. L. Sunithaa, M.B. Rajua, B.S. Srinivas, A comparative study between noisy data and outlier data in data mining. Int. J. Curr. Eng. Technol. (2013)
14. S. Lingwal, Noise reduction and content retrieval from web pages. Int. J. Comput. Appl. **73**(4) (2013). http://search.proquest.com/openview/ac440ddad43bcc282fe20f7e4256cba0/1?pq-origsite=gscholar&cbl=136216. Accessed 4 Apr 2017
15. S.S. Bhamare, B.V. Pawar, Survey on web page noise cleaning for web mining. Int. J. Comput. Sci. Inf. Technol. **4**(6), 766–770 (2013)
16. J. Kapusta, M. Munk, k MD, Cut-off time calculation for user session identification by reference length, in *2012 6th International Conference on Application of Information and Communication Technologies (AICT)* (2012), pp. 1–6
17. A. Dutta, S. Paria, T. Golui, D.K. Kole, Structural analysis and regular expressions based noise elimination from web pages for web content mining, in *2014 International Conference on Advances in Computing, Communications and Informatics (ICACCI)* (2014), pp. 1445–1451
18. P. Sivakumar, Effectual web content mining using noise removal from web pages. Wirel. Pers. Commun. **84**(1), 99–121 (2015)
19. T. Htwe, N.S.M. Kham, Extracting data region in web page by removing noise using DOM and neural network, in *3rd International Conference on Information and Financial Engineering* (2011), http://www.ipedr.com/vol12/22-C024.pdf. Accessed 4 Apr 2017
20. R.P. Velloso, C.F. Dorneles, Automatic web page segmentation and noise removal for structured extraction using tag path sequences. J. Inf. Data Manag. **4**(3), 173 (2013)
21. J. Onyancha, V. Plekhanova, D. Nelson, Noise web data learning from a web user profile: position paper, in *Proceedings of The World Congress on Engineering 2017*. Lecture Notes in Engineering and Computer Science, 5–7 July, 2017, London, U.K., pp. 608–611
22. M. John, J.S. Jayasudha, Methods for removing noise from web pages: a review (2016), https://www.irjet.net/archives/V3/i8/IRJET-V3I8359.pdf. Accessed 4 Apr 2017
23. M.E. Akpınar, Y. Yesilada, Vision based page segmentation algorithm: extended and perceived success, in *Revised Selected Papers of the ICWE 2013 International Workshops on Current Trends in Web Engineering—Volume 8295*. New York, NY, USA (Springer New York, Inc., 2013), pp. 238–252, http://dx.doi.org/10.1007/978-3-319-04244-2_22. Accessed 4 Apr 2017
24. A. Garg, B. Kaur, Enhancing performance of web page by removing noises using LRU. Int. J. Comput. Appl., **103**(6) (2014), http://search.proquest.com/openview/d15b65f1df9ab04d79abf1bcfa4de555/1?pq-origsite=gscholar&cbl=136216. Accessed 4 Apr 2017
25. N. Narwal, Improving web data extraction by noise removal, in *Fifth International Conference on Advances in Recent Technologies in Communication and Computing (ARTCom 2013)* (2013), pp. 388–395
26. P. Nithya, P. Sumathi, Novel pre-processing technique for web log mining by removing global noise and web robots, in *2012 National Conference on Computing and Communication Systems* (2012), pp. 1–5
27. A.K. Santra, S. Jayasudha, Classification of web log data to identify interested users using Naïve Bayesian classification. Int. J. Comput. Sci. Issues **9**(1), 381–387 (2012)

28. J. Sripriya, E.S. Samundeeswari, Comparison of Neural Networks and Support Vector Machines using PCA and ICA for Feature Reduction. Int J Comput Appl. 2012 Feb 29;40(16):31–6
29. H.K. Azad, R. Raj, R. Kumar, H. Ranjan, K. Abhishek, M.P. Singh, Removal of noisy information in web pages (ACM Press, 2014), pp. 1–5, http://dl.acm.org/citation.cfm?doid=2677855.2677943. 4 Apr 2017
30. S.P. Malarvizhi, B. Sathiyabhama, Enhanced reconfigurable weighted association rule mining for frequent patterns of web logs. Int. J. Comput. **13**(2), 97–105 (2014)
31. A. Nanda, R. Omanwar, B. Deshpande, Implicitly learning a user interest profile for personalization of web search using collaborative filtering, in *2014 IEEE/WIC/ACM International Joint Conferences on Web Intelligence (WI) and Intelligent Agent Technologies (IAT)* (2014), pp. 54–62
32. K.S. Rao, D.A.R. Babu, D.M. Krishnamurthy, Mining user interests from user search by using web log data. J. Web Dev. Web Des., **2**(1) (2017), http://matjournals.in/index.php/JoWDWD/article/view/1167. Accessed 8 Aug 2017
33. X. Wei, Y. Wang, Z. Li, T. Zou, G. Yang, Mining users interest navigation patterns using improved ant colony optimization. Intell. Autom. Soft Comput. **21**(3), 445–454 (2015)
34. M. Grčar, D. Mladenič, M. Grobelnik, User profiling for interest-focused browsing history, in *Proceedings of the Workshop on End User Aspects of the Semantic Web* (2005), pp. 99–109, http://ceur-ws.org/Vol-137/09_grcar_final.pdf. Accessed 9 Aug 2017
35. H. Kim, P.K. Chan, Implicit indicators for interesting web pages (2005), https://dspace-test.lib.fit.edu/handle/11141/162. Accessed 10 July 2017
36. X. Wu, P. Wang, M. Liu, A Method of mining user's interest in intelligent e-learning (2014)
37. K. Jiang, Y. Yang, Noise reduction of web pages via feature analysis, in *2015 2nd International Conference on Information Science and Control Engineering* (2015), pp. 345–348
38. X. Wang, B. Chen, F. Chang, A classification algorithm for noisy data streams with concept-drifting. J. Comput. Inf. Syst. **7**(12), 4392–4399 (2011)
39. H. Wang, Q. Xu, L. Zhou, Deep web search interface identification: a semi-supervised ensemble approach. Information. **5**(4), 634–651 (2014)
40. F. Hu, M. Li, Y.N. Zhang, T. Peng, Y. Lei, A non-template approach to purify web pages based on word density, in: *Proceedings of the International Conference on Information Engineering and Applications (IEA) 2012* (Springer, London, 2013), pp. 221–228, https://link.springer.com/chapter/10.1007/978-1-4471-4847-0_27. Accessed 4 Apr 2017
41. S.N. Das, M. Mathew, P.K. Vijayaraghavan, An efficient approach for finding near duplicate web pages using minimum weight overlapping method, in *2012 Ninth International Conference on Information Technology—New Generations* (2012), pp. 121–126
42. P. Sahoo, S.P. Rajagopalan, An efficient web search engine for noisy free information retrieval. Int. Arab. J. Inf. Technol. IAJIT. (2015)
43. I.-H. Ting, H.-J. Wu, in *Web Mining Applications in E-Commerce and E-Services* (Springer Science & Business Media, 2009), 181 p.
44. O. Hasan, B. Habegger, L. Brunie, N. Bennani, E. Damiani, A discussion of privacy challenges in user profiling with big data techniques: the excess use case, in *2013 IEEE International Congress on Big Data (BigData Congress)* (IEEE, 2013), pp. 25–30, http://ieeexplore.ieee.org/abstract/document/6597115/. Accessed 22 Aug 2017
45. P. Patel, M. Parmar, Improve heuristics for user session identification through web server log in web usage mining. Int. J. Comput. Sci. Inf. Technol. **5**(3), 3562–3565 (2014)
46. J.-C. Ou, C.-H. Lee, M.-S. Chen, Efficient algorithms for incremental Web log mining with dynamic thresholds. VLDB J. **17**(4), 827–845 (2008)

Leveraging Lexicon-Based Semantic Analysis to Automate the Recruitment Process

Cernian Alexandra, Sgarciu Valentin, Martin Bogdan and Anghel Magdalena

Abstract This paper presents the design and implementation of a semantic based system for automating the recruitment process, mainly by improving the identification of the best suited candidate for specific jobs. The process is based on using a skills and competencies lexicon, that we have developed specifically for this purpose, to provide a semantic processing of the resumes and match the candidates' skills with the requirements for each particular job description. The main objective is to reduce the recruiter's processing time by eliminating certain repetitive activities in the resume analysis procedure and to obtain a qualitative improvement by highlighting the competencies and qualities of candidates based on complex and customized semantic criteria.

Keywords Automate recruitment process · Competence-based search
Data mining · Lexicon · Recruitment process · Semi-automatic tool
Semantic technologies · Skills

1 Introduction in the Recruitment Context

People are the most valuable asset of an organization. One of the key roles of HR specialists is to find and hire the right people for the right job. But how do we find them? Their profile must match the organizational vision and mission, their set of

C. Alexandra (✉) · S. Valentin · M. Bogdan · A. Magdalena
Faculty of Automatic Control and Computers, University Politehnica of Bucharest,
313 Splaiul Independentei, 006042 Bucharest, Romania
e-mail: Alexandra.cernian@upb.ro

S. Valentin
e-mail: vsgarciu@aii.pub.ro

M. Bogdan
e-mail: Bogdan.martin7@gmail.com

A. Magdalena
e-mail: Magdalena.anghel@aii.pub.ro

S.-I. Ao et al. (eds.), *Transactions on Engineering Technologies*,
https://doi.org/10.1007/978-981-13-0746-1_15

values must be convergent with the organizational culture and their skills must match the job description and requirements. The evaluation process during recruitment is the key to finding the appropriate candidates in a timely manner.

At present, the recruitment process is mainly assisted by human experts, which can be a time-consuming task. The process is basically simple: HR specialists use various channels to collect resumes, then start filtering them based on specific background, skills and competencies, as required by the specific job they are recruiting for. According to [1], the most widely used engagement channels in the recruitment process are: print, job boards, agencies, job search aggregators, university recruiting, sites for professional networking, internal talents, talent communities and employee referrals. The authors of [1] identify 2 recruiting approaches:

- Traditional approach, which focuses more on the five engagement channels enumerated above
- Network recruiting, where decisions are based on social media analysis and on networking decision factors and which focuses more on the last five channels enumerated above

Therefore, we assist at a shift regarding the way recruiters reach their candidates, as well as in the mindset companies use throughout the recruitment process [2]. Companies currently create and use networks of leads, candidates, employees and alumni in order to leverage and increase the quality and performance of the recruitment process.

Improving the process of analyzing resumes and identifying the best suitable candidates for a job leads to an improvement in the recruitment process as a whole. This can be achieved by reducing working time with tools that eliminate certain repetitive activities when it comes to resumes' analysis and the qualitative improvement of this analysis by highlighting qualities based on more complex criteria. One such example is the aggregation of data from a large number of documents (resumes) using the organization where candidates have worked as a criterion.

It is important that this type of application contains one or more of the following components in order to appreciate the usefulness of such applications.

- Automatic processing of resumes regardless of their structure
- The ability to process documents for as many document types as possible (pdf, rtf, .doc etc.)
- The ability to perform complex searches in the candidates database
- The ability to generate a short list of candidates for a job description
- The ability to define and constantly update job profiles
- The possibility to display the reasoning used by the recruiter algorithm and to be able to permanently improve it

An automation we can bring into the process is a smarter way of processing candidates' resumes. If we build an improved and more intelligent parsing and skills matching algorithm, the entire recruiting process would benefit, and we would be able to efficiently identify the suitable candidates for a particular job. The final objective is to minimize the HR specialist input in the process and limit the time they spend in processing resumes by developing digital tools to automate the candidates' identification phase. Such a tool would bring two main benefits:

1. Eliminate certain repetitive activities in the resume analysis procedure, currently performed by HR specialists mostly empirically and
2. Improve the process from a qualitative point of view by identifying and highlighting the competencies and skills of candidates based on customized criteria [3].

This paper presents a semi-automatic tool for boosting the recruitment process based on a semantic approach for identifying candidates' skills and matching them with organizational requirements and job descriptions.

The rest of the paper is organized as follows: Sect. 2 presents and overview of the system, emphasizing the main objectives and benefits of this approach, Sect. 3 presents the design and implementation of the tool and Sect. 4 draws the conclusions of the hereby presented work.

2 Overview of the Semantic Based Tool for Automating the Recruitment Process

This paper presents the design and implementation of a semi-automatic tool able to assist HR specialists in the recruitment process by providing an automatic semantic-based analysis of the candidates resumes and identifying their key skills, by also matching their competencies against the requirements in the job description. The system will be able to recommend the best suited candidates for specific positions.

The main functional specifications of the tool are more exhaustive and include various actions conducted by the HR department, namely:

1. Provide the possibility to define and update job descriptions
2. Keep an updated database of candidates resumes
3. Perform searches to identify best suited candidates for specific jobs

The main **objectives** of the application are the following:

1. Implement a semi-automatic tool to assist HR specialist in the recruitment process
2. Implement an automatic module for parsing the resumes of candidates and identify their key skills and competencies
3. Develop an algorithm that automatically finds the best matches for a specific job profile based on candidates' skills

The process is based on a semantic processing of the candidates resumes, based on a lexicon/dictionary approach. The overview of the flow of the application is presented in Fig. 1.

In order to analyze a candidate's resume, it is necessary that the information be extracted so that it can be further processed. This process must take into account a

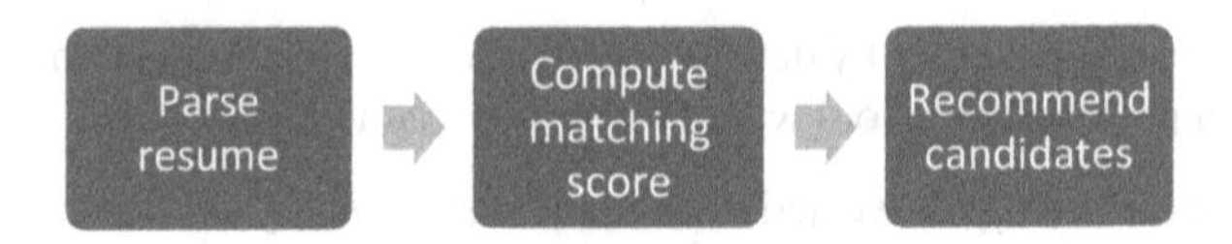

Fig. 1 The overview of the application flow

number of variables such as: how the resume is structured, any non-text items that it might contain, as well as the type of document that is being processed.

The process of analyzing candidates' profiles involves using the information extracted in the previous phase to determine whether a candidate is suitable for a particular job profile. At this stage, a recruiter is looking at three key elements:

1. The educational background of the candidate
2. The experience a candidate has in the field for which he has applied
3. The candidates' skills and competencies and their matching with the job profile.

Therefore, what we look for in the resume processing phase is to organize the information using the following classification criteria:

1. **Educational background**

We define a database of educational institutions and the abilities that students acquire after graduating them. This type of classification can provide a series of information based solely on the educational institution that a person has graduated from.

2. **Professional experience**

We define a database of organizations, job types and associated skills and competencies required. Thus, we can extract additional information about a person, based explicitly on the organizations in which they have worked.

3. **Skills and competencies**

We define a lexicon/dictionary that will contain a series of competencies and abilities. Further, the dictionary is used to identify and match candidates' skills and job profiles.

3 The Design and Implementation of the Semantic Based Recruitment Automation Tool

This section presents the design and implementation of the recruitment automation tool, based on a semantic approach for matching the best suited candidates for specific job profiles. The candidates' selection algorithm starts each search process based on the configurations made by the user and has the following flow:

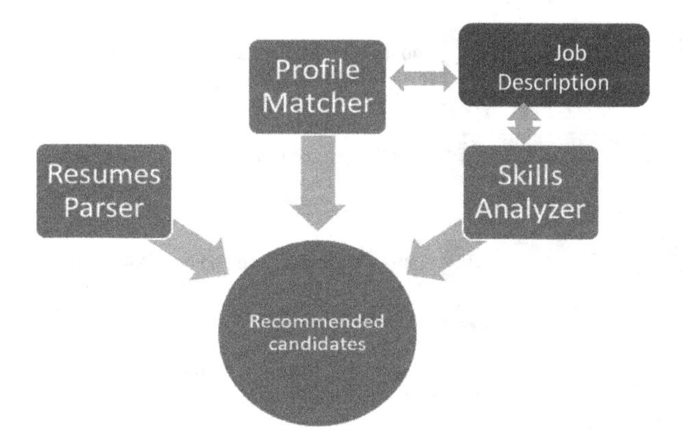

Fig. 2 The main components of the system

Candidates selection	The user configures the job description and candidate profile
	A special section is dedicated to the main competencies and skills required
	The candidates resumes are saved into the database
	The parsing algorithm is used to process the candidates resumes
	The candidates profile is matched against the job description and a similarity score is computed
	The list of best matches is generated by the application for each specific job description

The tool has three main components, each with specific functionalities within the flow of the automatic candidates' selection and recommendation process (Fig. 2).

3.1 The Resumes Parser Component

This component is responsible with processing the candidates' resumes in order to extract information about their education, work experience and different skills. The parser processes information in four main phases:

1. Get input data from the user (one or more resumes) and transform them into text files
2. Parse the resume in order to split the information into 4 individual sections: personal details, education, work experience and skills and competencies.

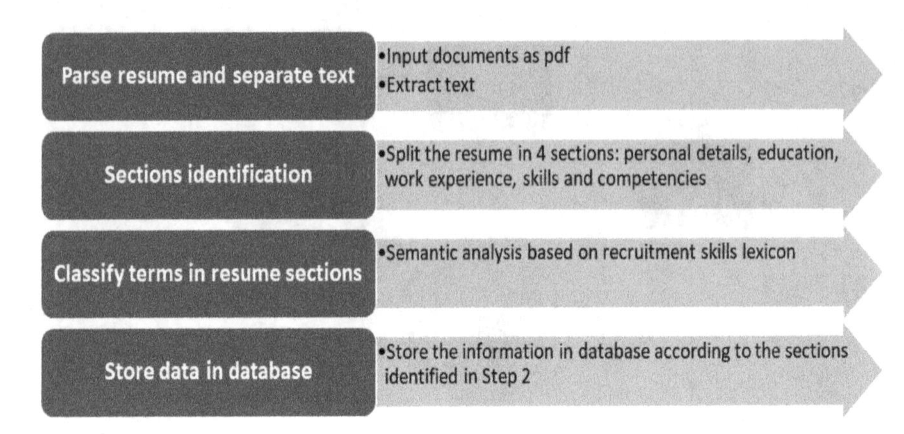

Fig. 3 The resumes parsing process

3. Classify specific words in each section in order to leverage the candidates' selection model
4. Store the information in a database (Fig. 3).

In the first step, CV text will be extracted from the provided pdf documents, namely the resumes of candidates. Text extraction is implemented using an open source library called "Apache PDFBox" [4, 5]. Apache PDFBox is a Java open source library which facilitates document management, providing functions to create documents and extract content from existing documents in different formats [4].

```
$converter = new PdfBox;
$converter->setPathToPdfBox('C:\wamp\www\quickstart\public\library\pdfbox-app-2.0.2.jar');
$text=$converter->textFromPdfFile($file);
```

Elements of a different type, other than text, such as images or graphical representations, will not be taken into account as they are not analyzed. Once the entire text has been extracted, the second step of identifying the corresponding text for each section of the resume.

In the second step, sections identification, we identify and split the sections of a resume and the corresponding text by the specific markups for the four mentioned sections of the resume. This process is automated and uses word categorization to identify sections, based on predefined lexicon containing these section-specific elements.

The first processing that we make on the text is to determine the parts of the speech in the sentences from the text source previously extracted. This processing is carried out by using the software developed by Stanford within the Stanford Natural Language Processing Group, "Pos Tagger", which is developed in Java [6, 7].

In order to extract only the relevant information, which is of interest to us, a brief processing of the obtained result is performed. Thus, a variable containing two

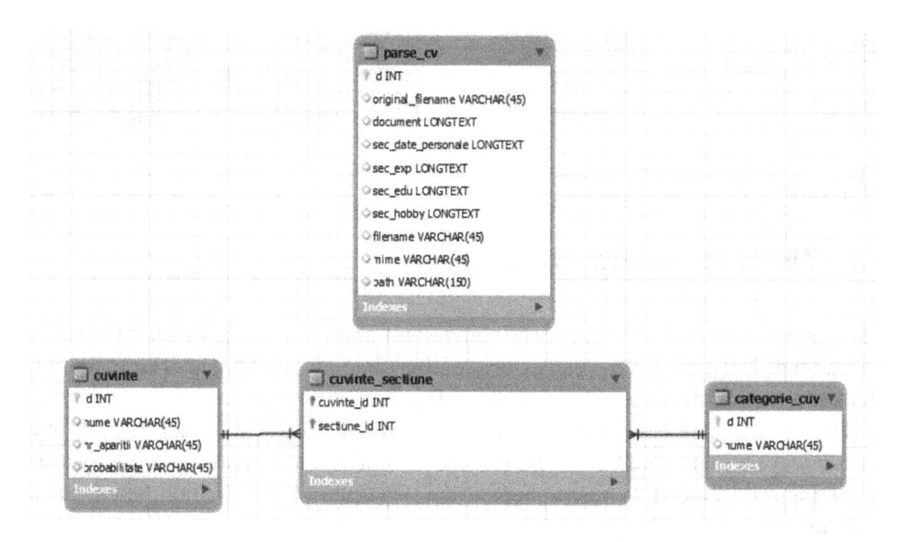

Fig. 4 Tables configuration for resumes parser

vectors was created in which we store both the extracted words and the part of speech identified for each section.

The calculation of the probability of occurrence of specific words in a particular section of the CV is made using the Naive Classifier Bayes [8].

$$S_j = \prod_{i=1}^{n} \frac{P(C_i|K_j)}{P(C_i)}$$

where Sj is the score for category j (education for example), Ci—a word in the new message, n—the number of words in the new message, K_j—one of the categories.

The keywords that will be searched for in each section of a resume are found in the skills and competency lexicon/dictionary, as well as in the professional experience and the education dictionaries, combined with the job title field.

In the last step, Store data in database, the text of the sections previously identified will be saved in the database for further processing of the information. Each resume will be stored in the database both in text form, after it has been parsed, and in PDF format for the convenience of resumes inspection once the candidate identification phase has been completed. Figure 4 depicts the main 4 tables used to store the information extracted in this phase of processing [5].

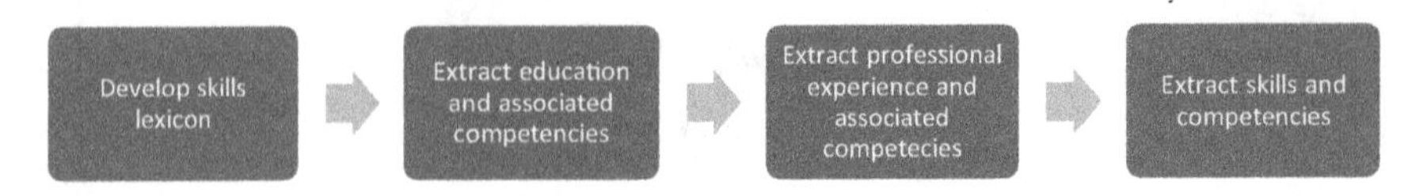

Fig. 5 The skills analyzer module architecture

3.2 *The Skills Analyzer Component*

This module is the most complex of the system and it uses a semantic processing based on a skills and competencies lexicon that we have developed. There are four steps involved in this process, as shown in Fig. 5.

The key element here is the skills and competency lexicon, which is built in a similar way to SentiWordNet [9]. SentiWordNet is a lexical resource used for opinion mining and developed by researchers at the "Institute of Information Science and Technology A. Faedo".

The skills and competencies lexicon contains 250 terms at present and we are constantly expending it. The design of the lexicon is flexible and extendable. It contains three classes, as follows:

- The skills dictionary
- The responsibilities dictionary—a list of competencies related to specific job profiles
- The educational institutions dictionary—which contains a list of educational institutions (mainly in Romania at present) and the competencies acquired by graduates of various fields of specialization.

3.2.1 The Skills Dictionary

The structure of the skills dictionary is the following:

- #TYPE is the column which identifies the type of the term defined as follows: 1—skill or 2—competency
- ID is the term id
- Reliability is the degree of trust associated with a skill or ability related to the description of the term in the dictionary. If a search returns the correct resume, reliability will be automatically increased and otherwise it will decrease. Reliability is defined as a percentage between 0 and 100%.
- Associated Terms includes a list of competencies/skills that are complementary to those specified in the Description column. The "#" marks the term that is associated more often. The scale starts from "#1" where "1" represents the most associated skill or competency.
- The Description column contains the description of skill/competency and various definitions of that competency. When searching for a competency in the resume, the algorithm will search for all these terms and definitions.

Table 1 Exemple of term in dictionary

# Type	ID	Reliability	Associated terms	Description
1	00006885	0.75	excel#1 pivottabels#2	The ability to use Microsoft Excel

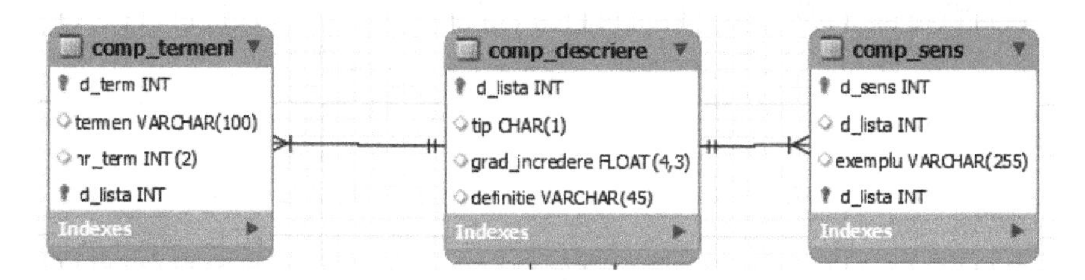

Fig. 6 Tables configuration for the skills and competencies dictionary

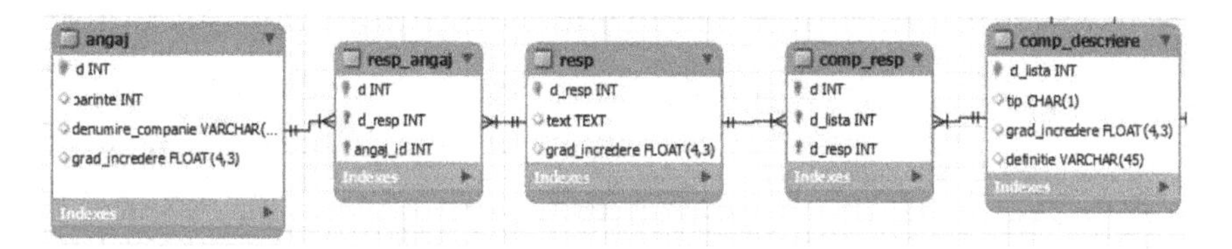

Fig. 7 Tables configuration for the responsibilities dictionary

Table 1 presents an example of a record from the skills dictionary.

Figure 6 shows the structure of the tables associated to the skills and competencies dictionary, as design in the system relational database.

3.2.2 The Responsibilities Dictionary

The responsibilities dictionary contains a list of all the skills and abilities developed by a person based on the responsibilities assigned to a job. It is defined similarly to the skills dictionary and contains a list of employers with an associated degree of trust for each employer. If the professional responsibilities associated with an employer are found in their employees, the degree of trust will be higher.

Since a professional responsibility element can be associated with several employers, it also has a degree of trust that can be decreased or increased depending on the number of resumes for which that particular element led to the successful identification of a resume.

Figure 7 shows the structure of the tables associated to the responsibilities dictionary, as design in the system relational database.

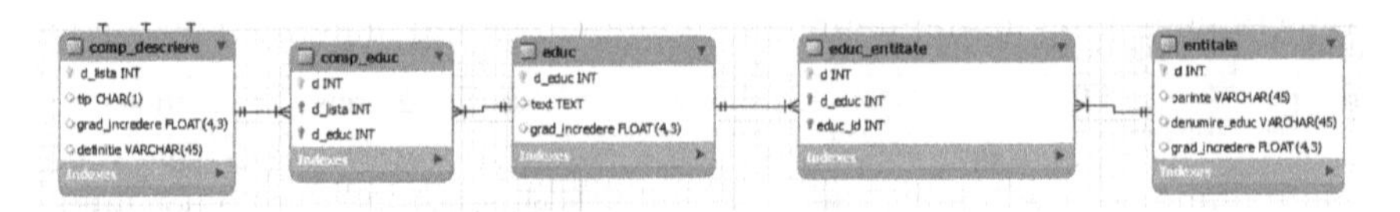

Fig. 8 Tables configuration for the educational institutions dictionary

Janus Identificare Parse CV CV-uri Competente Experienta prof Experienta educ Date Agregate Cuvinte Bogdan Martin

Probabilitate cuvinte

Show 10 entries Search:

ID	Tip	Grad	Desc	Termeni	Sensuri	ID	Resp	Angajator	ID	Educ	Entitate
1	competenta	2.2	Can use excel at an advanced level communication adaptable	excel#1.advanced#2.pivot#3.adaptable#4	(Can use all the features of the excel for day to day usage.).(Can use table design.)	2	Sharing information with peers.	Oracle	1	Learning Excel pivots and the use of advanced functions.	Politehnica
2	abilitate	1.6	Has proficient communication skills.	communication#1.proficient#2.pcm#3	(Has knowledge of PCM)	1	Creating reports using reporting tools like Microsoft Excel.	HP	2	Learning to work in a team.	Politehnica

Fig. 9 User interface for viewing skills and competencies

3.2.3 The Educational Institutions Dictionary

The educational institutions dictionary contains a list of educational institutions and competences and abilities developed after graduation. It is defined similarly to the competency dictionary and contains a list of educational institutions with an associated degree of trust for the educational institution. If the descriptions associated with a specific educational institution are found in their graduates, the degree of trust will be higher.

Since an element associated with an educational institute can be associated with several institutes, it also has a degree of confidence that can be decreased or increased depending on the number of resumes for which the element led to the successful identification of resumes.

Figure 8 shows the structure of the tables associated to the educational institutions dictionary, as design in the system relational database.

So, at this stage, the resumes have been split into the four sections previously defined, and each section has been semantically processed based on the skills and competencies lexicon. Each candidate is profiled according to their education, work experience and skills (Fig. 9).

An important aspect is that we also treat associated terms with specific skills, in order to find the appropriate correlations [10] between terms. The next step is

Fig. 10 Profile matcher process

to compute a matching score for each resume, in relation to the requirements for a specific position in a company.

3.3 *The Profile Matcher Module*

This module uses a score-based algorithm to determine the candidates that best match specific job descriptions, by taking into account the work experience, skills and educational background of applicants.

The score associated with a resume and its position between all the analyzed resumes is based on the number of identified terms that correspond to the search criteria, such as skills, job title, keywords or job responsibilities, as well as educational items related to employers and educational institutions in relation to the required skills.

The profile match phase has several steps, as follows (Fig. 10):

1. *Provide input as new job profile*
   The user introduces the job description with the following input: job title, professional competencies, skills, professional responsibilities, and educational profile. The HR specialists introducing the job can assign priority to specific elements, as percentage, which will be used in the formulas for computing the matching scores.
2. *Extract resumes from database*
   The resumes have been previously parsed into sections and split data.
3. *Compute matching score*
   Step 1. For each CV, a score is computed for each section. A section score is calculated by the number of appearances of elements that define the competencies or other required parameters

$$score_{section} = score_{section} + n * weight \tag{1}$$

   If the weight parameter is not customized, it will be considered equal to 1 by default.
   Step 2. Compute score for CV
   The score for the CV is calculated with the following formula:

$$score_{cv} = \sum score_{section} \tag{2}$$

4. *Generate the best matching candidates list*
   Step 1. Sort the candidates resumes by score.
   Step 2. Show the list of candidates based on the highest scores.

4 Conclusion

The application has been tested using a set of 375 resumes and 25 job descriptions and the results were 94% positive. The scores were accurately computed and the recommendations correctly matched the required profiles for specific jobs. The development of the skills and competencies lexicon is a key asset of the projects. The accuracy of the system will grow with the extension of the lexicon. The next step is to include a machine learning module in the system, which will automatically learn from the human expert's decisions and will be able to extend the lexicon with new sets of terms, as well as provide valuable input to the selection algorithms. Another envisaged development is to combine the semantic approach with classification methods [11] to improve accuracy.

Automating the recruitment process has gained more and more interest lately, in order to reduce the time spent on repetitive tasks. One of its main uses is during the resumes selection phase, which is intensively time consuming when performed manually. Therefore, this tool can be a relevant decision support instrument for HR and recruitment specialists in identifying the best suited candidates for specific jobs, based on their profile and experience.

References

1. J. Bersin, *Predictions for 2015 Redesigning the Organization for a Rapidly Changing World*, Deloitte (2015), http://www.cedma-europe.org/newsletter%20articles/misc/Predictions%20for%202015%20-%20Redesigning%20the%20Organization%20for%20a%20Rapidly%20Changing%20World%20(Jan%2015).pdf
2. A. Cernian, V. Sgarciu, Boosting the recruitment process through semi-automatic semantic skills identification, in *Proceedings of The World Congress on Engineering 2017*, London, U.K. Lecture Notes in Engineering and Computer Science (2017), pp. 605–607
3. H.-F. Hsieh, S. Shannon, Three approaches to qualitative content analysis. Qual. Health Res. **15**(9), 1277–1288 (2005)
4. PDFBox, https://pdfbox.apache.org/index.html. Accessed July 2016
5. A. Cernian, D. Carstoiu, B. Martin, Semi-automatic tool for parsing CVs and identifying candidates' abilities and competencies, in *International Conference on Education, Management and Applied Social Science EMASS*, 2016
6. O.R. Zaïane, in *Proceedings of Principles of Knowledge Discovery in Databases* (1999)
7. D.L. Olson, D. Delen, *Advanced Data Mining Techniques* (Springer, 2008)
8. N. Dracos, R. Moore, *Naïve Bayes Text Classification* (Cambridge University Press, 2008), https://nlp.stanford.edu/IR-book/html/htmledition/naive-bayes-text-classification-1.html
9. S. Baccianella, A. Esuli, F. Sebastian, in *SENTIWORDNET 3.0: An Enhanced Lexical Resource for Sentiment Analysis and Opinion Mining*, http://sentiwordnet.isti.cnr.it/. Accessed July 2016

10. X. Wei, H. Lin, L. Yang, Cross-domain sentiment classification via constructing semantic correlation. IAENG Int. J. Comput. Sci. **44**(2), 172–179 (2017)
11. K. Abdalgader, Clustering short text using a centroid-based lexical clustering algorithm. IAENG Int. J. Comput. Sci. **44**(4), 523–536 (2017)

The Universal Personalized Approach for Human Knowledge Processing

Stefan Svetsky and Oliver Moravcik

Abstract The processing of information for supporting human knowledge based processes still represents a big challenge. This is caused both by the abstract concept of the information and by missing a suitable information representation (how to represent human knowledge), which would enable a human—computer communication in natural language. However, any interdisciplinary definition of "information" or even "knowledge" does not exist until now. In the period 2010 to 2012, the authors presented their research approach to the automation of teaching processes as typical human knowledge based processes within the WCE's International Conference on Education and Information Technology. The following research showed that solving the automation of the knowledge processes requires one to solve not only a content knowledge processing but also issues of communication, transmission and feedback. Actually, the research covers some areas of computer science (e.g. computer applications, computing methodologies, theoretical computer science). The question of the information was solved by the design of a specific default data structure so called virtual knowledge which enables an individual to communicate with computer in natural language. In this context, the virtual knowledge unit symbolizes an information representation for human knowledge which is readable both by a human and machine. Such approach enables one to input any kind of information from human resources into the virtual knowledge and to process human knowledge as it is typical for computers. This contribution illustrates how various kind of information from human sources are processed by the in-house software BIKE. It also enables one to understand why the software works as an all-in-one tool when processing human knowledge which is inputted into the virtual knowledge. In addition, the information transmission in the knowledge based processes was solved by a specific utility model for processing unstructured data.

S. Svetsky (✉) · O. Moravcik
Faculty of Materials Science and Technology in Trnava,
Slovak University of Technology, Paulínska 16, 917 24 Trnava, Slovakia
e-mail: stefan.svetsky@stuba.sk

O. Moravcik
e-mail: oliver.moravcik@stuba.sk

S.-I. Ao et al. (eds.), *Transactions on Engineering Technologies*,
https://doi.org/10.1007/978-981-13-0746-1_16

Keywords Computer applications · Human knowledge processing
Human computer interaction · Information representation
Knowledge based processes · Personal informatics

1 Introduction

In the period 2010–2012 an empirical research on technology enhanced learning (TEL) was presented on the WCE's International Conference on Education and Information Technology [1–3]. As was presented on the conferences, the batch knowledge processing paradigm and in-house educational software BIKE (WPad) were developed and tested for design of teaching innovative methods within TEL. The software enables individuals to process large amount of data (the human information and knowledge). To program own software was needed because there was an absence of IT tools for TEL. As it is in [4] "a system design on the model basis has been widely ignored by the community" till now, and software engineering is missing in TEL system development.

After this initial phase, the following research was more systematic because the question of knowledge was needed to solve. There was another paradox yet. Despite the fact that each educational book, journal, research papers deals with knowledge, however, any interdisciplinary definition of knowledge exists. From terminologically point of view, the definition of information or knowledge has other meaning in computer science, information and communication technology, knowledge management, psychology, teaching and learning, or how laics understand it. Moreover, in computer science there is a big challenge, how to represent "human knowledge". So if one should solve any automation when human knowledge is the key parameter he must design a suitable representation for knowledge or information.

This challenge and considerations were reminiscent of the early period of Cybernetics. For example, in [5] teaching processes were considered as controlled "processes of acquisition, processing and storing information". Between two of its subsystem, i.e. teacher and student, the transmission of information exists on a basis of feedback loop. It is emphasized as well that the "cybernetization" enables one to model a control the teaching process.

In this context, the Cybernetics (within which computers were invented) represents the basic theoretic discipline focused on the automation (regulation, control) of such complex systems, i.e. on the information, information transfer and processing [6]. Teaching and learning processes are considered for cognitive processes, i.e. they are human knowledge based, thus logically connected with thinking and intelligence. Within psychology, behaviorists understand thinking as a process of solving problems (as an adaptation to the changes in environment), while in cybernetics it is a process of processing and using information, as in [7].

The Cybernetics as a science field deals in general with creation, transmission and processing of information [6]. According to the Shannon's information theory, as a part of the field, one bit represents the basic information unit (with the value 0 or 1). Computers are machines for information processing by using such sequences of numbers 0 or 1. A default block of the sequences then creates data types which enable the machine to process texts, images, video via standardized formats. In [6] Kotek mentions three category of information: syntactic, semantic and pragmatic. In the content of his book, there are chapters dedicated to machine learning, natural language recognition, artificial intelligence. Kotek emphasizes these topics should be considered as informatics constructs derived only from syntactic information, because at this time there was not enough research data affecting semantic information, especially pragmatic information (the pragmatic information should be an useful information within a report).

In other words, computers do not understand human natural language, so any "human information" must be converted to "machine information". It is important for understanding the terminological chaos in the current scientific literature. For instance, in the field of Information and Communication Technology, similarly as in Knowledge Management, there is other hierarchy common: *data* (strings, numbers), *information*, *knowledge* (how it is also understood by laics). This contribution shows on several examples from implemented case studies how the problem of human knowledge processing was solved based on designing a virtual knowledge unit which is readable both by human and machines (computers).

2 Challenges for Human Knowledge Processing

Other obstacles when programming human knowledge based processing are evident from the following schema.

Figure 1 illustrates that the human knowledge based processes are uncertain, unstructured, thus, an unification and solving algorithms are needed. Moreover, the application data from human resources are in natural language and heterogeneous. Therefore, an universal IT tool is needed which would work like an all-in-one software. Moreover, as was mentioned above, for the processing of information, which is collected from human resources, computers need specific rules and representations in order to process it as the sequences of numbers 0 and 1. In this view, e.g. clustering short texts by their meaning is considered as a challenging task in [8]. Authors even propose a semantic hashing approach for encoding the meaning of a text into a compact binary code. Other similar approaches can be found in literature concerning question of the syntactic or pragmatic information, see e.g. [9, 10]. These approaches are related to the Shannon's information theory, respectively as it was mentioned above in the context of the Cybernetics. However, from our point of view, none of the approaches relate the processing of human knowledge directly in the natural language. It must be emphasized, that all of them are dedicated to automatic information processing. These approaches try to solve how to find or input the meaning of "human information" into the syntactic, semantic or pragmatic "machine information". In other words, one could imagine that all activities are done by a machine.

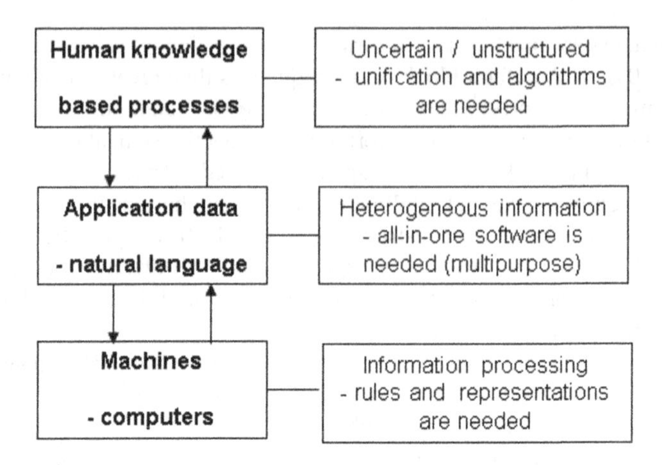

Fig. 1 Basic schema for using computers to process human knowledge

The same situation could be mentioned regarding Information Technology where ontologies "seem to be right way how to represent knowledge in machine readable form" [11]. It is emphasized that for information and knowledge sharing among users, the web-based systems need machine readable metadata, which define meaning of human readable-data. The so called Semantic Web uses for this purpose ontologies approaches to semantic representation (e.g. abstract representation of real objects, rules of a domain represented by knowledge graphs).

In contrast to these approaches, this contribution presents an entirely different approach which is based on the synergic collaboration between the human and the machine. For this purpose, a virtual unit knowledge was designed which simply works as a switch between human mental processes and machine (computer). This enables one to simulate and computerize mental processes, i.e. human knowledge based processes, by using this virtual knowledge. From a human point of view, the content of the virtual knowledge means simply any information or knowledge as it is understood by people, which is written in natural language. From a machine (computer) point of view, the machine understand the virtual knowledge unit as a "user data type", thus, as a kind of information representation which uses ASCII codes. An empty virtual knowledge has ca 10 KB. One could imagine that the empty default data structure is a primitive knowledge (knowledge cell/knowledge table). After inserting the human knowledge in natural language into the virtual knowledge one could measure in bytes how "wise" it is. From another point of view, it is important that the virtual knowledge enables a user to process unstructured data in a user friendly way. In other words, the content of the virtual knowledge is both human and machine readable. Thus such approach seems to be interdisciplinary understandable even from a layman's point of view.

332. ??? MYSQL - testy php
331. MYSQL - testy php
330. phpMyAdmin localhost PC svti
329. phpMyAdmin PC škola
328. KZP návrh
327. sessions 09.10.201615:53:08
326. MIX SME

Fig. 2 Screenshot of the WPad table column txtuni after numerating

2.1 Examples: The Batch Knowledge Paradigm

Visual FoxPro database platform seems to be one of the most user friendly software dedicated to Personal informatics. Especially, due to the specific programming language FoxPro which enables users to process even large data in batches. In other words, human information and knowledge can be processed in natural language even by users with lower informatics skills. For example, a user can use the following simply command to numerate hundred or thousand of records (virtual knowledge units) in the column *txtuni* ("universal text"):

Replace all txtuni with alltrim(str(RecNo()))+"."+alltrim(txtuni) for not "."$substr(txtuni,1,5)

Figure 2 shows the result, i.e. all records has been numbered with the exception of records that contain a dot.

The batch knowledge processing paradigm was developed by using thousands of such commands. The content of the knowledge tables is controlled and handled by the in-house WPad software (BIKE software is installed only on the computers of the author Svetsky).

2.2 Examples: The Personalized Learning Analytics

The state-of-the-art in the field of Learning Analytics is focusing on statistical monitoring and sampling of simple online data. It is based on the activities of students in online learning environments, but not on they knowledge. In our case, the research approach is focused directly on the knowledge flow because WPad works as well as a convertor of tacit knowledge into explicit knowledge (according to the terminology of the area of knowledge management), so once a student or teacher writes something into the WPad, then this represents the explicit knowledge. Therefore, such approach is closer to human mental processes, i.e. beyond the state-of-the art.

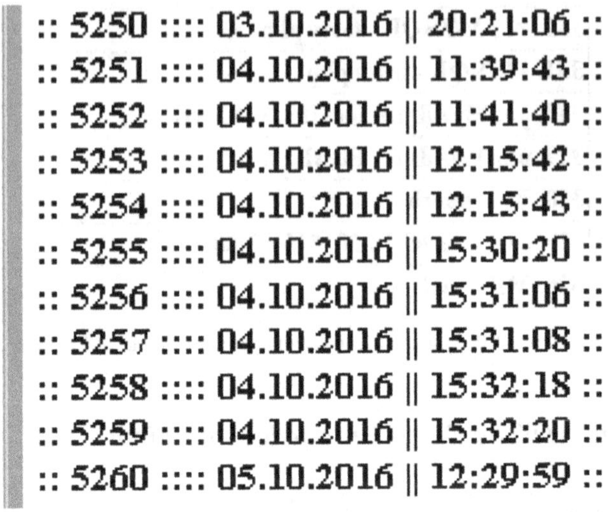

:: 5250 :::: 03.10.2016 || 20:21:06 ::
:: 5251 :::: 04.10.2016 || 11:39:43 ::
:: 5252 :::: 04.10.2016 || 11:41:40 ::
:: 5253 :::: 04.10.2016 || 12:15:42 ::
:: 5254 :::: 04.10.2016 || 12:15:43 ::
:: 5255 :::: 04.10.2016 || 15:30:20 ::
:: 5256 :::: 04.10.2016 || 15:31:06 ::
:: 5257 :::: 04.10.2016 || 15:31:08 ::
:: 5258 :::: 04.10.2016 || 15:32:18 ::
:: 5259 :::: 04.10.2016 || 15:32:20 ::
:: 5260 :::: 05.10.2016 || 12:29:59 ::

Fig. 3 Screenshot of the evidence of the personalised data

It can be demonstrated how the teacher—designer, monitored his own activities within the research on learning analytics. He wrote a specific programming code which indicated how many time the "knowledge" table was opened by him. In this case, such monitoring can indicate how the personalized data relate not only to his activities but also to the teaching content (knowledge), etc.

Figure 3 illustrates such personalized data from his home computer. For example, one can see that the knowledge table was opened by the teacher around five thousand times within ten months.

2.3 Examples: The Text Extraction from Large Files

The user had at his disposal proceedings with 250 pdf-files from a conference in Villach (Austria). He wanted to analyze research focus of contributions. Firstly he joined the 250 files into one file with Nuance pdf-software and saved as one plain text file. Then inserted the file into the knowledge table, extracted keywords and found the frequency of their occurrence. Figure 4 illustrates the result in the same knowledge table.

vocational	0.002641	11
virtual	0.005282	31
remote	0.005522	23
enhanced	0.004322	6
mobile	0.005282	22
language	0.002641	11
e-learning	0.002161	9
distance	0.002161	9
control	0.003361	14
component	0.004802	47
assessment	0.002641	11
management	0.002161	9
collaborative	0.002881	29

Fig. 4 Screenshot of the text extraction from 250 pdf-files

2.4 *Examples: The Use of Digital Pen*

The use of digital pen was also tested in order to automate transmission of handwritten schemas and chemical formulas to computer. Figure 5 illustrates personal notes of the designer made by the digital pen (in relation to the principle how human knowledge from various human sources is processed via WPad for knowledge based processes).

Despite the fact that the image seems to be quite good, it took more time when one was using a paper and pen. So it has been used only once, as a signature list of students (there was also a problem with the ink replacement, incl. the price).

2.5 *Examples: The Modelling Collaborative Activities*

The design of the universal virtual knowledge unit which is controlled by the in-house software BIKE (WPad) as a knowledge table (by using the Visual FoxPro database platform) was a first step to complex human knowledge processing. The second step

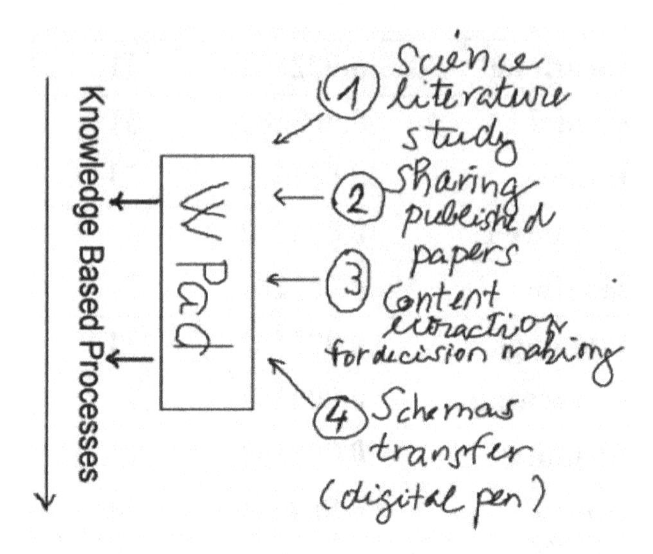

Fig. 5 Example of an image made by the digital pen

was a solution how to transmit the virtual knowledge between off-line and online environments in order to solve various collaborative activities. This was possible after implementation of the Utility Model 7340/2014 (Slovak Patent Office, registration on December 2015) which relates to processing and transmission of unstructured data. The question of the data transmission was important for solving collaborative activities on a shared virtual environment.

In this view, some applications were developed with focus on the so called Human Language Technology and Computer Supported Collaborative Learning (CSCL).

There was a project idea to create personal corpuses which would contain multilingual human expert knowledge from the field of machinery engineering [12]. The CSA proposal objectives were focused on cooperative benchmarking for the design of the pilot human processes-driven system of automated Machine Translation consisting of research infrastructure of recommended multilingual repositories, specialized corpora, ontological batch knowledge sets, aMT software and a federated trans-lingual pan-European portal. In other words, the expert human knowledge should be shared and exchanged via the shared virtual environment within the portal.

The developing CSCL application represents progress within the participation research on Technology-enhanced learning. The actual state of the research is illustrated by Fig. 6. Within the supportive IT infrastructure a shared virtual environment works for educational knowledge exchange.

The first results when modelling the CSCL-knowledge exchange and sharing shows, that the processing of the human knowledge in a knowledge based process requires one to solve three categories of algorithm—*pedagogical* (didactic methods), *informatics* (adaptation to the operating system) and *integrated* (CSCL—communi-

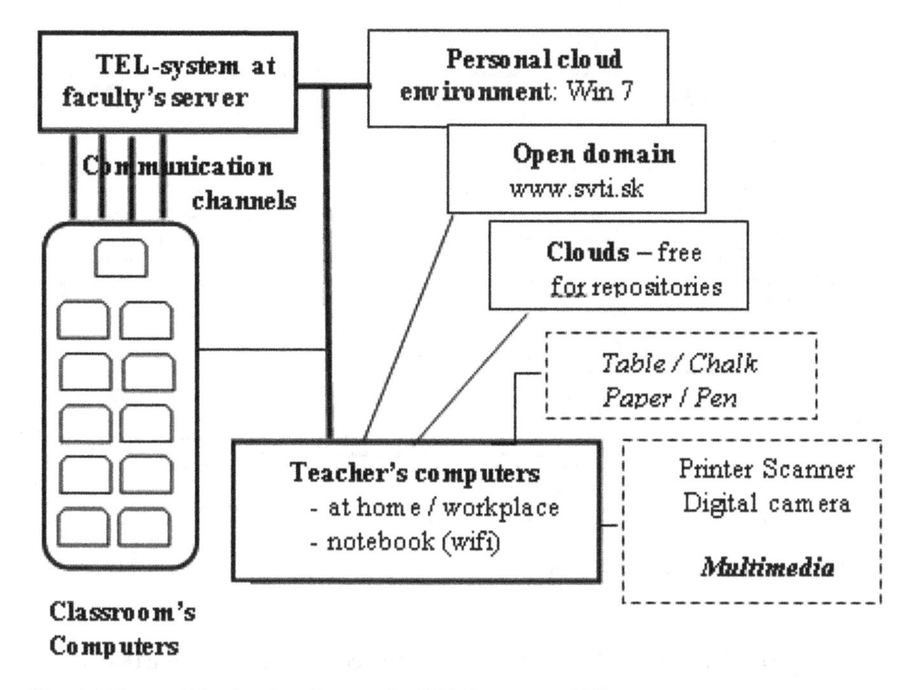

Fig. 6 Schema of the developed supportive IT infrastructure [13]

cation and human knowledge transmission). In this case, a design of templates for sharing the human knowledge is needed. Figure 7 shows such template for sharing annotation in virtual environment. The lower part of the figure represents annotation data in WPad and the upper part data after the conversion to html.

3 Some Aspects of Knowledge Processing in Engineering Education

The BIKE system was developed with the intention that a computer, like a user's partner, should expand one's quality of his life and sustainability. The representation of knowledge by using the virtual knowledge is analogous to how people think and act. People perform processes that use knowledge that needs to be worked into daily life. Knowledge is also a key element within teaching and learning.

The empirical research has showed that the work of university teachers in engineering knowledge is highly sophisticated. From an informatics perspective, this means that support of mental, learning activities and teaching is characterized by the use of unstructured information and knowledge. Thus, each type of teaching activity should be studied, analyzed regarding repeatability, elements, action sequences,

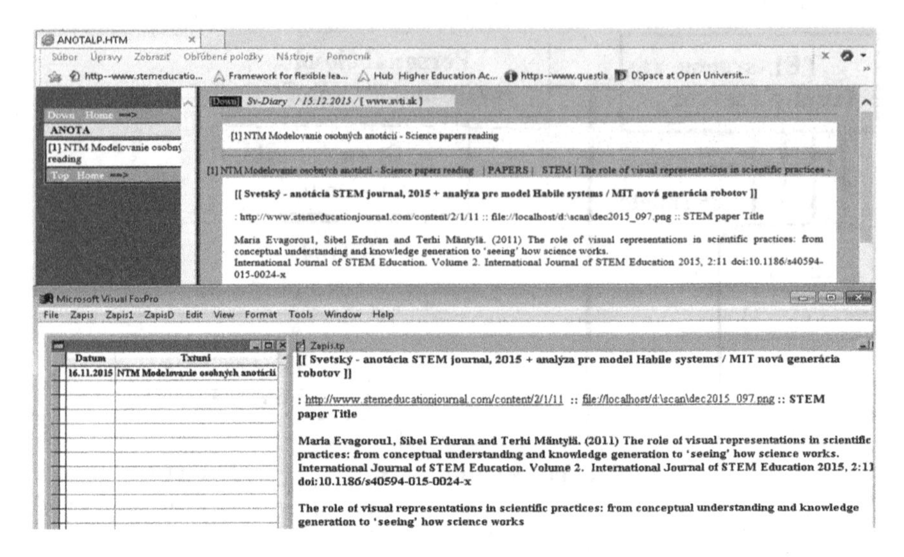

Fig. 7 Scrrenshot of the annotation template

teaching algorithms to enable writing the program codes, creating the BIKE's user menu and processing the knowledge flow.

The first stage of the above mentioned research on TEL was focused on the programming of "knowledge flow 1" between information sources and "DBMS—knowledge tables" (basically, content issues were computerized). This is illustrated by Fig. 8.

User outputs led both to the BIKE environment and HTML—format, which is readable by the common Internet browsers. Such browsable outputs supported a variety of educational activities (support for teacher's personal activities; for blended, informal, distance, active learning). In this context, a virtual learning space and personal learning environment with learning texts and libraries were created.

Despite the extensive progress in the first stage of research on TEL in bachelors' study, the practice revealed that any learning content or any excellent technology is "dead" if the knowledge flow is not working between the teacher and students in the classroom. Therefore, the further approach was focused more on the educational-driven approach of TEL. For example, a personal social network teacher—students was programmed within processing the "knowledge flow 2" (see Fig. 8), and feedback, with communication channels (an internet application), virtual calculation space and a set of informatics tools. The communication channels work as a personal network where information, knowledge and instructions "travel" between teacher and students to be shared. This approach also required to computerize various collaborative activities, so this research was focused on so called Computer Supported Collaborative Learning (CSCL).

In this context, the practice of the CSCL requires to solve not only the specific collaborative interactions of teachers and students, but also an additional design

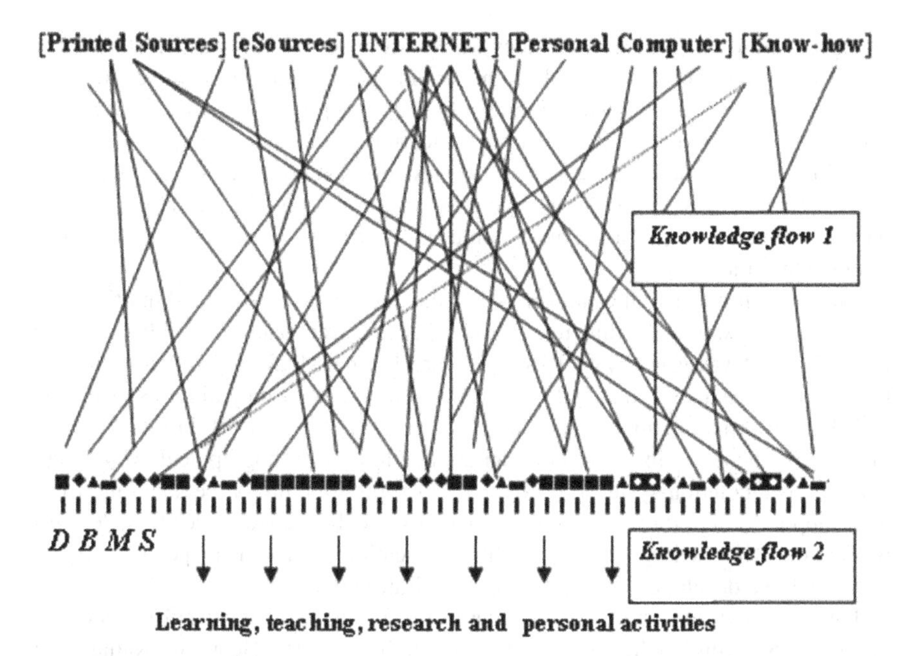

Fig. 8 Knowledge flow (Information sources—BIKE—Teacher's activities)

of the supporting IT background infrastructure for this kind of teaching. Because the practice of collaborative learning points to "collaboration" between the teacher and students, the didactics aspects (pedagogical methodologies), and informatics aspects (the supporting IT infrastructure) are more important in comparison with the traditional teaching and learning. So, in comparison with previous research, the collaborative shared space and personal cloud space were implemented.

The actual approach of authors' research on TEL is focused on a complex "automation" of educational processes, which requires an integration of three key areas: (1) didactics processes running in classrooms, (2) informatics processes running during teaching, and (3) the adaptation of individual's activities related to using the Windows operation system, Internet browsers, software, hardware, clouds, networks (this affects the Human-Computer-Interaction). Authors' long-term research showed that the computer support must be strictly tailor made to the teaching activities, so not vice versa (teachers usually applied only a general software, e.g. office packages).

4 The Future Works

The above mentioned examples demonstrated universal "all-in-one" function of the in-house software BIKE (WPad) when processing human knowledge. Batch internet retrieving, extraction data from WEB databases, text processing, collaboration with internet search or translation services (these are also used by bachelor students) can also be mentioned as examples. One could speak about a system "evolution". Namely, each software has a user menu with ten to twenty items divided into certain categories. By solving a tailor made modification of the in-house software, when modelling the support of knowledge based processes, a combination of menus had to be used. This resulted in a new category of applications. In the future, each of them could be developed as an autonomous software, or solved as classes in object oriented programming.

These all-in-one activities are similar to the ones made by so called multi-agents or robots. For example, at this time a possibility to extract data from bibliographical MARC format was tested by the BIKE software. Format MARC (MAchine-Readable Cataloging), is a data format and set of related standards used by libraries to encode and share information about books and other material they collect. In future works, there would be a possibility to build an application for bibliographical purposes (a citation or annotation system). In this case also programming a personal multilingual support system for writing science papers could be planned (even this idea was worked out when writing this paper). In an education environment specific applications could be developed as well (see for inspiration in [14]).

In this context, authors of this contribution see a certain analogy to Nilsson's vision of "Habile Systems" in the field of Artificial Intelligence [15]. He wrote "Rather than work toward this goal of automation by building special-purpose systems, I argue for the development of general-purpose, educable systems that can learn and be taught to perform any of the thousands of jobs that humans can perform". Figure 9 illustrates how authors were inspired by the Nilsson's vision. In other words, the BIKE(E) should work as a black box for human knowledge processing. The outputs would be thousands of applications, which are useful for various human "jobs". So future works leads to a vision of an "educational robot" or "multipurpose intelligent system" (Fig. 9).

5 Conclusion

In this contribution, the universal personalized approach for human knowledge processing within knowledge based processes was described. A specific focus was given on explaining this approach to the information representation of human knowledge by a specific default data structure—so called virtual knowledge unit. Such approach seems to be related to "pragmatic" information because the human knowledge is useful from the users point of view (e.g. a user can make decisions).

This enables one to simulate and computerize mental processes, i.e. human knowledge based processes, by using the virtual knowledge. From a human point of view, the content of the virtual knowledge means simply any information or knowledge as it is understood by people, which is written in natural language. From a machine (computer) point of view, the machine understands the virtual knowledge unit as an "user data type", thus, as a kind of information representation which uses ASCII codes. In addition, the virtual knowledge enables the user to process unstructured

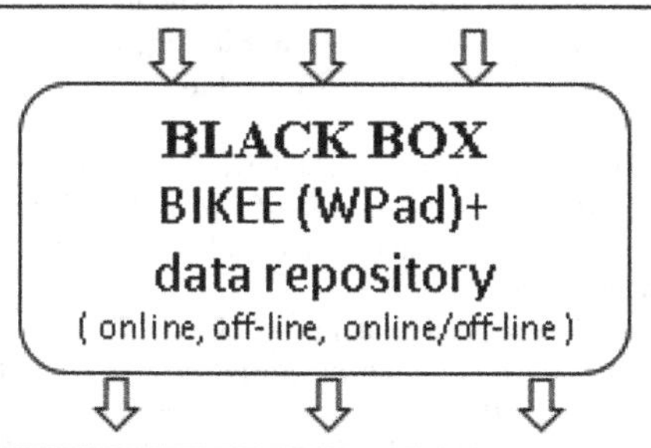

Fig. 9 Schema of function in-house software as a black box

data in a user friendly way. In other words, the content of the virtual knowledge is both human readable and machine readable. Any other rules, representations, languages, semantics graphs, etc. are not needed. It all works as an all-in-one IT tool, a black box, because if the user finds any way how to write human knowledge into the virtual knowledge, it automatically means, that he embedded into the data structure his own semantics and ontology.

This was illustrated via several examples from human knowledge processing (in each screenshot, a different kind of human knowledge was used). Thus, the inputs were human knowledge (information) in natural language with user's embedded semantics and ontology. The outputs, which are produced by the computer, were also in natural language. This enables user to "automate" his mental processes.

Moreover, if one defined a set of algorithms for processes where the human knowledge is a basic parameter, many applications could be developed in the future. This seems to be a certain analogy to Nilsson's vision of "Habile" systems. The analogy lies in the idea that the applications for human knowledge processing could be performed by a "Virtual Educational Robot" functioning as a "multi-purpose intelligent system" (see the Partner Search section within the IDEAL IST web-page: http://www.ideal-ist.eu/ps-sk-99947).

References

1. S. Svetsky, O. Moravcik, J. Stefankova, P. Schreiber, The implementation of the personalized approach for technology enhanced learning, in *Proceedings of The World Congress on Engineering and Computer Science 2010*, WCE 2010, San Francisco, USA. Lecture Notes in Engineering and Computer Science, pp. 321–323
2. S. Svetsky, O. Moravcik, P. Schreiber, J. Stefankova, The informatics tools development and testing for active learning, in *Proceedings of The World Congress on Engineering and Computer Science 2011*, WCE 2011, San Francisco, USA. Lecture Notes in Engineering and Computer Science, pp. 265–268
3. S. Svetsky, O. Moravcik, P. Schreiber, J. Stefankova, The educational—driven approach for technology enhanced learning, in *Proceedings of The World Congress on Engineering and Computer Science 2012*, WCE 2012, San Francisco, USA. Lecture Notes in Engineering and Computer Science, pp. 290–296
4. A. Martens, Software engineering and modeling in TEL, in *The New Development of Technology Enhanced Learning Concept, Research and Best Practices*, ed. by R. Huang, K. Nian-Shing Chen (Springer, 2014), pp. 27–40
5. L. Habinak, Využitie výpočtovej techniky pri riadení výchovno-vzdelávacieho procesu (The use of computation technique for the controlling educational processes), in *Conference: The Use of Computation Technique in Education*, DT CSVTS Bratislava, 1984, pp. 11–21
6. Z. Kotek, P. Vysoky, Z. Zdrahal, *Kybernetika (Cybernetics)* (SNTL, Praha, 1990)
7. J. Boros, *Zaklady psychologie (Basics of Psychology)* (SPN, Bratislava, 1976)
8. Z. Yu, H. Wang, M. Wang, Understanding short texts through semantic enrichment and hashing, IEEE Trans. Knowl. Data Eng. **28**(2), (2016)
9. F.Y.Y. Choi, P. Wiemer-Hastings, J. Moore, Latent semantic analysis for text segmentation, in *Proceedings of The Conference on Empirical Methods in Natural Language Processing*, Carnegie Mellon University, Pittsburgh, PA, 2016, pp. 109–117
10. E. Weinberger, A theory of pragmatic information and its application to the quasi-species model of biological evolution, Max Planck Institute for Biophysical Chemistry, https://pdfs.semanticscholar.org/3ade/f57e8f5c34566fc7f1e56079b7114631238f.pdf
11. M. Bieliková et al., Personalized conveying of information and knowledge, Studies in informatics and information technology, Research Project Workshop Smolenice (SUT Press, Bratislava, 2012), pp. 53–86
12. S. Svetsky et al., Pan-European language network 21 + 3 for designing the infrastructure of human processes—driven System of machine translation for specialized high-quality translation, Proposal No.: 643961 submitted to H2020-ICT-2014–1 Call
13. S. Svetsky, O. Moravcik, The implementation of digital technology for automation of teaching processes, in *The FTC 2016—Future Technologies Conference*, San Francisco, USA, http://ieeexplore.ieee.org/stamp/stamp.jsp?tp=&arnumber=7821632
14. D.G. Moursund, A faculty member's guide to computers in higher education, (April 2007), http://uoregon.edu/~moursund/Books/Faculty/Faculty.html
15. N.J. Nilson, Human-level artificial intelligence? Be serious!, AI Magazine (2005), http://ai.stanford.edu/~nilsson/OnlinePubs-Nils/General%20Essays/AIMag26-04-HLAI.pdf

The Use of Scalability of Calculations to Engineering Simulation of Solidification

Elzbieta Gawronska, Robert Dyja, Andrzej Grosser, Piotr Jeruszka and Norbert Sczygiol

Abstract The paper focuses on the use of scalability of available development tools to engineering simulation of solidification in the mold. An essential aspect of the considerations are the ways parallelization of computations taking into account the contact between the two materials. The implementation uses a TalyFEM and PETSc library. A problem solved with finite element method (FEM) can be parallelized in two ways: the parallelization on the mathematical formulas level and the division of tasks into smaller subtasks—assignment of nodes and elements into specific computational units. Both methods can be used in the TalyFEM library if the input files loading module is modified. We have designed our own parallel input module (finite element mesh) providing a division of loaded nodes and elements into individual computational units. Our solutions enable the full potential of parallel computing available in the TalyFEM library using the MPI protocol. This implemented software can be run on any computer system with distributed memory.

Keywords Distributed memory · FEM · Parallel computing · Relative speed up Solidification · Strong scalability

---

E. Gawronska (✉) · R. Dyja · A. Grosser · P. Jeruszka · N. Sczygiol
Czestochowa University of Technology, Dabrowskiego 69, 42-201 Częstochowa, Poland
e-mail: gawronska@icis.pcz.pl

R. Dyja
e-mail: dyja@icis.pcz.pl

A. Grosser
e-mail: grosser@icis.pcz.pl

P. Jeruszka
e-mail: jeruszka@icis.pcz.pl

N. Sczygiol
e-mail: sczygiol@icis.pcz.pl

S.-I. Ao et al. (eds.), *Transactions on Engineering Technologies*,
https://doi.org/10.1007/978-981-13-0746-1_17

1 Introduction

The development of the computing power of personal computers increased the possibility of carrying out numerical calculations. Large engineering simulations, in which billion of unknowns has to be estimated, can be calculated an ever shorter period on High-Performance Computing (HPC) systems. However, today's PCs have sufficiently powerful computational units which can solve small problems. Nowadays, due to technological limitations, the possibility of accelerating the calculations is focused on the use of multiple computational units in the same numeric process. Instead of accelerating the cycle frequency of one unit, a higher number of such units is used, especially on HPC systems. Software should be adapted to such hardware. Also, the use of supercomputers, consisting of several thousands of nodes operating in parallel, requires scalable software project.

The distinguishment of elements which can be calculated independently is required to avert widespread problems in scalable software at the design stage of the algorithms. Each part that depends on the other desires the use of costly time-synchronization task to avoid errors resulting from the use of the same variables. This paper shows a method that uses splitting of tasks into subtasks which can be calculated separately on each computing unit.

The numerical method presented in this paper uses a system of linear equations which is usually kept in the form of a coefficients matrix and additional vectors. They are well described, implemented in the existing components and ready to be used in the source code. Vectors, matrices, and solvers—objects responsible for solving the system of equations—can conduct parallelly. Our engineering software exploited ready-made data structure used to the scalable calculations of the linear equations systems.

The presented problems such as the division of tasks and parallel calculation had to be implemented in the form of packaged engineering software. From the software engineering's viewpoint, this process can be accelerated by using the core adjusted to the required applications. It reduces the time needed for the design, creation, and implementation of software.

This task was carried out in the original software using appropriate frameworks. As a numerical example, the two-component alloy was used in the solidification. The finite element method (FEM) [11] is used as a domain discretization method [7]. That technique is prevalent in resolving the scientific problems connected with a system of equation obtained from the partial differential equations.

Consideration of the problem of two-component alloy solidification with the usage of the FEM may lead to the need to calculate the system of equations with millions of unknowns in which the coefficient matrix is the sparse matrix. In addition, the symmetry of this matrix can be impaired with the introduction of boundary conditions. This situation introduces laborious numerical calculations which—due to technical limitations—is practically impossible to speed up with the consideration of a single computational unit. Because of this reason, using parallel and distributed processing to resolve this issue becomes the interesting algorithmic problem. An important aspect of the considerations are the ways parallelization of computations taking into account the fourth-type boundary condition—the contact between the two materials. The rest of this article presents the numerical model, the use of the preprocessor, the parallel methods of implementation of the model in the software, and the results [3].

2 Solidification Model

Engineering calculations were performed on the example of the solidification process of an alloy. The model of solidification process is built on the basis of the equation of heat conduction with the source member [5]:

$$\nabla \cdot (\lambda \nabla T) = c\rho \frac{\partial T}{\partial t} - \rho_s L \frac{\partial f_s}{\partial t} \tag{1}$$

wherein: T—temperature t—time, λ—thermal conductivity, c—specific heat, ρ—density, L—latent heat of solidification, f_s—the solid phase fraction.

2.1 *Discretization of the Area and of the Time*

Equation (1) is converted in the model using enthalpy formulation (basic capacitive formulation). As a result, it gives one equation that describes the liquid phase and the solid area of solidification. The apparent heat capacity formulation keeps the temperature as the unknown in the equation [2].

As a result of this transformation and the implementation of the discretization of the area by the finite element and the discretization of the time with the finite difference (Euler Backward modified scheme [10]), the linear system of equations was achieved. It has a suitable form to be solved by the one of the algorithms equations above:

$$(\mathbf{M} + \Delta t \mathbf{K})\mathbf{T}_{t+1} = \mathbf{M}\mathbf{T}_t + \Delta t \mathbf{b} \tag{2}$$

where:

$$\mathbf{K} = \int_{\Omega} \lambda \nabla \mathbf{N} \nabla \mathbf{N} d\Omega$$

$$\mathbf{M} = \int_{\Omega} c^*(T) \mathbf{N}\mathbf{N} d\Omega \tag{3}$$

$$\mathbf{b} = \int_{\Gamma} \mathbf{N} \lambda T_{,i} n_i d\Gamma$$

wherein **K**—the conductivity matrix, **M**—the mass matrix, **b**—the boundary conditions vector, **N**—the shape function vector, i—spatial coordinates, Ω—the computational domain, c^*—effective heat capacity, Γ—boundary.

2.2 Boundary Conditions

The model uses two types of boundary conditions: the Newton's boundary condition which models heat transfer between volume and environment, and the boundary condition of contact, which reflects the flow of heat between the two domains (cast and mold) taking into account to the separation layer. Both of these boundary conditions are natural boundary conditions.

The introduction of the natural boundary conditions is carried out by means of elements which are the boundary's discretization. In the case of the three-dimensional mesh consisting of the tetrahedral elements, boundary elements are the triangular elements and introducing the boundary conditions is made with the following equations:

$$\begin{Bmatrix} b_1 \\ b_2 \\ b_3 \end{Bmatrix} = \frac{A}{12} \begin{bmatrix} 2 & 1 & 1 \\ 1 & 2 & 1 \\ 1 & 1 & 2 \end{bmatrix} \begin{Bmatrix} q_1 \\ q_2 \\ q_3 \end{Bmatrix} \tag{4}$$

This system is a solution of the integral in the expression defining **b** in the formula (Eq. 3), wherein A is the area of the boundary element and q's are the flows of heat at the boundary of a given vertex. The value of heat flux occurring in the formula is calculated in accordance with the type of boundary condition. For Newton's boundary condition the following formula is used:

$$\Gamma: \quad q = \alpha(T - T_{ot}) \tag{5}$$

wherein α—the coefficient of heat exchange with the environment, T—the temperature of the body on the boundary Γ and T_{ot}—the ambient temperature. In contrast, the following formula shows the exchange of heat by the boundary layer separation:

$$\Gamma: \quad \begin{cases} q = \kappa(T^{(1)} - T^{(2)}) \\ T^{(1)} \neq T^{(2)} \end{cases} \tag{6}$$

wherein κ—a heat transfer of a separation layer, $T^{(1)}$ and $T^{(2)}$—the temperature of two areas on contact point.

As it has been shown, the introduction of above boundary conditions requires modifying the coefficient matrix because the boundary conditions consist of a temperature of the current time step.

3 The Implementation of the Required Functionality

Implementation of the computing module using the TalyFEM library [4, 9] (with PETSc library, developed by [1]) requires the implementation of several classes, presented in Fig. 1. The basic class is `SolidEquation`. It provides an implementation of assembling the global matrix and boundary conditions by means of overridden `Integrands` methods and `Integrands4Side`. Another `SolidInputData` class manages loaded properties of the material from the text file. The next class `SolidGridField`, thanks to overriding the `SetIC` methods with the `GridField` method, allows the initial conditions to be introduced. On the other hand, `SolidNodeData` objects store the results in nodes. As noted in the introduction, one of the implementation problems was the inclusion of the boundary condition of the fourth type with the separation layer. To generate the Finite Element mesh, the GMSH preprocessor was used. That preprocessor allows entering many physical areas by setting appropriate flags for the material. This makes it possible to distinguish the cast, mold or external surfaces during the loading stage of the grid. TalyFEM was not compatible with files generated by the GMSH, therefore, implementation of several classes in order to load and distribute the parallel distribution of the data between all processes was required.

The second problem occurred with the limitations of the TalyFEM library. Boundary conditions may be imposed on elements with smaller dimension than the dimension of the task. For example, the boundary conditions of two-dimensional plates

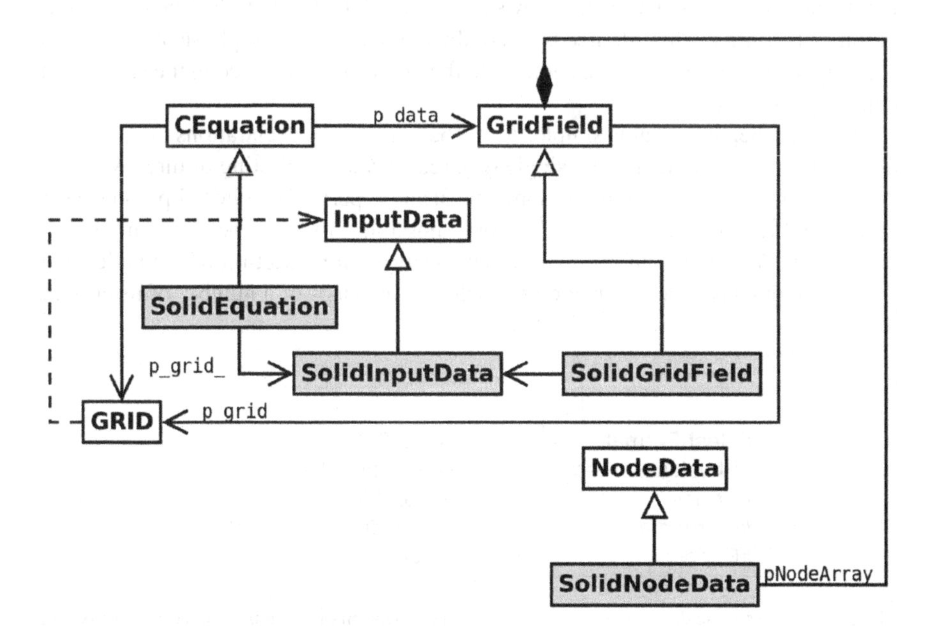

Fig. 1 Class diagram of solidification module

(divided into two-dimensional elements) refer to the one-dimensional edge which can be regarded as one-dimensional finite elements. Similarly, with a cube (3D finite elements)—boundary conditions apply to the 2D walls. With the used library, all elements must be the same type which unfortunately, makes it difficult to use the data boundary (for example: to select only one surface element located on the edge of the area). This situation had to be resolved by the appropriate use of configuration data.

Another required functionality was to ensure proper operation of the fourth boundary condition. In the solidification model, the fourth condition requires a physical data derived from adjacent elements. For example, considering the heat transfer between the mold and the casting, the thermal conductivity of the two areas should have been taken into account. At the same time, it is apparent that the knowledge about the connections between nodes (a node of one area corresponding to the node of second area) is required.

3.1 Implementation of the GMSH Format

One of the essential elements of the simulation is to load input data representing the considered problem. The problem presented earlier loaded input data generated by the GMSH preprocessor. Due to the scalability of the TalyFEM library, the very process of loading must take into account the distribution of the data (nodes, elements) into processes. The Fig. 2 shows a diagram of the GMSH file. The implementation of loading of this file needs to combine a declaration of physical conditions (here in the #PhysicalNames section) with the properties described in the simulation configuration file.

`*.msh` scheme format is simple. At the beginning, the declarations of a physical object group—components of a single type (eg. surface consisting of three walls)—are given. These objects combine some physical property. The GMSH preprocessor is not used to set specific value. It is only possible to select specific components to connect into a physical group. The next sections are a section of the nodes and the finite elements. The description of each node consists of a number of its current

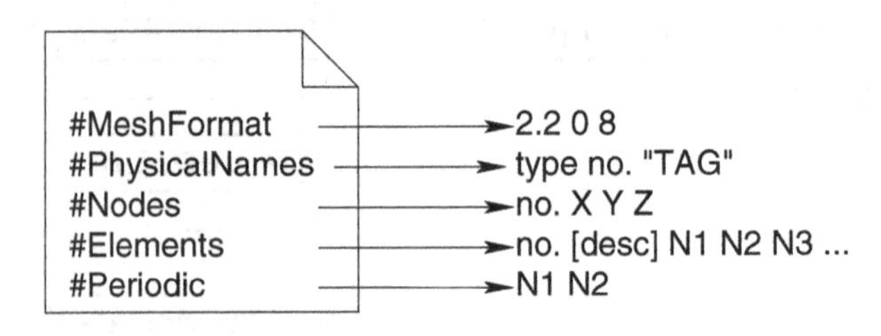

Fig. 2 The `*.msh` file data scheme. *[desc]* means a description of the item which differs in syntax depending on the pre-processor and the element

and subsequent coordinates. It should be noted that regardless of the dimension of the task, the GMSH saves the three-dimensional coordinates—the z coordinate is stored with a value of 0 for the two-dimensional tasks which had to be taken into account in implementation. The description of an element consists of the item index, a description of geometric and physical properties, and the serial indices of nodes that make up the item. A description of property, in the case of meshes generated for solidification, consists of four integers—the type of an element, unused value (usually a value of 2), the index of the physical object, and the geometry of the object. The type of an item and physical object of the group are important for the TalyFEM Library.

Loading nodes did not cause major implementation issues but loading elements required following a few rules. First of all, for the TalyFEM library, the main information about the physical properties (a boundary condition; they are called indicators in the library) are stored in objects representing nodes, and the `.msh` format stores this information in the elements. The library retains the image of a mesh consisting only of the elements of one type; GMSH also saves boundary elements. For the three-dimensional grid for calculations, tetrahedral elements should be only read; for boundary conditions, recorded triangular elements located on the surfaces of volume must be analysed. Loading nodes and elements can be divided into the following steps:

- load the number of nodes and move the position of the next character into the elements section;
- load the total number of elements and analyse the element. The loading module analyses the boundary elements and the space, creating a map of the node index of the physical property. After this step, the number of spatial elements is known, and this number determines the number of all the items used in the calculations;
- load spatial elements;
- go back to nodes section and load nodes data using information of the constructed map of physical properties of nodes.

The last interesting piece of the file is the information about nodes adjacent to each other (Periodic section). Basically, this section consists of indices pairs, which represent nodes having the same coordinates. This information is used during the boundary condition of the fourth type with the contact of two boundaries.

The loading was implemented for two scenarios for the program based on TalyFEM:

- the calculation with the decomposition of the area (the grid division into smaller sub-areas stored on each process) which required the implementation of data communications;
- calculation without decomposition of the area; all data is stored in one process only.

The first scenario is interesting from the programming point of view. The library supports parallel data storage with distributed memory but low-level data, such as node coordinates or description of the elements, which had to be sent using the MPI interface [8]. Figure 3 shows a scheme of loading the `.msh` file using decomposition area for the main process. The main process (process 0) reads information but does not keep it—the individual data is transferred to the separate processes. At the beginning, physical properties are loaded and broadcasted to all processes. It has been recognized that there is no need for the division of information into the individual processes. Such information requires very little memory resources compared with information about the mesh.

Then the analysis and loading of elements, including map of the physical properties of nodes is carried out. At this stage, information is sent in the form of a $N \times M$ matrix where N is the number of elements for the n-th process, and M is the number of data describing an element—physically (for simplicity of communication) it is an array of $N \times M$ size. Similarly, the information about the nodes is sent, wherein, additional information about physical group is transmitted (the physical properties of a given node) and information on the adjacent node (if the node does not have a boundary condition of the fourth kind, a value of -1 is sent). The task of the remaining processes is to take the data and assemble it in a simple dynamic arrays, what is presented in Fig. 4. The TalyFEM library converts information from the arrays into its

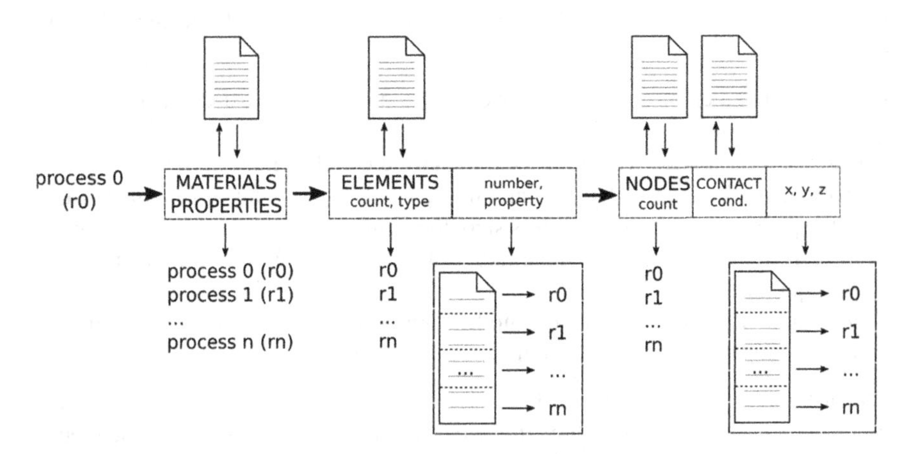

Fig. 3 Scheme of loading and decomposing data by the main process (process 0)

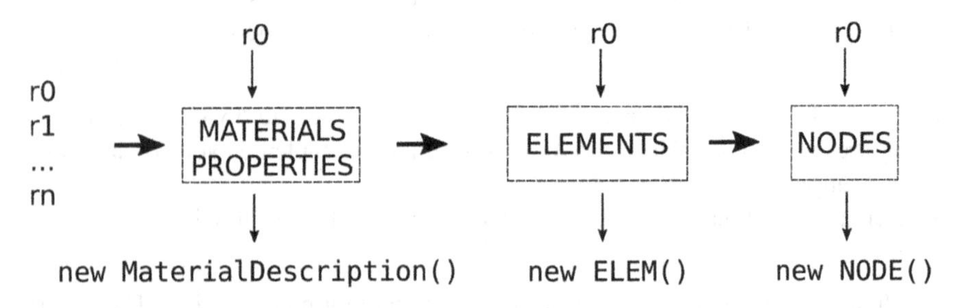

Fig. 4 The diagram of aggregation of data received from the main process

object-oriented counterpart. There is no need for manual communications between different processes nor even involve other processes in the process of sending the data—excluding the information on the connected areas.

3.2 *The Adjacent Nodes in Different Processes*

The problem with parallel processing of the fourth condition mode arises due to the necessity of modifying a local matrix (in each process) using the physical parameters and a local matrix of neighbouring element. To properly assemble the system of equations, the knowledge of a pair of nodes that have the same coordinates (but belong to different areas) is required. These nodes can belong to different processes. Moreover, one process does not have information on the location (process rank) of its adjacent node. As shown in Fig. 5, communication during the loading of the mesh is limited to the communication of the main process with other processes. Another difficulty is the renumbering nodes while performing of the parMETIS tools—the numbers of nodes from a `.msh` file are converted to the numbers optimal in parallel communication. Each process keeps the map of the physical index of the node (specified in the `.msh` file) in solution index but this map is limited solely to the ones stored in the process. The library and the modifications (shown in the next section) require knowledge about the number of nearly processes (contain neighbouring nodes). To

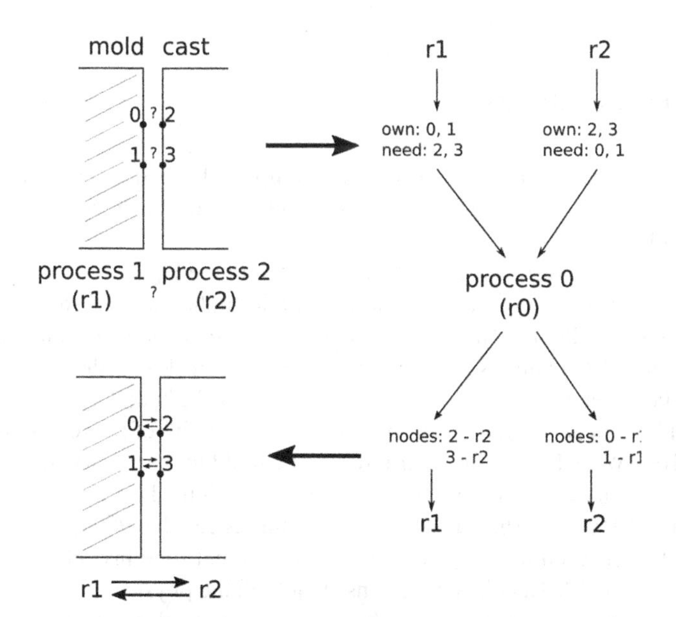

Fig. 5 The scheme of exchanging information on adjacent nodes (in the boundary condition of the fourth type)

avoid troublesome point-to-point (peer-to-peer) communication, the data repository seeking for all adjacent nodes was used in the 0 process.

Any process obtaining the nodes and converting them into an object, analyses the proximity of the node. As shown in the previous section, information about a node consists of coordinates, the index of physical groups, and the index (according to the numbering of the preprocessor) of adjacent node. For process 0, the following information is prepared:

- map of the index from `.msh` file into the solution index of nodes which are adjacent to another node (and, therefore, they will be adjacent to the another node);
- the physical indices of those nodes for which information about the process is required.

Process 0 takes information from each of the processes and prepares the map reflecting the physical indices of nodes of the two values: solution index and the rank of the process that the node has. The last step is to re-communicate with each of the processes which receive the missing information on the adjacent nodes—solution index and the rank of the process having the adjacent node. Communication between individual processes in the course of calculating the unknown value is provided by TalyFEM so there is no need for manual communications between processes.

Above solution is not scalable (parallelization is not used to speed up calculations), but is necessary for the proper implementation of the fourth-type boundary condition. Further, the finding of adjacent nodes is executed only once, while loading task properties.

4 Numerical Results

Performance tests were done on the CyEnce computer located at Iowa State University. CyEnce is a cluster consisting of 248 computing nodes each with two eight-core Intel Xeon E5 and 128 GB RAM (Fig. 6).

The calculations were made for the area shown in the Fig. 7.

This area consists of a cast (dark gray) and mold (light gray). As shown, the mold is a cube with a side of 0.2 m. The cast also has the shape of a cube but with a side of 0.1 m. The Al_2Cu alloy is the material for the cast; steel—for the mold. The heat exchange between the cast and the mold was held using the boundary condition of 4th type including the separation layer. A thermal conductivity of the release layer was equal to 1000 W/m^2K. Heat transfer through the mold to the environment was held with the boundary condition of the third type in which the heat transfer coefficient was equal to 10 W/mK. The initial cast temperature is equal to 960 K; mold—660 K.

Exemplary distributions of temperature for different time instants are shown in the Figs. 8, 9 and 10. The results are consistent with the physics.

Performance tests were made for three sizes of finite elements mesh: small (of 3.5 million elements, see Fig. 11), bigger (of 25 million elements, see Fig. 12), and the biggest of 75 million elements.

Fig. 6 CyEnce—a cluster used in computing, located at Iowa State University

Fig. 7 The relevant area. Cast has dark gray color, mold—light gray

Fig. 8 Temperatures in the relevant area after 5 s ($T_{min} = 655.9$ K, $T_{max} = 960$ K)

Fig. 9 Temperatures in the relevant area after 25 s ($T_{min} = 658.4$ K, $T_{max} = 949.4$ K)

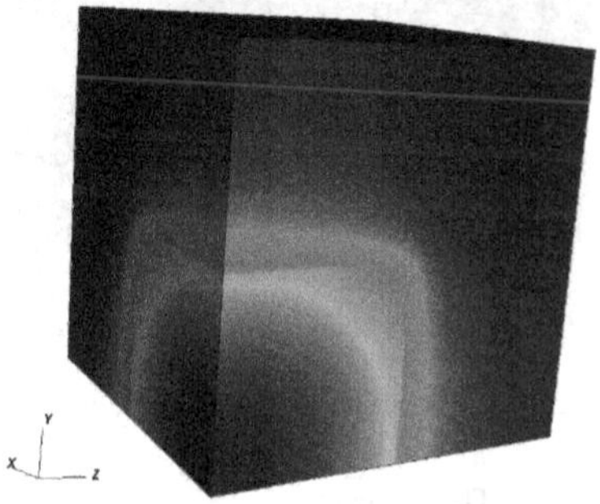

Fig. 10 Temperatures in the relevant area after 50 s ($T_{min} = 668.4$ K, $T_{max} = 925.9$ K)

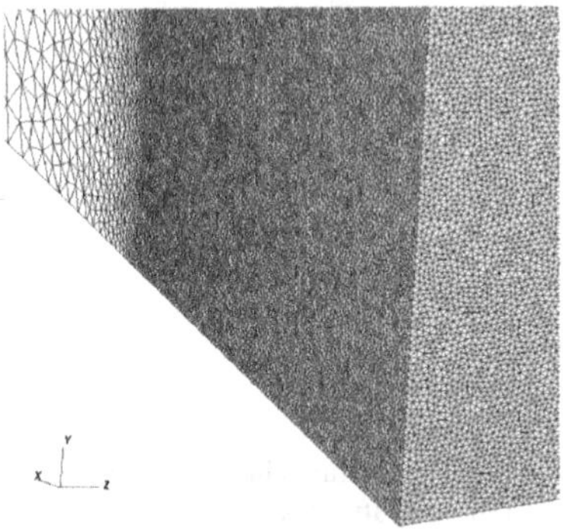

Fig. 11 Mesh details with 3.5 million elements

The time required to perform 100 steps of simulation (where the size of the time step was equal to 0.05 s) is shown in the charts on Figs. 13, 15, and 17 (Figs. 14 and 16).

The Figs. 14, 16 and 18 show the scalability, that is, the acceleration of calculations while adding further processors. This is called strong scalability. The basis for comparison is the computation time using 16 processors, for which it was assumed that they give a 16-fold increase in time in relation to a single processor. Thus, the graphs show relative speed up. The ideal increment is linear and is also marked on the chart.

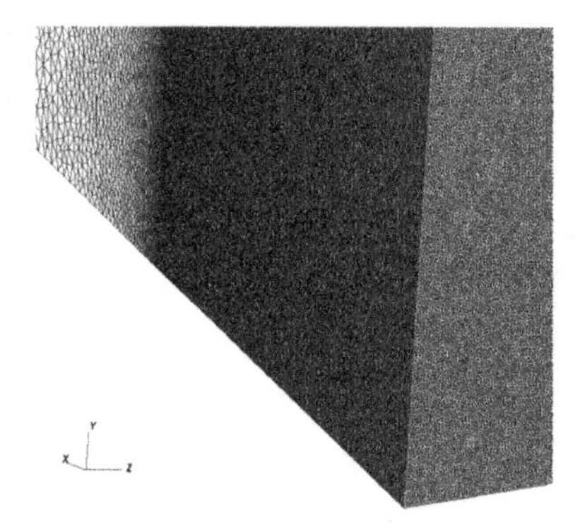

Fig. 12 Mesh details with 25 million elements

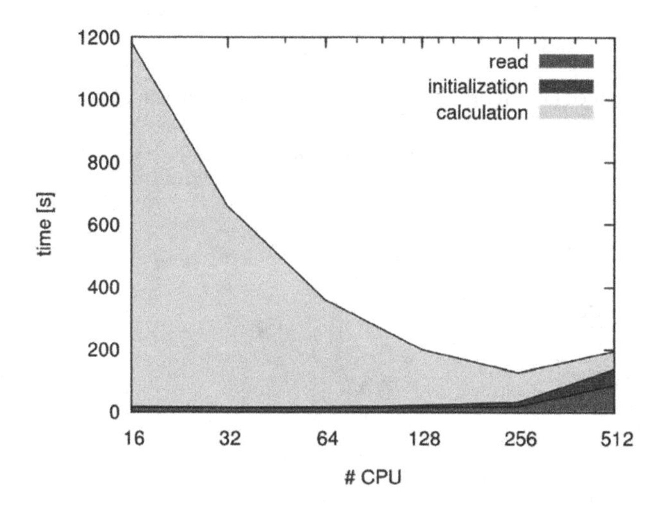

Fig. 13 Time required to perform 100 steps of simulation in mesh with 3.5 million elements

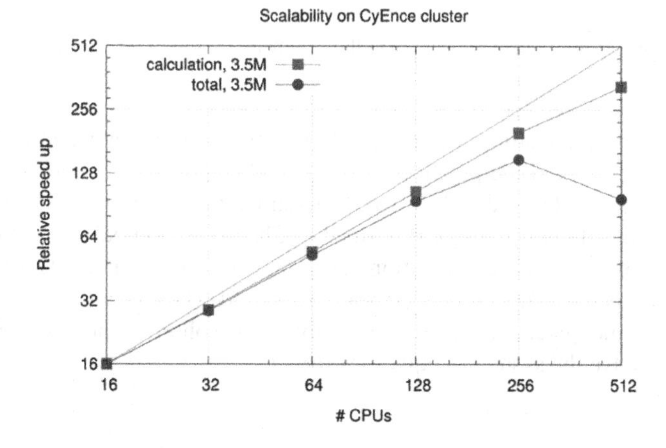

Fig. 14 Strong scalability in mesh with 3.5 million elements

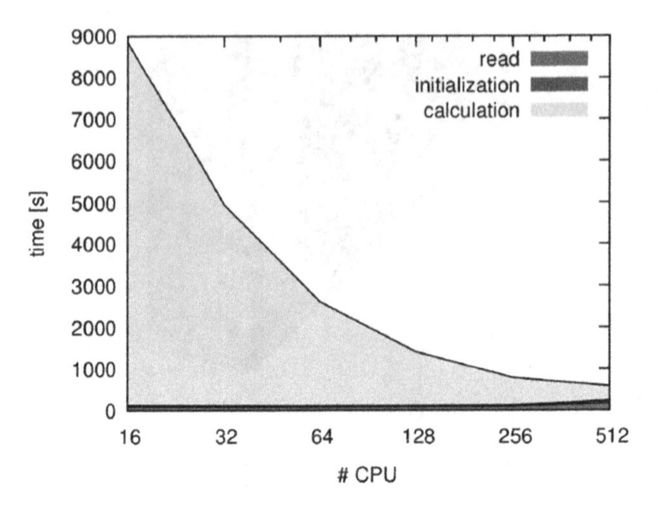

Fig. 15 Time required to perform 100 steps of simulation in mesh with 25 million elements

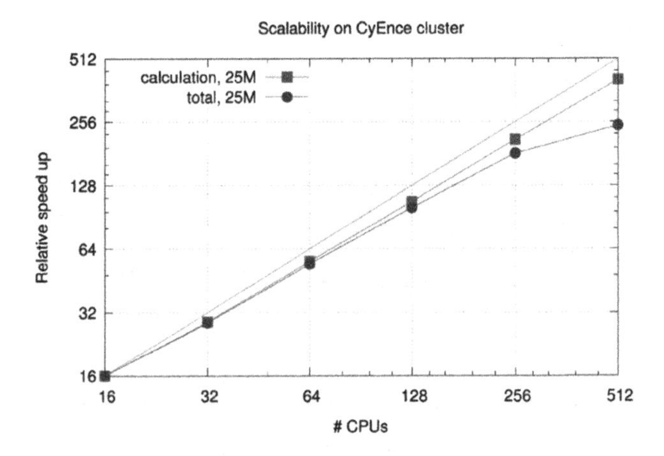

Fig. 16 Strong scalability in mesh with 25 million elements

For performance testing, the output file saving has been switched off because the analysed times and the preparation of the calculation could be affected by the writing operation. The charts show summarised loading time of retrieve data from files operation and the operations described in Chap. 3. The initialization time is the time to perform optimal decomposition of the area into the sub-areas. The number of sub-areas was equal to the number of processors used. The calculation time consists of assembling the system of equations and subsequent solution. The GMRES method was used to solve the system of equations [6].

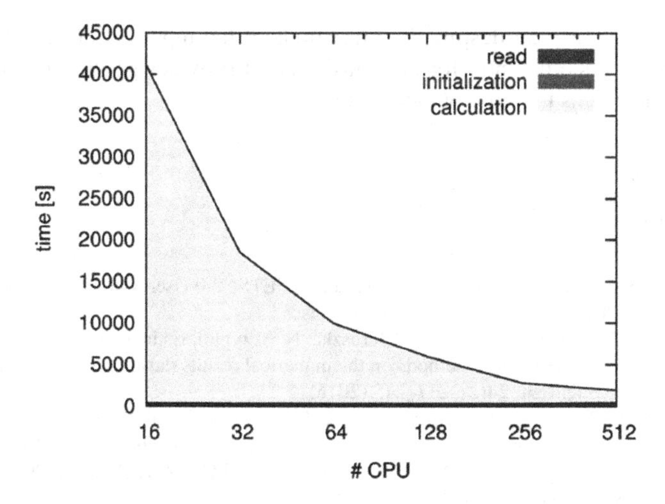

Fig. 17 Time required to perform 100 steps of simulation in mesh with 75 million elements

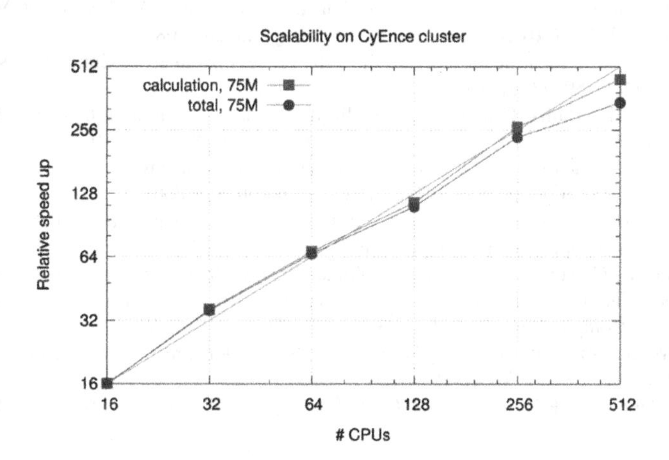

Fig. 18 Strong scalability in mesh with 75 million elements

5 Conclusions

The paper presents the use of libraries that support calculations which use the finite element method applied to the solidification simulation. Although the TalyFEM library had already been tested on computers of high power, its use in solving a specific problem required adding essential pieces of code. The essence of the research presented in the paper was to answer the question of how changes affect the performance and to determine the ability of using computers of distributed memory for calculations.

The results show that despite the limitations of the input/output operations, the example performs efficiently for large tasks and it provides good scalability while the load time equals the time of calculation.

References

1. S. Balay, W.D. Gropp, L.C. McInnes, B.F. Smith, PETSc 2.0 User Manual. Argonne National Laboratory (2014), http://www.mcs.anl.gov/petsc/
2. R. Dyja, E. Gawronska, A. Grosser, P. Jeruszka, N. Sczygiol, Estimate the impact of different heat capacity approximation methods on the numerical results during computer simulation of solidification. Eng. Lett. **24**(2), 237–245 (2016)
3. E. Gawronska, R. Dyja, A. Grosser, P. Jeruszka, N. Sczygiol, Scalable engineering calculations on the example of two component alloy solidification, in *Proceedings of the World Congress on Engineering 2017*. Lecture Notes in Engineering and Computer Science, 5–7 July 2017, London, U.K. (2017), pp. 226–231
4. H.K. Kodali, B. Ganapathysubramanian, A computational framework to investigate charge transport in heterogeneous organic photovoltaic devices. Comput. Methods Appl. Mech. Eng. **247–248**, 113–129 (2012). https://doi.org/10.1016/j.cma.2012.08.012
5. J. Mendakiewicz, Identification of solidification process parameters. Comput. Assist. Mech. Eng. Sci. **17**(1), 59–73 (2010)
6. Y. Saad, M.H. Schultz, GMRES: a generalized minimal residual algorithm for solving non-symmetric linear systems. SIAM J. Sci. Stat. Comput. **7**(3), 856–869 (1986)
7. N. Sczygiol, Modelowanie numeryczne zjawisk termomechanicznych w krzepncym odlewie i formie odlewniczej. Politechnika Czstochowska (in Polish) (2000)
8. University of Tennessee, Knoxville, MPI: a message passing interface (2016), http://mpi-forum.org/docs/mpi-3.0/mpi30-report.pdf
9. O. Wodo, B. Ganapathysubramanian, Computationally efficient solution to the cahnhilliard equation: adaptive implicit time schemes, mesh sensitivity analysis and the 3D isoperimetric problem. J. Comput. Phys. **230**(15), 6037–6060 (2011)
10. W.L. Wood, *Practical Time-Stepping Schemes* (Clarendon Press, Oxford University Press, Oxford [England], New York) (1990)
11. O.C. Zienkiewicz, R.L. Taylor, *The Finite Element Method*, 5th edn. (Butterworth, Oxford, 2000)

Vision-Based Collision Avoidance for Service Robot

Mateus Mendes, A. Paulo Coimbra, Manuel M. Crisóstomo and Manuel Cruz

Abstract ASSIS is a service robot which uses a camera, sonars and infra-red sensors for navigation. It uses images stored into a Sparse Distributed Memory, implemented in parallel in a Graphics Processing Unit, as a method for robot localization and navigation. It is controlled from a web-based interface. Algorithms for following previously learnt paths using visual and odometric information are described. A stack-based method for avoiding random obstacles, using visual information, is proposed. The results show the algorithms are adequate for indoors robot localization and navigation.

Keywords Autonomous navigation · Obstacle avoidance · SDM · Service robot Vision-based navigation · Vision-based localization

1 Introduction

Development of intelligent robots is an area of intense and accelerating research. Different models for localization and navigation have been proposed. The present approach uses a parallel implementation of a Sparse Distributed Memory (SDM) as the support for vision and memory-based robot localization and navigation. The SDM is a type of associative memory suitable to work with high-dimensional Boolean

M. Mendes (✉)
Polytechnic Institute of Coimbra - ESTGOH and Institute of Systems and Robotics, Coimbra, Portugal
e-mail: mmendes@isr.uc.pt

A. P. Coimbra · M. M. Crisóstomo · M. Cruz
Institute of Systems and Robotics, University of Coimbra, Coimbra, Portugal
e-mail: acoimbra@deec.uc.pt

M. M. Crisóstomo
e-mail: mcris@isr.uc.pt

M. Cruz
e-mail: M.Joao-1991@hotmail.com

S.-I. Ao et al. (eds.), *Transactions on Engineering Technologies*,
https://doi.org/10.1007/978-981-13-0746-1_18

vectors. It was proposed in the 1980s by Kanerva [3] and has successfully been used before for vision-based robot navigation [9, 13]. Simple vision-based methods, such as implemented by Matsumoto [6], although sufficient for many environments, in monotonous environments such as corridors may present a large number of errors.

In the present approach, ASSIS is a robot that learns new paths during a supervised learning stage. While learning, the robot captures views of the surrounding environment and stores them in the SDM, along with some odometric and additional information. During autonomous navigation, the robot captures updated views and uses the memory's algorithms to search for the closest image, using the retrieved image's additional information as basis for localization and navigation. Memory search is performed in parallel in a Graphics Processing Unit (GPU). Still during the autonomous navigation mode, the environment is scanned using sonar and infra-red sensors (IR). If obstacles are detected in the robots path, an obstacle-avoidance algorithm takes control of navigation until the obstacle is overcome. In straight paths, the algorithm creates a stack of odometric data that is used later to return to the original heading. Afterward, vision-based navigation is resumed [10].

Related approaches for service robots include Rhino [1], Indigo [5], REEM [16] and Care-O-bot [4].

Section 2 briefly describes the SDM. Section 3 presents the key features of the experimental platform used. The principles for visionbased navigation are explained in Sect. 4. Section 5 describes two of the navigation algorithms implemented in the robot. It also presents the results of the tests performed with those algorithms. In Sect. 6, the obstacle avoidance algorithms are described, along with the validation tests and results. Conclusions and future work are presented in Sect. 7.

2 Sparse Distributed Memory

The properties of the SDM are inherited from the properties of high-dimensional binary spaces, as originally described by P. Kanerva [3]. Kanerva proves that high-dimensional binary spaces exhibit properties in many aspects related to those of the

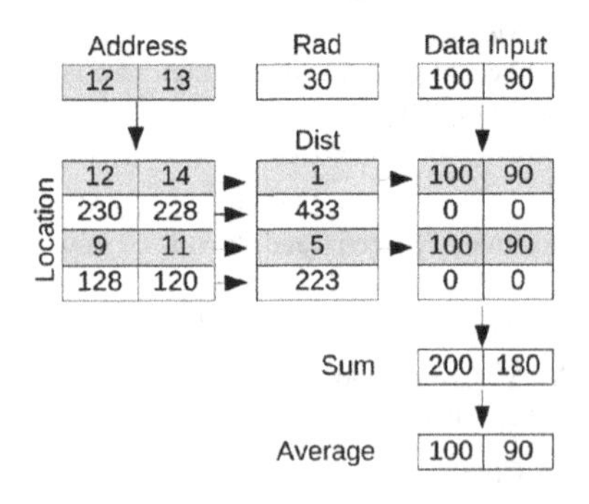

Fig. 1 Arithmetic SDM model, which operates using integer values

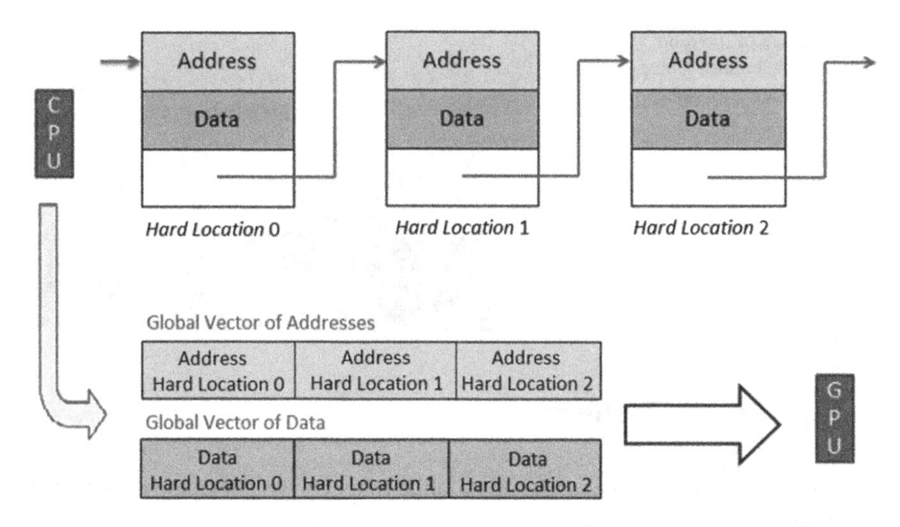

Fig. 2 Block diagram of the SDM implemented in the CPU and in the GPU

human brain, such as naturally learning (one-short learning, reinforcement of an idea), naturally forgetting over time, ability to work with incomplete information and large tolerance to noisy data.

The original SDM model, using bit counters, has some disadvantages, such as low storage rate, or sensitivity to data using the binary code [12]. Figure 1 shows a model based on Ratitch et al.'s approach [14], which groups data bits as integers and uses the sum of absolute differences instead of the Hamming distance to compute the distance between an input address and location addresses in the memory. That has been the model used in the present work. It was named arithmetic SDM and its performance has been superior to the original model [8].

The SDM model was implemented first in a CPU, using linked lists as depicted in the upper part of Fig. 2. As the volume of data available to process in real time increased, it became clear that the system could benefit from a parallel implementation of the search procedure. The SDM was then implemented in parallel, in a GPU, using CUDA architecture, as shown in the lower part of Fig. 2. During the learning stage of the robot, the linked list is built using only the CPU and central RAM memory. When learning is over, the list contents are copied to an array of addresses and an array of data in the GPU memory. Later, when necessary to retrieve any information from the SDM, multiple GPU kernels are launched in parallel to check all memory locations and get a quick prediction. The parallel implementation is described in more detail in [15].

Fig. 3 Picture of the robot used, with the control laptop, while executing a mission

3 Experimental Platform

The experimental platform consists of a mobile robot carrying a laptop running the control and interface software.

3.1 Hardware

The robot used was an X80Pro1,[1] as shown in Fig. 3. It is a differential drive vehicle equipped with two driven wheels, each with a DC motor, and a rear swivel caster wheel, for stability. The robot has a built-in digital video camera, an integrated WiFi system (802.11g), and offers the possibility of communication by USB or serial port. For additional support to obstacle avoidance it has six ultrasound sensors (three facing forward and three facing back) and seven infra-red sensors with a sensing range of respectively 2.55 m and 0.80 m.

The robot was controlled in real time from a laptop with a 2.40 GHz Intel Core i7 processor, 6 Gb RAM and a NVIDIA GPU with 2 Gb memory and 96 CUDA cores.

3.2 Software Modules

Figure 4 shows a block diagram of the main software modules, as well as the interactions between the different software modules and the robot. The laptop was carried on board and linked to the robot through the serial port. Besides the SDM module, where the navigation information is stored, different modules were developed to interact with the robot, control the supervised learning process and the autonomous run mode, which required the obstacle avoidance module.

[1]From Dr Robot Inc., www.drrobot.com.

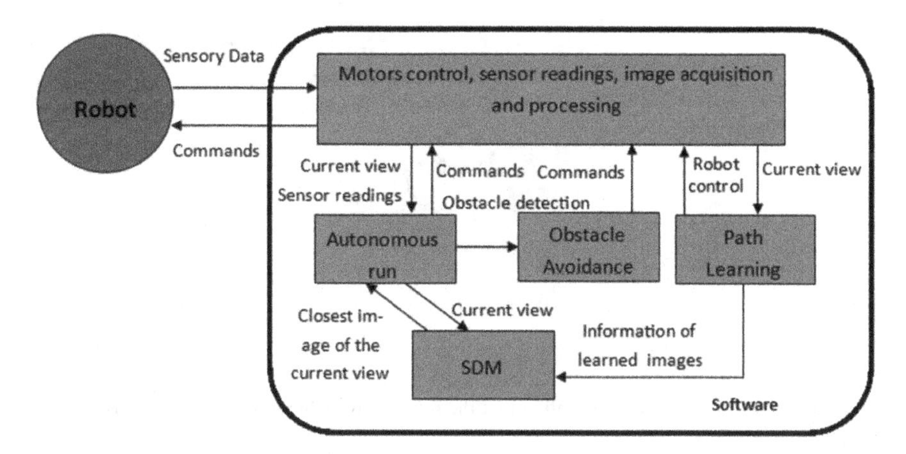

Fig. 4 Interactions between software modules and the user robot

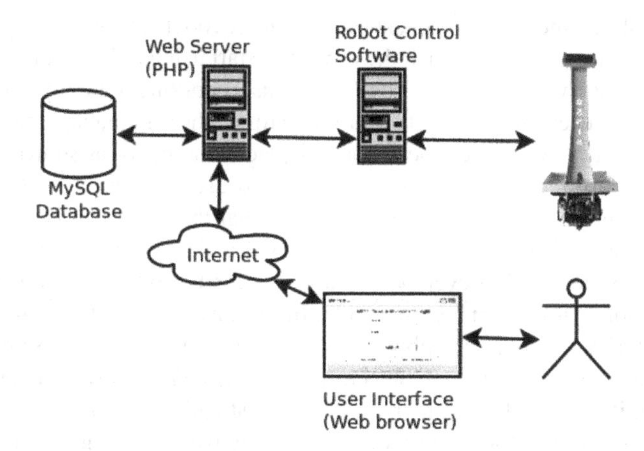

Fig. 5 Interactions between all system components and the user

3.3 *Mission Control and User Interface*

To facilitate interaction with users, a web-based interface was developed. Figure 5 is a high-level representation of the system. As the figure shows, there is a web server which communicates with the robot control through a socket connection. The web server also stores a database. In the database there is information about the users who can manage and use the robot. There is also information about each mission that the robot can perform and the paths that it can follow. As for users, three levels of privileges were implemented: (1) administrators, who can teach the robot and manage other users; (2) standard users, who can use the robot and also store their own missions; and (3) anonymous users, who can only use the robot to perform missions.

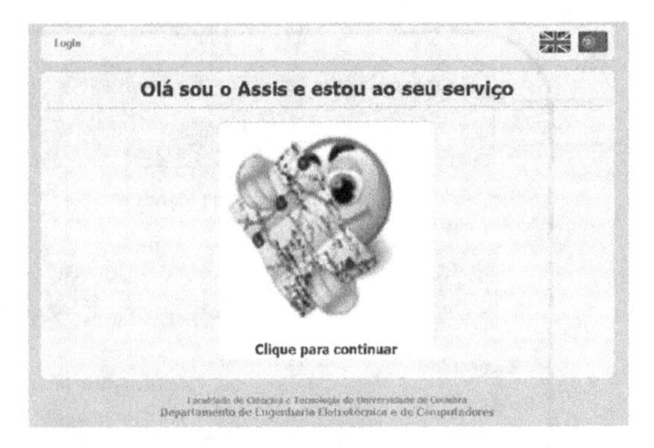

Fig. 6 Image of the login screen, stating that the robot is available to perform missions or learn new paths

Figure 6 shows the login screen, shown when the robot is waiting for new orders. In this state, any user can approach the robot and start using it. Users that log in with username and password will see different buttons, according to their privileges. The maximum privileges are granted to administrators, who can manage the privileges of other users, as well as teach the robot new places and paths, as shown in Fig. 7.

Registered users or anonymous users can also use the robot to carry on some missions, as shown in Fig. 8. The first mission that the robot can be used to is to guide the person to a known place. So far, only this mission was implemented, but the interface was already developed to account for the other mission types. The second mission is to perform a tour guide. In this case the robot can stop at points of interest during the tour and show additional information about those points. The third mission type is to deliver an object. In this case the robot invites the user to put the object in its tray, and when it arrives at the destination it will invite the receiver to grab the object. The fourth mission type is to deliver a message. In this case the robot can record a text or voice message and reproduce it at the destination point. The fifth option is Manual control, in which case the robot accepts to be manually driven to another location. When one of the first four options is chosen, the robot will let the user choose one of the known destinations, as shown in Fig. 9. The robot automatically determines which locations can be reached starting from its present localization, so that the list shown contains only places that can be reached through one of the known paths.

4 Vision-Based Navigation

Robot localization and navigation are based on visual memories, which are stored into the SDM during a learning stage and later used in the autonomous run mode.

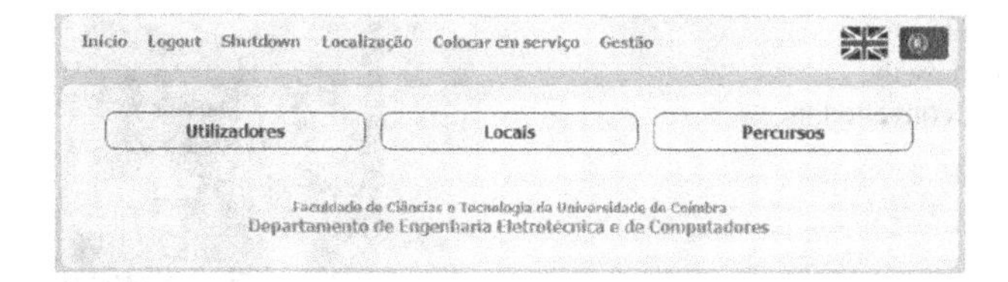

Fig. 7 Image of the administrator menu, where the logged administrator has the options to manage other users, places and paths

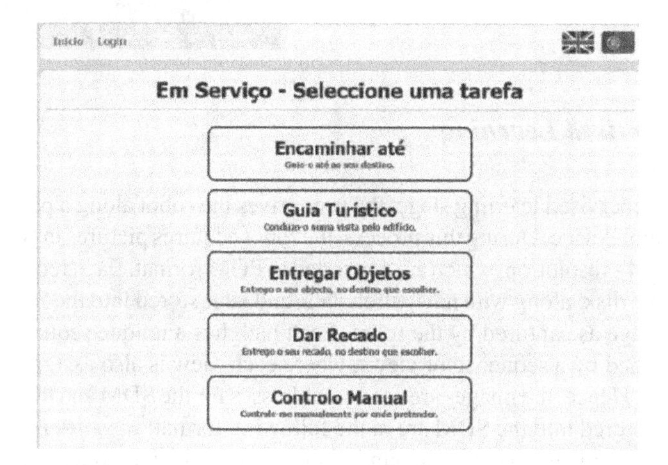

Fig. 8 Image of the missions menu, where the user can choose a type of mission. The missions available are (i) Guide to a place; (ii) Touristic tour; (iii) Deliver an object; (iv) Deliver a message; and (v) Manual control

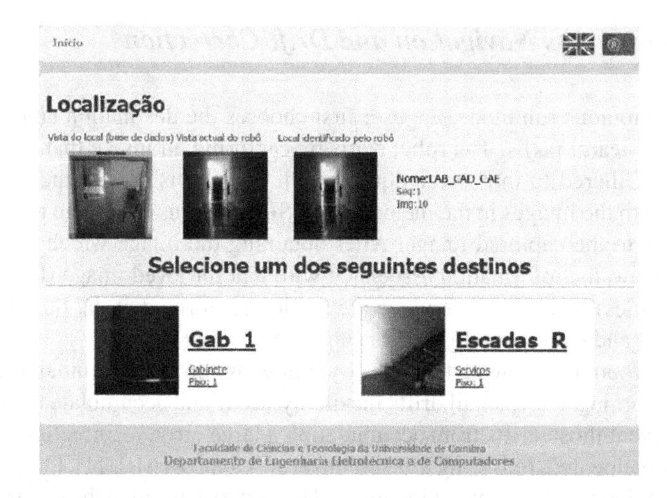

Fig. 9 Screen showing known paths, so that the user can choose where to go

Fig. 10 Example of a picture captured by the robot's camera, converted to PGM

4.1 Supervised Learning

During the supervised learning stage, the user drives the robot along a path using the mouse as input device. During this process, the robot acquires pictures in BMP format with 176×144 resolution, which are converted to PGM format. Each recorded image is saved in the disk, along with navigation data, and later stored into the SDM. Fig. 10 shows an image as captured by the robot. Each path has a unique sequence number and is described by a sequence of views, where each view is also assigned a unique view number. Hence, the images are used as addresses for the SDM and the navigation data vectors stored into the SDM are in the following format: $< seq\#, im\#, vr, vl >$ The sequence number is $seq\#$. The number of the image in the sequence is $im\#$. The velocities of the right and left wheels are, respectively, vr and vl.

4.2 Autonomous Navigation and Drift Correction

For the autonomous run mode, the user first chooses the destination among the set of previously learnt paths. The robot starts by capturing an image that is converted to PGM and filtered to improve its quality [7]. Then, it is sent to the SDM to be compared with the images in the memory. The SDM returns the image ranked as the most similar to the captured image. After obtaining the image which is closest to its current view, the information associated with that retrieved image (left and right wheel velocities) is used to reproduce the motion performed at the time the image was captured and stored during the learning stage.

Since the robot is following the paths by following the same commands executed during the learning stage, small drifts inevitably occur and accumulate over time. In order to prevent those drifts from accumulating a large error, a correction algorithm was also implemented, following Matsumoto et al.'s approach [6, 11]. Once an image has been predicted, a block-matching method is used to determine the horizontal drift between the robot's current view and the view retrieved from the memory which was stored during the learning stage. If a relevant drift is found, the robot's heading is adjusted by proportionally decreasing *vr* or *vl*, in order to compensate the lateral drift.

4.3 Sequence Disambiguation

During the autonomous run mode, the data associated with each image that is retrieved from the memory is checked to determine if the image belongs to the sequence (path) that is being followed. Under normal circumstances, the robot is not expected to skip from one sequence to another. Nonetheless, under exceptional circumstances it may happen that the robot is actually moved from one path to another, for unknown reasons. Possible reasons include slippage, manual displacement, mismatch of the original location, among many others. Such problem is commonly known as the "kidnapped robot," for it is like the robot is kidnapped from one point and abandoned at another point, which can be known or unknown.

To deal with the "kidnapped robot" problem and similar difficulties, the robot uses a short term memory of n entries (50 was used in the experiments). This short term memory is used to store up to n of the last sequence number S_j that were retrieved from the memory. If the robot is following sequence S_j, then S_j should always be the most popular in the short term memory. If an image from sequence S_k is retrieved, it is ignored, and another image is retrieved from the SDM, narrowing the search to just entries of sequence S_j. Nonetheless, S_k is still pushed onto the short term memory, and if at some point S_k becomes more popular in the short term memory than S_j, the robot's probable location will be updated to S_k. This disambiguation method showed to filter out many spurious predictions while still solving kidnapped robot-like problems.

4.4 Use of a Sliding Window

When the robot is following a path, it is also expected to retrieve images only within a limited range. For example, if it is at the middle of a long path, it is not expected to get back to the beginning or right to the end of the path. Therefore, the search space can be truncated to a moving "sliding window," within the sequence that is being followed. For a sequence containing a total of z images, using a sliding window of width w, the search space for the SDM image im_k is limited to images with image number in the interval $\{max(0, k - \frac{w}{2}), min(k + \frac{w}{2}, z)\}$. The sliding window in the SDM is implemented by truncating the search space, a method similar to Jaeckel's selected coordinate design [2]. In Jaeckel's method, coordinates which are deemed irrelevant to the final result are disregarded in the process of calculating the distance between the input address and each memory item, so computation can be many times faster. In the present implementation, however, the selected coordinates are used just to select a subset of the whole space. The subset is then used for computing the distance using all the coordinates.

The sliding window may prevent the robot from solving the kidnapped robot problem. To overcome the limitation, an all-memory search is performed first and the short-term memory retains whether the last n images were predicted within the sliding window or not, as described in section "Sequence Disambiguation". This means that the sliding window actually does not decrease the total search time, since an all-memory search is still required in order to solve the kidnapped robot problem. The sliding window, however, greatly reduces the number of Momentary Localisation Errors (MLE). A momentary localisation error is counted when the robot retrieves from the memory a wrong image, such as an image from a wrong sequence or from the wrong place in the same sequence. When a limited number of MLE occur the robot does not get lost, due to the use of the sliding window and the sequence disambiguation procedures, as described in [7].

5 Comparison of Navigation Algorithms

Different navigation algorithms were implemented and tested. The two most relevant of them are described in the following subsections: the simplest and the most robust.

5.1 *Basic Algorithm*

The first navigation algorithm is called "basic," for it performs just the simplest search and navigation tasks, as described in sections "Autonomous Navigation and Drift Correction"and "Sequence Disambiguation" above, and a very basic filtering technique to filter out possibly wrong predictions, described next.

Image search, for robot localisation, is performed in all the memory. Its performance is evaluated counting the number of MLEs. Detection of possible MLEs is based on the number of the image, balanced by the total size of the sequence. If the distance between image im_t, predicted at time t, and image $im_{t\pm1}$, predicted at time $t \pm 1$, is more than $\frac{1}{3}z$, for a path described by z images, $im_{t\pm1}$ is ignored and the robot continues performing the same motion it was doing before the prediction. The fraction $\frac{1}{3}z$ was empirically found for the basic algorithm.

The performance of this basic algorithm was tested indoors in the corridors of the Institute of Systems and Robotics of the University of Coimbra, Portugal. The robot was taught a path about 22 m long, from a laboratory to an office, and then instructed to follow that path 5 times. Figure 11 shows the sequence numbers of the images that were taught and retrieved each time. The robot never got lost and always reached a point very close to the goal point.

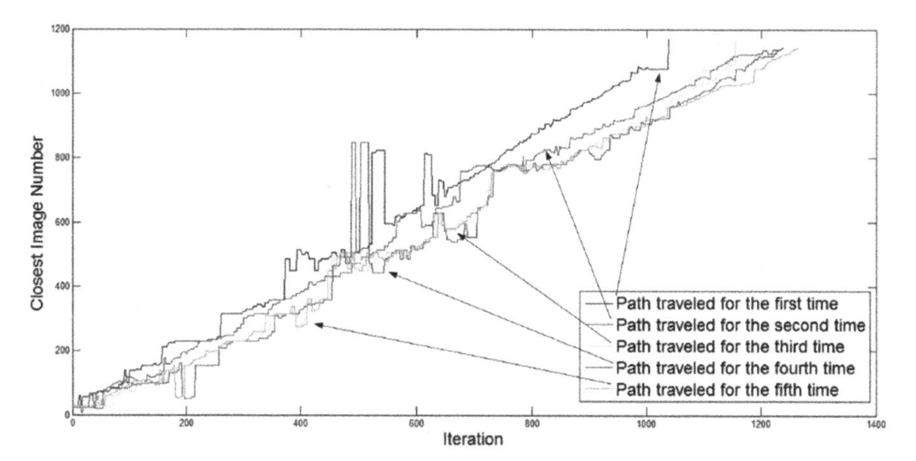

Fig. 11 Image sequence numbers of the images predicted by the SDM following the path from a laboratory to an office (approx. 22 m). The graph shows the number of the images that are retrieved from the memory as the robot progresses towards the goal

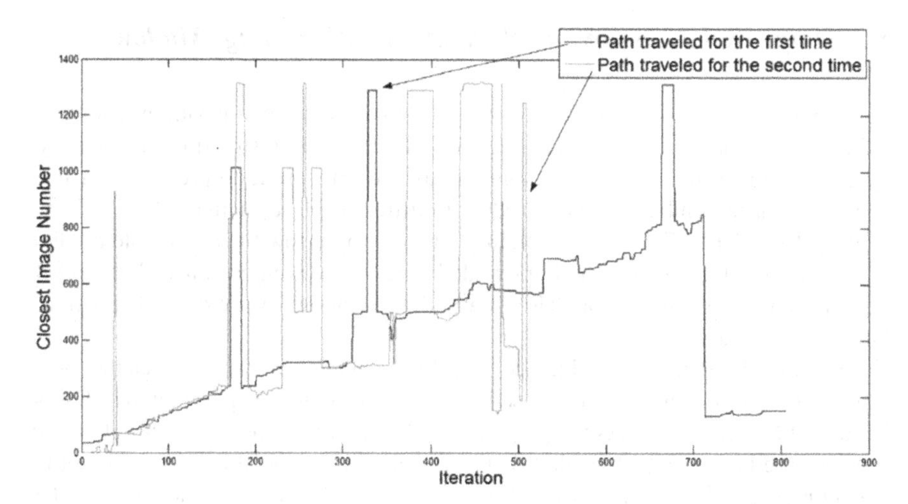

Fig. 12 Images predicted by the SDM following the second test path (approx. 47 m) using the basic algorithm

In a second test, the robot was taught a path about 47 m long. The results are shown in Fig. 12. As the figure shows, the robot was not able to reach the goal, it got lost at about the 810th prediction in the first run and at the 520th in the second run.

The results obtained in the second test show that the basic algorithm is not robust enough, at least for navigating in long and monotonous environments such as corridors in the interior of large buildings.

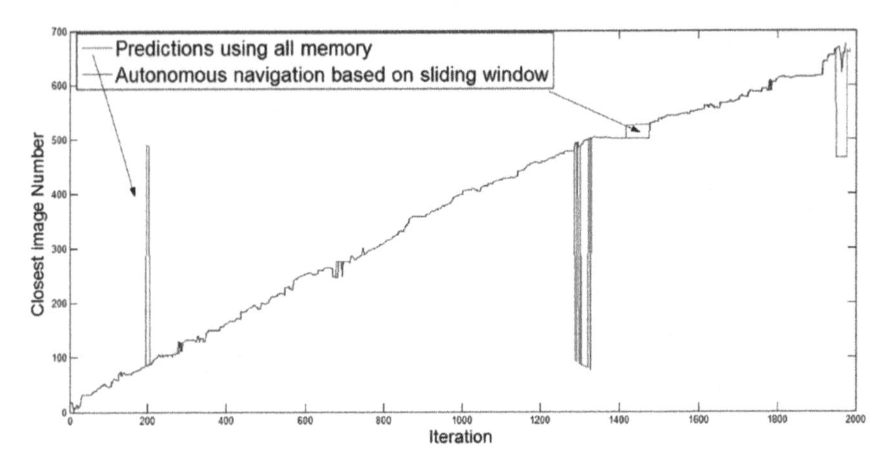

Fig. 13 Images predicted by the SDM following the second test path using the sliding window algorithm

5.2 Improved Autonomous Navigation with Sliding Window

Many of the prediction errors happen where the images are poor in patterns and there are many similar views in the same path or other paths also stored in the memory. Corridors, for example, are very monotonous and thus prone to prediction errors, because a large number of images taken in different places actually have a large degree of similarity. The use of a sliding window to narrow the acceptable predictions improves the results. The improved sliding window algorithm with the other principles described in section "Use of a Sliding Window" worked in all situations that it was tested.

Figure 13 shows the result obtained with this navigation algorithm when the robot was made to follow the same path used for Fig. 12 (second test path), with a sliding window 40 images wide. As the graph shows, the sliding window filters out all the spurious predictions which could otherwise compromise the ability of the robot to reach the goal. Close to iterations number 200 and 1300, some of the most similar images retrieved from the memory are out of the sliding window, but those images were not used for retrieving control information because they were out of the sliding window. This shows the use of the sliding window correctly filtered wrong predictions out and conferred stability to the navigation process.

6 Obstacle Avoidance

In order for the robot to navigate in real environments, it is necessary that the navigation process in autonomous mode is robust enough to detect and avoid possible obstacles in the way. Two algorithms were implemented, one for obstacles which appear in straight line paths and another for obstacles that appear in curves.

6.1 Obstacles in Straight Paths

In the autonomous navigation mode, the front sonar sensors are activated. All objects that are detected at less than 1 m from the robot are considered obstacles and trigger the obstacle avoidance algorithm. A median of 3 filter was implemented to filter out possible outliers in the sonar readings. When an obstacle is detected, the robot suspends memory-based navigation and changes direction to the side of the obstacle that seems more free. The robot chooses the side by reading the two lateral front sonar sensors. The side of the sensor that gives the higher distance to an object is considered the best side to go. If both sensors read less than 50 cm the robot stops to guarantee its safety. Providing the robot senses enough space, it starts circumventing the obstacle by the safest side. While circumventing, it logs the wheel movements in a stack. When the obstacle is no longer detected by the sensors, the wheel movements logged are then performed in reverse order emptying the stack. This process returns the robot to its original heading. When the stack is emptied, the robot tries to localize itself again based on visual memories and resume memory-based navigation. If the robot cannot localise itself after the stack is empty, then it assumes it is lost and navigation is stopped. In the future this behaviour may be improved to an active search method.

Figure 14 shows an example of a straight path previously taught and later followed with two obstacles placed in that path. After avoiding collision with the first obstacle, the robot resumes memory-based navigation maintaining its original heading, performing lateral drift correction for a while. It then detects and avoids the second obstacle and later resumes memory-based navigation maintaining its original head-

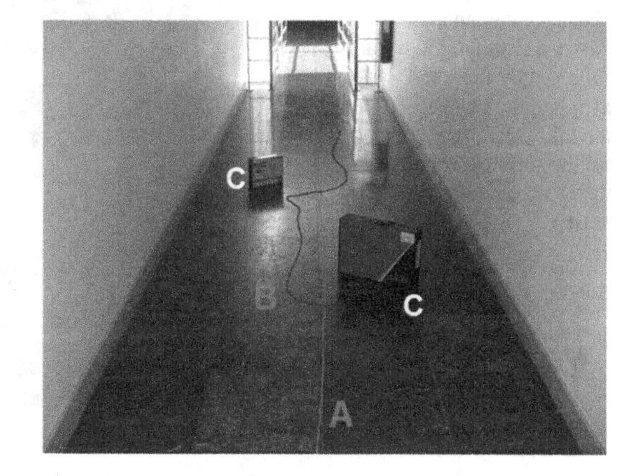

Fig. 14 Examples of obstacle avoidance in a straight path (the lines were drawn using a pen attached to the rear of the robot, so they actually mark the motion of its rear, not its centre of mass). **a** Path followed without obstacles. **b** Path followed with obstacles. **c** Obstacles

ing. Note that in the figure, because the lines were drawn using a pen attached to the rear of the robot, when the robot turns to the left it draws an arc of a line to the right.

6.2 Obstacles in Curves

The stack method described in the previous subsection works correctly if the obstacle is detected when the robot is navigating in a straight line. If the obstacle is detected while the robot is performing a curve, that method may not work, because the expected robot's heading after circumventing the obstacle cannot be determined in advance with a high degree of certainty. If an obstacle is detected when the robot is changing its heading, then the stack method is not used. The robot still circumvents the obstacle choosing the clearer side of the obstacle. But in that case it only keeps record of which side was chosen to circumvent the obstacle and what was the previous heading. Then, when the obstacle is no longer detected, the robot uses vision to localise itself and fine tune the drift using the algorithm described in section "Autonomous Navigation and Drift Correction". In curves, the probability of confusion of images is not very high, even if the images are captured at different distances. The heading of the camera often has a more important impact on the image than the distance to the objects. Therefore, after the robot has circumvented the obstacle it will have a very high probability of being close to the correct path and still at a point where it will be able to localise itself and determine the right direction to follow.

Figure 15 shows examples where the robot avoided obstacles placed in a curve. In path B (blue) the wall corner at the left is also detected as an obstacle, hence there is actually a second heading adjustment. The image shows the robot was still able to localise itself after the obstacle and proceed in the right direction, effectively getting back to the right path.

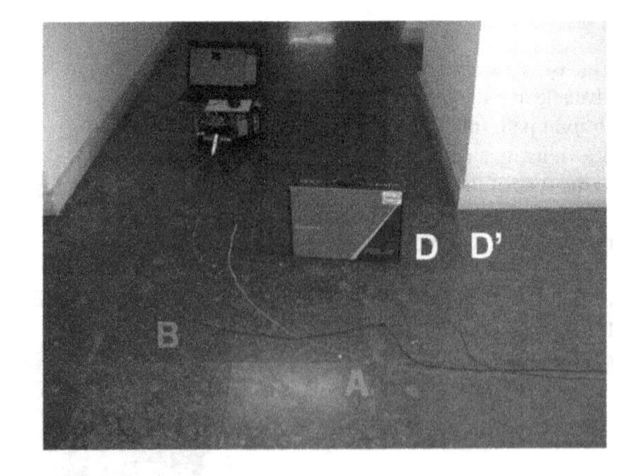

Fig. 15 Examples of obstacle avoidance in a curve. **a** Path taught. **b** Path followed by the robot avoiding the obstacle (**d**) and the left wall corner. **c** Path followed by the robot when the obstacle was positioned at (**d'**)

6.3 Summary

The experimental results show that the algorithms proposed can safely be used to navigate the robot, even in the presence of obstacles. If the robot finds obstacles it will either circumvent them or stop at safe distance to avoid collisions. The architecture proposed is easily scalable, so that the robot can perform the already implemented missions or other missions.

7 Conclusion

A service robot which uses vision to navigate and circumvent obstacles was described. The visual navigation memories are stored into an SDM with the retrieving process implemented in parallel in a GPU for better performance.

The use of a stack to store the robot motions when circumventing obstacles showed good performance in straight paths. For curves, after circumventing the obstacle the robot adjusts its heading faster using only memory-based navigation for localization and the drift-correction algorithm for heading adjustment.

Experimental results show the method was effective for navigation in corridors, even with corners.

The service robot currently can work as a cicerone, but the web-based interface and the architecture chosen are easily scalable, so that future work will include development of higher level modules to implement other missions and behaviours.

Acknowledgements The authors acknowledge Fundacão para a Ciência e a Tecnologia (FCT) and COMPETE 2020 program for the financial support to the project UID-EEA-00048-2013.

References

1. W. Burgard, D. Fox, G. Lakemeyer, D. Haehnel, D. Schulz, W. Steiner, S. Thrun, A. Cremers, Real robots for the real world—the rhino museum tour-guide project, in *Proceedings of the 1998 AAAI Spring Symposium* (1998)
2. L.A. Jaeckel, An alternative design for a sparse distributed memory, in *Technical Report* (Research Institute for Advanced Computer Science, NASA Ames Research Center, 1989)
3. P. Kanerva, *Sparse Distributed Memory* (MIT Press, Cambridge, 1988)
4. R. Kittmann, T. Fröhlich, J. Schäfer, U. Reiser, F. Weißhardt, A. Haug, Let me introduce myself: i am care-o-bot 4, a gentleman robot. Mensch und computer 2015–proceedings (2015)
5. S. Konstantopoulos, I. Androutsopoulos, H. Baltzakis, V. Karkaletsis, C. Matheson, A. Tegos, P. Trahanias, Indigo: Interaction with personality and dialogue enabled robots (2008)
6. Y. Matsumoto, K. Ikeda, M. Inaba, H. Inoue, Exploration and map acquisition for view-based navigation in corridor environment, in *Proceedings of the International Conference on Field and Service Robotics* (1999), pp. 341–346
7. M. Mendes, A.P. Coimbra, M.M. Crisóstomo, Improved vision-based robot navigation using a sdm and sliding window search. Eng. Lett. **19**(4) (2011)
8. M. Mendes, A.P. Coimbra, M.M. Crisóstomo, Intelligent robot navigation using view sequences and a sparse distributed memory. Paladyn J. Behav. Robot. **1**(4) (2011). https://doi.org/10.2478/s13230-011-0010-z

9. M. Mendes, A.P. Coimbra, M.M. Crisóstomo, Robot navigation based on view sequences stored in a sparse distributed memory. Robotica (2011)
10. M. Mendes, A.P. Coimbra, M.M. Crisóstomo, Circumventing obstacles for visual robot navigation using a stack of odometric data, in *2017 Lecture Notes in Engineering and Computer Science: Proceedings of The World Congress on Engineering*, London, UK (2017), pp. 172–177
11. M. Mendes, M.M. Crisóstomo, A.P. Coimbra, Robot navigation using a sparse distributed memory, in Proceedings of the IEEE International Conference on Robotics and Automation. Pasadena, California, USA (2008)
12. M. Mendes, M.M. Crisóstomo, A.P. Coimbra, Assessing a sparse distributed memory using different encoding methods, in *Proceedings of the World Congress on Engineering (WCE)* London, UK (2009)
13. R.P.N. Rao, D.H. Ballard, Object indexing using an iconic sparse distributed memory. Technical Report 559, The University of Rochester, Computer Science Department, Rochester, New York (1995)
14. B. Ratitch, D. Precup, Sparse distributed memories for on-line value-based reinforcement learning, in *ECML* (2004)
15. A. Rodrigues, A. Brandão, M. Mendes, A.P. Coimbra, F. Barros, M. Crisóstomo, Parallel implementation of a sdm for vision-based robot navigation, in *13th Spanish-Portuguese Conference on Electrical Engineering (13CHLIE)*, Valência, Spain (2013)
16. R. Tellez, F. Ferro, S. Garcia, E. Gomez, E. Jorge, D. Mora, D. Pinyol, J. Oliver, O. Torres, J. Velazquez, et al., Reem-b: an autonomous lightweight human-size humanoid robot, in 2008 8th IEEE-RAS International Conference on Humanoid Robots (IEEE, 2008), pp. 462–468

Identity and Enterprise Level Security

William R. Simpson and Kevin E. Foltz

Abstract Intrusions to enterprise computing systems have led to a formulation that put in place steel gates to prevent hostile entities from entering the enterprise domain. The current complexity level has made the fortress approach to security implemented throughout the defense, banking, and other high-trust industries unworkable. The alternative security approach, called Enterprise Level Security (ELS), is the result of a concentrated 15-year program of pilots and research. The primary identity credential for ELS is the PKI certificate, issued to the individual who is provided with a Personal Identity Verification (PIV) card with a hardware chip for storing the private key. This process provides a high enough identity assurance to proceed. However, in some instances the PIV card is not available or in a compromised position and a compatible approach at a higher level of assurance is needed. This chapter discusses a multi-level authentication approach designed to satisfy the level of identity assurance specified by the data owner, add assurance to derived credentials, and to be compatible with the ELS approach for security.

Keywords Assurance · Authentication · Enterprise level security · Identity
Multi-factor authentication · Personal identification verification

1 Introduction

Adversaries continue to penetrate, and in many cases, already exist within, our network perimeter, i.e., they have infiltrated the online environment, jeopardizing the confidentiality, integrity, and availability of enterprise information systems. The fortress model—hard on the outside, soft on the inside—assumes that the boundary can prevent all types of penetration [1], but this assumption has been proven wrong

W. R. Simpson (✉) · K. E. Foltz
Institute for Defense Analyses, 4850 Mark Center Drive, Alexandria, VA 22311, USA
e-mail: rsimpson@ida.org

K. E. Foltz
e-mail: kfoltz@ida.org

S.-I. Ao et al. (eds.), *Transactions on Engineering Technologies*,
https://doi.org/10.1007/978-981-13-0746-1_19

by a multitude of reported network-related incidents. The previous statements are no longer controversial but a wise assumption for data and information security practitioners. Network attacks are pervasive, and nefarious code is present even in the face of system sweeps to discover and clean readily apparent malware. The focus of this chapter is on the security aspects of countering existing known and unknown threats based on robust identity and access management (IdAM) and on how this access control system can dynamically support mission information requirements. A working prototype has been developed and evaluated for security, functionality, and scaling issues.

Enterprise Level Security (ELS) is a capability designed to counter adversarial threats by protecting applications and data with a dynamic claims-based access control (CBAC) solution. ELS helps provide a high-assurance environment in which information can be generated, exchanged, processed, and used. It is important to note that the ELS design is based on a set of high-level tenets that are the overarching guidance for every decision made, from protocol selection to product configuration and use [2]. From there, a set of enterprise-level requirements are formulated that conforms to the tenets and any high-level guidance, policies, and requirements.

The basic tenets, used at the outset of the ELS security model, are the following:

0. Malicious entities are present	8. Need-to-share as overriding need-to-know
1. Simplicity	9. Separation of function
2. Extensibility	10. Reliability
3. Information hiding	11. Trust but verify (and validate)
4. Accountability	12. Minimum attack surface
5. Specify Minimal detail	13. Handle exceptions and errors
6. Service-driven rather than a product-driven solution	14. Use proven solutions
7. Lines of authority should be preserved	15. Do not repeat old mistakes

The current paper-laden access control processes for an enterprise operation are plagued with ineffectiveness and inefficiencies. In a number of enterprises, tens of thousands of personnel transfer locations and duties annually, which on a daily basis introduce delays and security vulnerabilities into their operations. ELS mitigates security risks while eliminating much of the system administration required to manually grant and remove user/group permissions to specific applications/systems. Early calculations show that for government and defense, 90–95% of recurring man-hours will be saved and up to 3 weeks in delay for access request processing will be eliminated by ELS-enabled applications [3]. The fortress architecture assumes that threats are stopped at the front gates, but ELS does not accept this precondition and is designed to mitigate many of the primary vulnerability points at the application using a distributed security architecture, shown in Fig. 1. This chapter is based in part on a paper published by WCE 2017 [4].

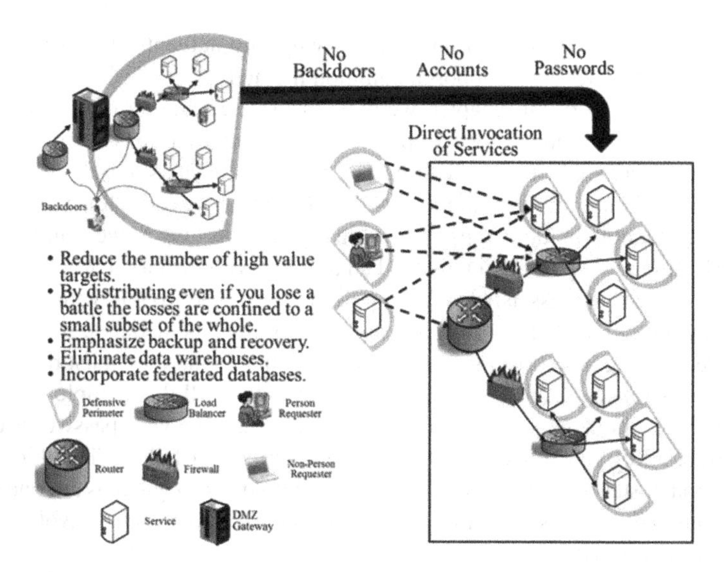

Fig. 1 Distributed security architecture

2 Background

When intercommunications between computers began, there were no security concerns, just a bunch of academics experimenting. Making this happen, we found it useful on many levels as did the hackers. Once organized into enterprises, real resources were at risk.

Network operating systems generally require that a user be authenticated in order to log onto the network [5]. This can be done by entering a password, inserting a smart card and entering the associated PIN, proximity of a nearfield device, providing a biometric verification or using some other means to prove to the system that you are who you claim to be. The network may provide an identity token that provides identity to applications and providers of service on the network. This token (sometimes called single-sign-on (SSO)), is vulnerable and subject to theft or forgery and may be replayed for nefarious activity. When logging on to a web site, you are not generally aware of whether you are logging into an application or a network because the magic is all behind the scenes. With ELS, the network is the medium and authentication is to the application or provider of services. This of course, must be bi-lateral so that both entities have an assurance of their communicating partner's identity.

For some networks, a message integrity process is invoked called Internet Protocol Security (IPSec). IPSec transmissions can use a variety of authentication methods, including ticket or token based approaches, public key certificates issued by a trusted certificate authority (CA), or a simple pre-shared secret key (a string of characters known to both the sender and the recipient). This type of authentication assumes the fortress model previously discussed. As with any authentication process, the requester and provider must support a common method.

For the remainder of this chapter, we will discuss the requester/provider authentication process. The earliest form of protection was the use of passwords. We have gotten much more sophisticated with passwords over the years, regularly changing them, making them complex, and not using the same passwords for multiple purposes. As an identity method, these passwords were ok as long as we could keep the secret. The username/password unlocked an account with the target and the account had my privileges and assets, etc. Passwords are easy. The trouble with passwords is that they are not safe. They must be strong, they must be updated, they must be kept secret by several entities, they must be transmitted, and they must be stored at multiple sites for logistical as well as security reasons. All this complexity means you either use simple passwords over and over, or you write your passwords down, or you trust all of your passwords to some single point of failure. The thieves are getting good at stealing them. Passwords falling into the wrong hands are one of the biggest causes of network vulnerability: 63% of known data breaches involved weak or stolen passwords [6].

Attempts to resolve the password issues include extended passwords [7] by using special characters or adding additional factors to the password. Servers also store hashes of passwords instead of the raw passwords. To prevent someone with this hash list from breaking the passwords through brute-force hash trials, a common practice is the use of salt (random data that is used as an additional input to a one-way function that "hashes" a password or passphrase). Salt forces an attacker to guess one password at a time instead of all at once. The salt process is an attempt to thwart a password cracking method using rainbow tables [8]. Passwords and Accounts are part of the fortress approach where we constantly renew the mantra of adding more and better software to sort and identify and deny the unwanted from out enterprises. It turns out that the strong password is still a weak identity credential.

Attempts to thwart programs from trying to "guess" passwords include two or three-factor authentication (discussed later in this chapter) and the captcha [9] which is supposed to be a test (visual, sound or other) that humans can easily pass but programs have difficulties passing. Most of these tests have been cracked [10]. If the enterprise uses its knowledge of the user to supplement the password with secondary factors the strength of authentication is greatly increased, but asking simple questions does not appear to be the solution. Many of these questions may be answered with modest research [11]. Better is the out-of-band query, where a message is sent to a phone number or e-mail, and a correct response indicates that the presenter of the password at least has the out-of-band device [12]. Mobile devices have recently begun to use the camera function to provide biometric authentication to the device. These devices use a fingerprint, face print or iris scan to verify that the person who is in control of the device is the one registered to that device [13]. At least authentication to the device does not have to be transmitted and the digital representation is less subject to theft. The jury is still out on these attempts to increase security [14].

A strong credential would be one where you alone keep the secret locked away where only you can get to it, and you identify yourself by proving that you have this secret locked away (not necessarily producing the secret).

One identity credential that partially satisfies these criteria is the One-Time Password (OTP) [15]. OTP is the provision of a single-use password that is provided at the time of use.

Simple forms of the OTP include distributed lists (where a sequential list of user passwords are provided to the user—bookkeeping is a bit tedious here, and the lists may be stolen or intercepted).

Somewhat more sophisticated forms include algorithmically produced codes (usually based upon some shared values between the requester and the provider). These suffer from control of secrecy issues in both the shared values and the algorithm.

A more satisfactory OTP solution would include hardware provision of password generators tied to the user and synchronization between the user and the target of communication. Some Personal Identification Verification (PIV) (see next paragraph) include OTP generators included in their functionality. These have the ability to be registered, verified and revoked if lost, or stolen. These suffer from needing to be transmitted, making them available to a Man-in-the-Middle, and algorithm cracking or theft. Theft is less of a concern since the password is one time use. Algorithm theft is another issue. This latter has already been accomplished for one provider [16, 17].

Another Identity credential that meets this criterion is the PIV card [18] or equivalent. This is the preferred credential for ELS. The PIV card uses PKI credentials and has a public certificate, issued by a recognized certificate issuing authority, with a public key and a private key locked into tamper-proof hardware. Identity is established when the holder of the PIV card can decrypt a message encrypted by his/her public key. The proof is called Holder-of-Key (HOK). If the user or the card is compromised, the certificate may be revoked [19]. This does not solve all identity problems; it does provide a strong credential that is more difficult to steal.

3 Enterprise Level Security

The ELS design addresses five security principles that are derived from the basic tenets:

- Know the Players—this is done by enforcing bi-lateral end-to-end authentication;
- Maintain Confidentiality—this entails end-to-end unbroken encryption (no in-transit decryption/payload inspection);
- Separate Access and Privilege from Identity—this is done by an authorization credential;
- Maintain Integrity—know that you received exactly what was sent;
- Require Explicit Accountability—monitor and log transactions.

By abiding with the tenets and principles discussed above, ELS allows users access without accounts by computing targeted claims for enterprise applications (using enterprise attribute stores and asset-owner-defined claims for access and privilege). ELS has been shown to be a viable, scalable alternative to current access control schemas [5]. A complete description of ELS basics is provided in [19].

4 Identity Issues

Identity in the enterprise is a unique representation of an entity. For users, it begins with the human resources who maintain their files. The assigned identity is called the Distinguished Name (DN) and it must be unique over space and time. There may be five John Smiths in the enterprise, but only one John.Smith2534, UID = Finance, HID = Chicago. These and PKI information are normally encoded into a Personal Identity Verification (PIV) card for network access and provided to the entity for its use. Certain pieces of the information may be tagged as verifiable by DN for identity purposes, such as wife's middle name.

There is a need to allow users without PIVs some degree of access based on alternative authentication methods. PIVs may not be available to all, but also, the user device may not be capable of reading and using a PIV. Additional use cases include lost PIV, waiting for issuance of a PIV, or a user being unable to get a PIV compatible with the ELS certificate authority trust. Additionally, there are federation partners, contractors, and other vetted external individuals with short-term needs.

Each application ultimately decides what kind of authentication is strong enough (through a registration process with Enterprise Attribute Ecosystem (EAE)).

The creation of a non-PIV identity comprises three separate stages. The first stage is creation of a proposed identity. This value is provided by the user. The goal is to correlate this with the enterprise files. It may be an email, a common name, or simply a name. The second stage is creation of a candidate identity (starting point for identity determination), in which the proposed identity is paired with an enterprise identity, and a DN is determined. As we will discuss, the process also takes steps to verify that the pairing between the proposed identity and the DN is owned by the individual making the request. The last stage is creation of the assured identity. The candidate identity becomes the assured identity when enough correlated information and personal verification about the candidate identity has a sufficient level of pairing with the enterprise identity that it can be trusted with access to an application using his/her claims that have been computed for his/her use.

5 Scale of Identity Assurance

If you search the literature for multi-factor authentication, you will find a predominance of processes based upon account-based systems and starting with username—password [20–30]. These systems intertwine the security issues of authentication and

Table 1 Multifactor authentication identity assurance

Method	Comment—strength	Id assurance
1. AUTHN hard	Standard ELS—strong	0.80
2. ATHN soft	Closest to ELS	0.70
3. OOB	A start—minimal	0.25
4. OOB Bio	Solid	0.50
5. OOB Bio + 1mf.	Strong	0.80
6. OOB + 1mf	Moderate	0.60
7. OOB + 3mf	Strong	0.70
Greater than Normal ID Assurance directed by Web Application		
8. Hard token +	Very strong	0.85
9. Hard token ++	Very strong	0.90
10. Hard token +++	Highest value	0.95

authorization. In fact, the popular definition of multi-factor authentication merges the two:

"Multi-factor authentication (MFA) is a method of computer access control in which a user is only granted access after successfully presenting several separate pieces of evidence to an authentication mechanism—typically at least two of the following categories: knowledge (something they know); possession (something they have), and inherence (something they are)." [31].

ELS separates the identity and access/privilege security issues. Thus there are no accounts and no usernames with passwords. Further, ELS uses no proxies and limits access to the enterprise attribute system, thus reducing the threat surface.

Each data owner will decide what the requirements for access and privilege to their data are, and this includes the level of assurance that is acceptable. ELS represents a strong identity assurance and will be assigned a value of 0.80 (values are arbitrary and subject to revision). It is assumed that if the data owner wishes strong identity assurance he will specify 0.70 or 0.75 as the identity assurance value (from the collection below, the value of 0.75 requires biometric information in the absence of PIV). This will allow all enterprise users with a PIV to actually present access and privilege claims to the application. The lowest level of identity assurance would come from self-assertion; however, we will require several additional factors for this minimum, including a presence in the enterprise catalog, verification by an out-of-band (OOB—phone or e-mail) method; and of course for authorization, claims must be available for the individual. This lowest level will be described as User Asserted Identity with OOB verification and assigned a value of 0.2, which should also be the minimum specified by a data owner. A total of seven identity cases were developed, as follows, with strengths shown in Table 1:

1. Bi-lateral AUTHN (Hard Token)—AUTHN Hard
2. Bi-lateral AUTHN (prior issued Soft Token) in protected store—ATHN Soft
3. User Asserted Identity with Out-of-Band (OOB) verification—OOB

4. User Asserted Identity with OOB verification and with any Biometric factor—OOB Bio
5. User Asserted Identity with OOB verification and with any Biometric factor and with any non-biometric multi-factor verification—OOB Bio + 1mf.
6. User Asserted Identity with OOB verification and with any non-biometric multi-factor verification—OOB + 1mf
7. User Asserted Identity with OOB verification and with three non-biometric multi-factor verifications—OOB + 3mf.
   Enhanced Identity Assurance:
8. Hard token plus one non-biometric multi-factor verification—Hard token + 1mf
9. Hard token plus one biometrics authentication—Hard token + 1bio
10. Hard token plus one biometric and one non-biometric multifactor verification—Hard token +1bio + 1mf.

6 A Token Server with Certificate Authority

In order to preserve the ELS paradigm, a temporary soft certificate needs to be provided and the user claims must be provided with a SAML credential through TLS. The user needs to be in the attribute system with claims for services sought.

6.1 Non PIV STS/CA Issued X.509

Non-PIV owners go to a security token server with certificate issuance authority (STS/CA) and provide a proposed identity. This may be email or full name, etc. The STS/CA calls a service that scans the Enterprise Attribute Store (EAS) and rejects any identity that it cannot find in EAS. The STS/CA then confirms that the requester is not an automated system (via Captcha, etc.). This avoids a number of threat vulnerabilities. The STS/CA then asks questions of the non-PIV user to resolve ambiguity (if present). For example, there are five Jon Smiths in the enterprise, but only one works in Finance. The STS/CA then establishes the DN. To this point, the identity is still a proposed identity. The STS/CA saves the DN attributes in separate temporary store and sets up a server side TLS. The next step is a requirement, and non-PIV users must maintain an OOB contact for this. This OOB (one or more) is provided to the human resources for inclusion in the user's enterprise data. The token server resolves OOB (email, phone voice, phone text, etc.) communication methods for DN. We note that OOB means not on the network, and if the enterprise desk phone is part of the enterprise network, it does not work as OOB. Anyone without at least one OOB is rejected.

At this point the token server sends a one-time token (10 min or less life) to the OOB and requests input. No input or improper input will be rejected. A successful exchange results in the identity moving to a candidate identity.

The STS/CA will attempt to identify if the user is using a managed device (looking for bio capability like face or fingerprints). The STS/CA retrieves the claims from the enterprise claims store for the established DN, presents a choice from among the services the user has claims to, and asks for a selection. This establishes the application for later SAML transmission. The STS/CA chooses the maximum and minimum identity assurance needed for claims. The minimum identity assurance may not be achievable with the device, and a polite rejection is issued if so. Otherwise, the token server begins a multifactor verification, including biological, if applicable. Any multi-level failure leads to exit. If the multi-factor maximum achievable authentication for the identity assurance is successful, the identity becomes an assured identity. The STS/CA then creates and issues a temporary certificate, in the name of the assured identity DN, and sends this certificate and separately the private key to a specially configured application on the user's device for installation. The temporary certificate contains the identity assurance and has a life of 90 min or less. Comments in the temporary certificate specify the assurance level and the method for the application's use as appropriate. The temporary certificate may be reused for the life of the certificate by selecting any application (this will go to the normal STS for claims).

When the user selects an application, the token server posts a SAML through the browser to the application. The SAML is specifically for the audience (selected application). The temporary certificate is used for authentication to the application, and all else works as with normal ELS for an application. The interaction between the STS/CA and the attribute system is shown in Fig. 2.

6.2 *PIV Usage of the STS/CA*

A PIV user may be redirected to the STS/CA when the identity assurance requirement for the web application exceeds 0.80. The post will include the identity assurance value of the user (0.80), the identity assurance value sought, and the audience for the multi-factor authentication. The STS will use the user's PIV to authenticate, and the STS/CA will try to increase the identity assurance to the level sought by the application using the methods shown in Table 1.

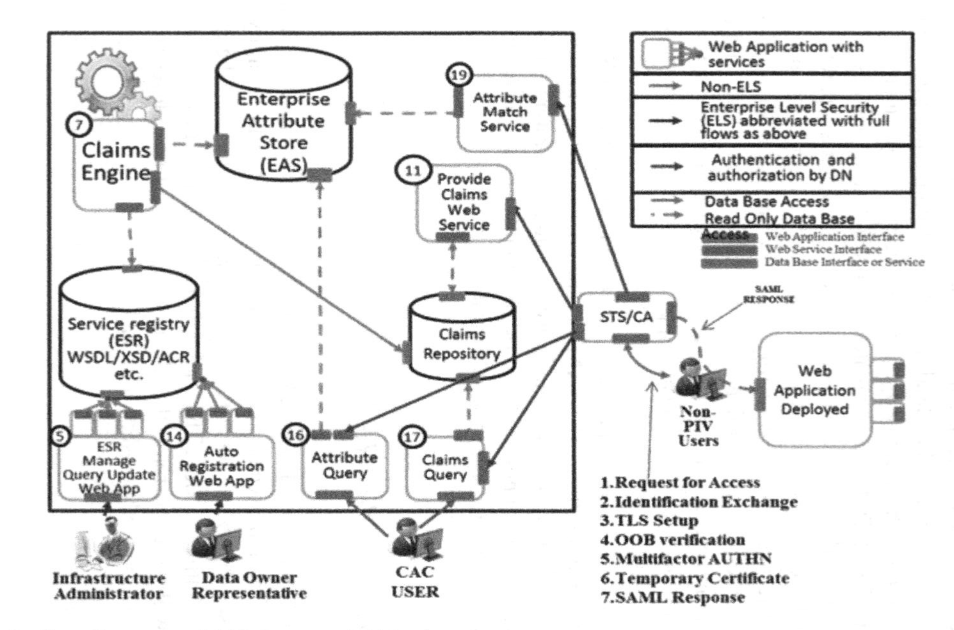

Fig. 2 Partial Enterprise Attribute Ecosystem (EAE) for extended identity assurance

7 Required Additional Elements

From an ELS standpoint, accommodation of non-PIV users adds the following requirements:

- Data Owners must specify the level of assurance on applications when specifying requirements for access and privilege in the enterprise service registry.
- STS/CA for non-PIV Users needs to be developed.
- An additional service must be placed in the EAE for comparison of attributes in DN retrieval.
- STS/CA must have full crypto and key management capability (generating asymmetric key pairs).
- Device software is needed to install temporary certificates on the end user device.
- The application must recognize temporary certificates generated by the STS/CA (STS/CA must be placed in the trust store).
- The application must recognize SAML certificates provided by the STS/CA.
- The application must check signatures and timestamp, but there is no need for revocation checking of the temporary certificate.

7.1 Advantages of the New Additions

- The derived process in this chapter is not username/password—there are no accounts and no storage of user data at individual applications.
- The process will handle retirees, contractors, and temporary employees if they are included in EAS.
- The process will handle missing or forgotten PIV cards.
- Since DN is in EAS claims are computed for each DN in the enterprise stores.
- Claims may be from Delegation (recommend non-PIV cannot delegate)
- All of the ELS software and handlers work without modification.
- The EAS has same attack surface as before.
- Temporary certificates expire out of system quickly.

7.2 However, the Following Disadvantages Are Noted

- Only covers person entities (not for Non-Person Entities (NPE)—but an adaption may be possible for NPEs).
- Extraction and sharing of temporary credentials—short duration is a mitigation.
- Manipulation of identities is possible (OOB requires the threat to have an OOB device in EAS that is really not part of the network).
- The threat's ability to initiate exchange with STS/CA (takes on all comers—reconnaissance by threat entities is facilitated under these circumstances).

- Intercept of temporary credentials (transmission is in TLS—some mitigation).
- On-device recovery of temporary credential (short duration provides mitigation).
- Credential forging (signatures and timeouts are some mitigations).
- The current identity assurance process treats all biometric identifications the same. For future versions, we may wish to distinguish between the types of biometric.
- The current identity assurance process treats all multi-factor queries as the same. For future versions, we may wish to distinguish between the types of multi-factor queries.

8 Conclusion and Future Work

We have reviewed the identity issues in a high-assurance security system. We have also described an approach that relies on high-assurance architectures and the protection elements they provide through PKI. The basic approach becomes compromised when identity is not verified by a strong credential for unique identification (such as holder-of-key in a PKI). The PKI usage is so fundamental to this approach that we have provided non-certificated users a way to obtain a temporary PKI certificate based on their enterprise need and the level of identity assurance needed to provide access and privilege to applications. The process is fully compatible with ELS and works as a complement to existing infrastructure. This work is part of a body of work for high-assurance enterprise computing using web services. Elements of this work are described in [19, 30–43].

Acknowledgements This work was supported in part by the U.S. Secretary of the Air Force and The Institute for Defense Analyses (IDA). The publication of this chapter does not indicate endorsement by any organization in the Department of Defense or IDA, nor should the contents be construed as reflecting the official position of these organizations.

References

1. F. Konieczny, E. Trias, N. Taylor, SEADE: countering the futility of network security. Air Space Power J. **29**(5), 4 (2015)
2. Technical Profiles for the Consolidated Enterprise IT Baseline, release 3.0. Available at (CAC required) (currently working 4.0): https://intelshare.intelink.gov/sites/afceit/TB
3. Email from Rudy Rihani, Project Manager, Accenture Corporation, dated 6 Mar 2016, Subject: "manpower savings with ELS."
4. W.R. Simpson, K.E. Foltz, Assured identity for enterprise level security, in *Proceedings of the World Congress on Engineering 2017*, 5–7 July 2017, London, U.K. Lecture Notes in Engineering and Computer Science, pp. 440–445
5. Email from Michael Leonard, MITRE Organization on behalf of USAF AFMC ESC/HNCDDD, dated 10 May 2012, Subject: "Performance/Scalability."
6. TechRepublic, McAfee, Understanding and selecting authentication methods, https://www.techrepublic.com/article/understanding-and-selecting-authentication-methods/. Accessed 27 Nov 2017

7. Verizon Communications, Verizon 2016 Data Breach Investigations Report, http://www.verizonenterprise.com/resources/reports/rp_DBIR_2016_Report_en_xg.pdf. Accessed 22 Nov 2017
8. Open Web Application Security Project (OWASP) Foundation, https://www.owasp.org/index.php/Password_special_characters, Apr 2013. Accessed 23 Nov 2017
9. Learn Cryptography, Password salting, https://learncryptography.com/hash-functions/password-salting, copyright 2017. Accessed 23 Nov 2017
10. StackExchange, Information Security, 2-Factor Authentication vs Security Questions, https://security.stackexchange.com/questions/96884/2-factor-authentication-vs-security-questions. Accessed 23 Nov 2017
11. House, Margaret, TechTarget, CAPTCHA (Completely Automated Public Turing Test to Tell Computers and Humans Apart), http://searchsecurity.techtarget.com/definition/CAPTCHA, August 2017. Accessed 23 Nov 2017
12. A. Griffin, UK Independent News, Google Kills off the Captcha…, http://www.independent.co.uk/life-style/gadgets-and-tech/news/google-captcha-re-robot-image-recognition-artificial-intelligence-website-a7627331.html, Mar 2017. Accessed 23 Nov 2017
13. K. Hickey, GCN Magazine, Biometric authentication growing for mobile devices, but security needs work, https://gcn.com/articles/2016/12/07/biometrics-maturity.aspx?admgarea=TC_Mobile, Dec 2016. Accessed 23 Nov 2017
14. L.M. Mayron, Arizona State University, Biometric authentication on mobile devices. IEEE Secur. Priv. **13**(3) (2015)
15. IBM Corporation, Upgrade Your Security with Mobile Multi-Factor Authentication, https://www-01.ibm.com/common/ssi/cgi-bin/ssialias?htmlfid=WGW03242USEN&. Accessed 23 Nov 2017
16. Gemalto, One Time Password (OTP), https://www.gemalto.com/companyinfo/digital-security/techno/otp. Accessed 23 Nov 2017
17. D. Goodin, Security Editor at Ars Technica, RSA SecurID software token cloning, https://arstechnica.com/information-technology/2012/05/rsa-securid-software-token-cloning-attack/, May 2012. Accessed 23 Nov 2017
18. National Institute of Technology and Standards, Computer Security Division, Applied Cybersecurity Division, Best Practices for Privileged User PIV Authentication, 21 Apr 2016, https://csrc.nist.gov/publications/detail/white-paper/2016/04/21/best-practices-for-privileged-user-piv-authentication/final. Accessed 22 Nov 2017
19. W.R. Simpson, CRC Press, *Enterprise Level Security—Securing Information Systems in an Uncertain World* (Auerbach Publications), ISBN 9781498764452, May 2016, 397 pp.
20. A.P. Sabzevar, A. Stavrou, Universal multi-factor authentication using graphical passwords, in *IEEE International Conference on Signal Image Technology and Internet Based Systems, 2008. SITIS '08*. (IEEE, 2008)
21. W. Gordon, Two-Factor Authentication: The Big List Of Everywhere You Should Enable It Right Now (3 Sept 2012), LifeHacker, Australia. Retrieved 1 Nov 2012
22. L. Lamport, Password authentication with insecure communication. Commun. ACM **24**(11), 770–772 (1981)
23. D.T. Bauckman, N.P. Johnson, D.J. Robertson, Multi-Factor Authentication, U.S. Patent No. 20, 130, 055, 368, 28 Feb 2013
24. A. Bhargav-Spantzel et al., Privacy preserving multifactor authentication with biometrics. J. Comput. Secur. **15**(5), 529–560 (2007)
25. F. Aloul, S. Zahidi, W. El-Hajj, Two factor authentication using mobile phones, in *AICCSA 2009, IEEE/ACS International Conference on Computer Systems and Applications, 2009* (IEEE, 2009)
26. S. Bruce, The Failure of Two-Factor Authentication, Mar 2005. https://www.schneier.com/blog/archives/2012/02/the_failure_of_2.html
27. M. Alzomai, B. AlFayyadh, A. Josang, Display security for online transactions: SMS-based authentication scheme, in *2010 International Conference on Internet Technology and Secured Transactions (ICITST)*

28. J.-C. Liou, S. Bhashyam, A feasible and cost effective two-factor authentication for online transactions, in *2010 2nd International Conference on Software Engineering and Data Mining (SEDM)* (IEEE, 2010)
29. Multi-factor authentication—Wikipedia, the free encyclopedia, https://en.wikipedia.org/wiki/Multi-factor_authentication
30. W.R. Simpson, C. Chandersekaran, A. Trice, A persona-based framework for flexible delegation and least privilege, in *Electronic Digest of the 2008 System and Software Technology Conference*, Las Vegas, Nevada, May 2008
31. W.R. Simpson, C. Chandersekaran, A. Trice, Cross-domain solutions in an era of information sharing, in *The 1st International Multi-Conference on Engineering and Technological Innovation: IMET2008*, vol I, Orlando, FL, June 2008, pp. 313–318
32. C. Chandersekaran, W.R. Simpson, The case for bi-lateral end-to-end strong authentication, in *World Wide Web Consortium (W3C) Workshop on Security Models for Device APIs*, 4 pp., London, England, Dec 2008
33. W.R. Simpson, C. Chandersekaran, Information sharing and federation, in *The 2nd International Multi-Conference on Engineering and Technological Innovation: IMETI2009*, vol. I, Orlando, FL, July 2009, pp. 300–305
34. C. Chandersekaran, W.R. Simpson, A SAML framework for delegation, attribution and least privilege, in *The 3rd International Multi-Conference on Engineering and Technological Innovation: IMETI2010*, vol. 2, Orlando, FL, July 2010, pp. 303–308
35. W.R. Simpson, C. Chandersekaran, Use case based access control, in *The 3rd International Multi-Conference on Engineering and Technological Innovation: IMETI2010*, vol. 2, Orlando, FL, July 2010, pp. 297–302
36. C. Chandersekaran, W.R. Simpson, A model for delegation based on authentication and authorization, in *The First International Conference on Computer Science and Information Technology (CCSIT-2011)*. Lecture Notes in Computer Science (Springer, Berlin, Heidelberg), 20 pp.
37. W.R. Simpson, C. Chandersekaran, An agent based monitoring system for web services, in *The 16th International Command and Control Research and Technology Symposium: CCT2011*, vol. II, Orlando, FL, Apr 2011, pp. 84–89
38. W.R. Simpson, C. Chandersekaran, An agent-based web-services monitoring system. Int. J. Comput. Technol. Appl. (IJCTA) **2**(9), 675–685 (2011)
39. W.R. Simpson, C. Chandersekaran, R. Wagner, High assurance challenges for cloud computing, in *Proceedings World Congress on Engineering and Computer Science 2011, WCECS 2011*, San Francisco, USA, 19–21 Oct 2011. Lecture Notes in Engineering and Computer Science, pp. 61–66
40. C. Chandersekaran, W.R. Simpson, Claims-based enterprise-wide access control, in *Proceedings World Congress on Engineering 2012, WCE 2012*, London, U. K., 4–6 July 2012. Lecture Notes in Engineering and Computer Science, pp. 524–529
41. W.R. Simpson, C. Chandersekaran, Assured content delivery in the enterprise, in *Proceedings World Congress on Engineering 2012, WCE 2012*, London, U. K., 4–6 July 2012. Lecture Notes in Engineering and Computer Science, pp. 555–560
42. W.R. Simpson, C. Chandersekaran, Enterprise high assurance scale-up, in *Proceedings World Congress on Engineering and Computer Science 2012, WCECS 2012*, San Francisco, USA, 24–26 Oct 2012. Lecture Notes in Engineering and Computer Science, pp. 54–59
43. C. Chandersekaran, W.R. Simpson, A uniform claims-based access control for the enterprise. Int. J. Sci. Comput. **6**(2), 1–23 (2012). ISSN: 0973-578X

Comprehensive Study for a Rail Power Conditioner Based on a Single–Phase Full–Bridge Back–to–Back Indirect Modular Multilevel Converter

Mohamed Tanta, José A. Afonso, António P. Martins, Adriano S. Carvalho and João L. Afonso

Abstract This chapter presents a rail power conditioner (RPC) system based on an indirect AC/DC/AC modular multilevel converter (MMC) where a V/V power transformer is used to feed the main catenary line and the locomotives. The proposed control strategy for this system has been introduced to guarantee a good compensating performance of negative sequence currents (NSCs) and harmonics on the public grid side. This control strategy has also the ability to achieve balanced and equal voltage between the MMC's submodules (SMs) capacitors. Simulation results for this RPC based on an indirect MMC are presented in this chapter to show the main advantages of using this topology. The results show how the proposed system is able to compensate NSCs and harmonics on the public grid side when the V/V power transformer feeds two unequal load sections.

Keywords Current harmonics · Electric locomotives · Modular multilevel converter (MMC) · Negative sequence currents (NSCs) · Power quality Rail power conditioner (RPC) · Submodules (SMs) · V/V power transformer

M. Tanta (✉) · J. L. Afonso
GEPE - Centro Algoritmi, University of Minho,
Campus of Azurém, 4800-058 Guimarães, Portugal
e-mail: mtanta@dei.uminho.pt

J. L. Afonso
e-mail: jla@dei.uminho.pt

J. A. Afonso
CMEMS, University of Minho, Campus of Azurém, 4800-058 Guimarães, Portugal
e-mail: jose.afonso@dei.uminho.pt

A. P. Martins · A. S. Carvalho
SYSTEC, University of Porto, Rua Roberto Frias, 4200-465 Porto, Portugal
e-mail: ajm@fe.up.pt

A. S. Carvalho
e-mail: asc@fe.up.pt

S.-I. Ao et al. (eds.), *Transactions on Engineering Technologies*,
https://doi.org/10.1007/978-981-13-0746-1_20

1 Introduction

The last decades have witnessed a rapid growth of high-speed and high power railways systems, with the appearing of new technologies to improve the efficiency and reduce the effects of AC railways traction grids on the power quality of the public grid side [1], especially when the railways traction grids are interconnected and have the same frequency of the public grid side (50 Hz), such as the case of Portugal and Finland. On the other hand, some countries, such as Germany and Austria, use different frequencies for the public grid (50 Hz) and the railway traction grid (16.7 Hz) [2]. The particular scopes of this chapter are the harmonic distortion produced by the electric locomotives and the currents imbalance created by the single-phase traction loads [3].

These matters are normally associated with the AC supply traction networks and they have been noticed from the early use of the AC electrified railways. These drawbacks have the ability to create an adverse effect on the electric devices and they can threat the safety and the economic operation of the high voltage public grids [1, 3].

Harmonic distortion in railway traction grids is mainly resultant from the electric locomotives and their old converter components, especially the old half controlled ones [4]. The second factor that affects the power quality is the unbalanced loads, which cause negative sequence currents (NSCs) [1, 5]. Eliminating the effects of NSCs requires to obtain a balanced load on the public grid, and that is achievable by using different techniques, like the static VAR compensators (SVCs) or the static synchronous compensators (STATCOMs) [6].

The main disadvantage of SVCs is the production of harmonics, which are caused by the low switching frequency of power electronics devices [3]. Besides, a SVC may decrease the total power factor of the system when it is designed to compensate NSCs, because there is still a trade-off between power factor correction and NSCs compensation when using SVCs in railways traction systems [7]. STATCOM devices are used with the medium voltage levels to compensate NSCs, and when this device is used in conjunction with a Scott power transformer or a V/V power transformer at the medium voltage levels of railways traction grid, it can be called as a rail power conditioner (RPC) [8].

Modular multilevel converters (MMCs) have been widely used with the medium voltage levels because of their flexibility and expandability to the required power value. Therefore, they have been enhanced recently to operate as RPCs in railways systems with different topologies, as in [9, 10]. The RPC presented in [9] combines a direct AC/AC MMC and a Scott power transformer, while the one in [10] presents the two–phase three–wires MMC with a V/V power transformer. The RPC in [11] combines an indirect AC/DC/AC MMC and a Scott power transformer. Consequently, the direct AC/AC conversion with MMC requires to use full bridge submodules (SMs) that have positive, negative and zero output voltages, while the indirect AC/DC/AC conversion needs only half bridge SMs with only positive and zero output voltages [3].

This chapter focuses on a RPC based on the indirect AC/DC/AC MMC with half bridge SMs and a V/V traction power transformer. This RPC system consists of a back-to-back single-phase MMC connected directly on parallel to the secondary side of a V/V power transformer and two load sections, as shown in Fig. 1. This system is designed to compensate NSCs and harmonics on the public grid side.

This chapter presents a comprehensive study about the RPC based on a full–bridge back–to–back indirect MMC and a V/V power transformer. The RPC is connected directly on parallel to the secondary side of the V/V power transformer as shown in the Fig. 1. The chapter has been organized as follows: Sect. 2 demonstrates the RPC system structure and the MMC characteristics. Section 3 explains the RPC operation principle and presents the mathematical analysis. Section 4 describes the RPC control algorithm. Section 5 presents the simulation results. Finally, Sect. 6 outlines the final conclusions.

2 RPC System Description

This section presents a brief description of the proposed RPC system based on a full–bridge indirect MMC and a V/V power transformer. The next paragraphs demonstrate the main advantages and the benefits of using the RPC based on a full–bridge indirect MMC, as well as the V/V power transformer characteristics.

2.1 Indirect MMC Characteristics

This converter has the ability to be linked directly to medium voltage levels without using transformers and its structure depends on a cascade connection of half-bridge SMs. The DC-link of this converter has an inherent performance and is mainly formed from individual SMs capacitors. Therefore, there is no need to use a special DC-link capacitor like in a normal two-level back-to-back converter. Hence, capacitors voltage balancing control between SMs is important to guarantee a stable DC-link voltage [12]. The main motivations and features of using an indirect MMC with high and medium voltage levels are [13]:

- When using the two–level power converter, it is necessary to connect many power switches in series to stand the high voltage level. This is not necessary in the MMC, since normal power switches can be used.
- It is easy to increase the converter's total power by adding more SMs.
- Increasing the voltage level or the number of SMs reduces the total switching frequency. Therefore, there is no need to use filters at the AC power grid side.
- The MMC has a simple construction compared to other multilevel power converters topologies (e.g., the flying capacitors multilevel converter).

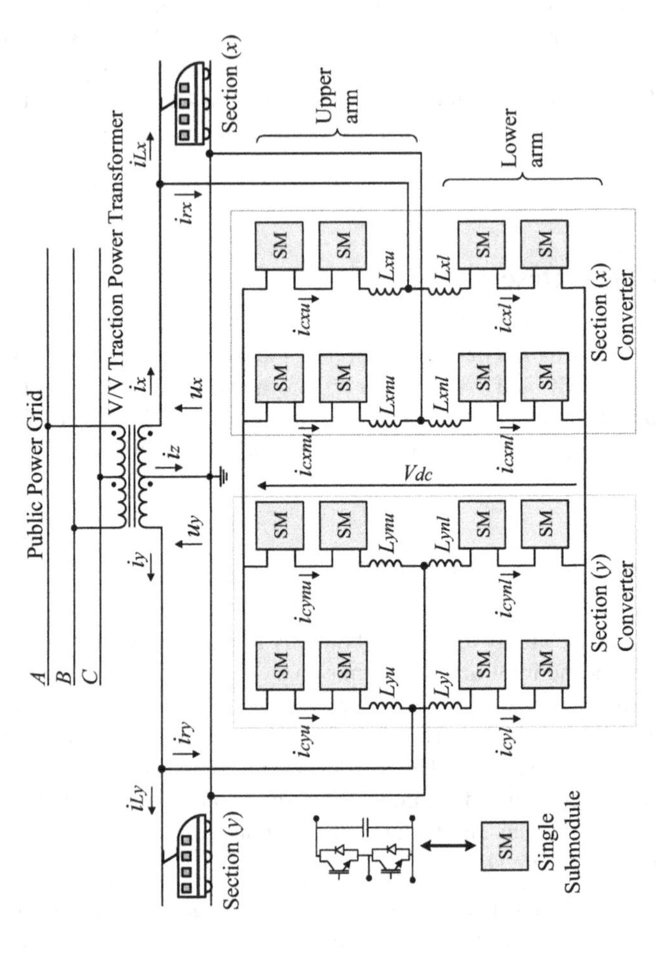

Fig. 1 RPC based on an indirect AC/DC/AC MMC connected to a V/V power transformer

- The power is divided equally between all SMs, thus reducing the power ratings of the power electronics components.

A single-phase indirect MMC consists of two back-to-back converters and each converter connected directly to the catenary-ground lines of both load sections (x) and (y), as shown in Fig. 1. Using the coupled inductor is very important to limit the current during voltage steps and to limit the circulating current between legs, especially after knowing that this current has high harmonics content, mainly the second harmonic; thus the inductances work like a converter's inner filter [13]. In addition, the coupled inductors are connected in series with SMs capacitors, so they have the ability to suppress any fault currents which could be resulting from the collapse of one or more capacitors [14].

2.2 V/V Power Transformer Characteristics

V/V transformers are commonly used in the high speed railway traction grids, because of reasons such as their simple structure, low price and high overload capability, when they are compared with other power transformers [15]. It is very important to know that, without applying the compensation strategy of the RPC, and when an unbalanced V/V transformer is in use, the NSCs injected to the grid are half of the fundamental positive sequence when both load sections consume the same power. However, when a balanced transformer is in use (such as a Scott transformer or a Woodbridge transformer), no NSCs are injected to the public grid side when both load sections consume the same power [16, 17].

As a result and regardless to the used power transformer, equalization between two load sections is almost impossible in practical applications because the locomotives are operating asynchronously (e.g., one locomotive may be accelerating while the other one is braking), and the RPC devices in such situations are required to compensate NSCs on the public grid side.

3 RPC Operation Principle

By assuming that the power losses in the RPC power electronics devices are negligible, the main RPC operation principle is to shift half of the load currents difference from the highly loaded section to the lightly loaded one [16]. In the normal case, and without applying any compensation strategy, the voltages of two load sections u_x and u_y are in phase with the line voltages of u_{AC} and u_{BC} respectively as shown in the Fig. 2a. Consequently, and by considering a unity power factor, currents on public grid side are given by the Eq. (1), where K_V is the turns ratio of the V/V power transformer and I_{Lx}, I_{Ly} are the RMS values of the two load sections currents.

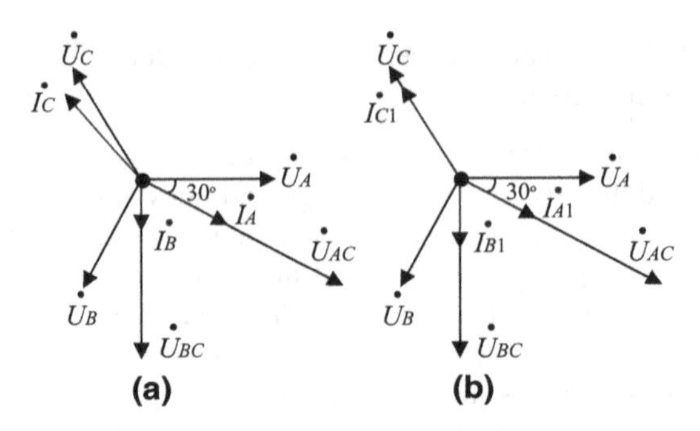

Fig. 2 Phasor diagrams of public grid currents: **a** When both load sections are loaded unequally without any compensation; **b** When shifting the active power difference

$$\left.\begin{aligned} \dot{I}_A &= (I_{Lx}/K_V)e^{-j30} \\ \dot{I}_B &= \left(I_{Ly}/K_V\right)e^{-j90} \\ \dot{I}_C &= -(I_{Lx}/K_V)e^{-j30} - \left(I_{Ly}/K_V\right)e^{-j90} \end{aligned}\right\} \tag{1}$$

As in the RPC operation principle, the converter can shift half of the load currents difference, so Eqs. (2) and (3) can be applied [16]. Equation (3) show that phase A and phase B currents have now the same RMS value.

$$\Delta I = \frac{1}{2}(I_{Lx} - I_{Ly}) \tag{2}$$

$$\left.\begin{aligned} \dot{I}_{A1} &= \dot{I}_A - \frac{\Delta I}{K_V}e^{-j30} = \frac{1}{2K_V}(I_{Lx} + I_{Ly})e^{-j30} \\ \dot{I}_{B1} &= \dot{I}_B + \frac{\Delta I}{K_V}e^{-j90} = \frac{1}{2K_V}(I_{Lx} + I_{Ly})e^{-j90} \end{aligned}\right\} \tag{3}$$

Phase C does not have the same RMS value that is presented in Eq. (3), because the angles between phase A and phase B currents are not equal to 120°, as shown in Fig. 2b. Moreover, phase C current is in phase now with its voltage, while the other two phases are still shifted by a 30° angle with the corresponded phase voltages because the reactive power is not compensated yet [15, 16, 18].

In order to make the three-phase currents balanced, it is important to add a certain reactive current to phase x and phase y, as shown in Fig. 3b. The RPC system has the responsibility of injecting the compensation currents irx and iry. Phase x generates reactive power because the reactive current component $irxr$ that is injected by the section (x) converter leads the voltage vector of phase x. Phase y consumes reactive power because the reactive current component $iryr$ that is absorbed by the section (y) converter lags the voltage vector of phase y [19], as shown in Fig. 3b. Furthermore, the phase currents on the public grid side i_{A2}, i_{B2}, i_{C2} have the same magnitude and

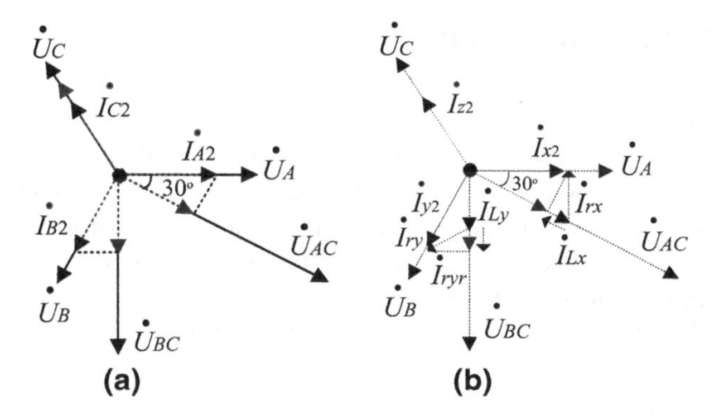

Fig. 3 Phasor diagram after shifting the active power difference and compensating the reactive power: **a** Public grid side; **b** Secondary side of the V/V power transformer

are balanced [15]. From Fig. 3b and Eq. (3), it is possible to obtain the Eq. (4), which present the same RMS values of the reactive compensation currents components.

Equation (5) give the section (x) current phasor after compensation. The same previous steps are also applied for the section (y). The instantaneous values of the three–phase currents after compensation are presented in Eq. (6). These are the final balanced currents after shifting the active power difference and after compensating the reactive power [15, 16]. The total compensation currents (reference currents) that contain the active and the reactive components can be obtained from Fig. 3b and Eq. (7).

$$\left.\begin{aligned} I_{rxr} &= I_{x1}\tan\frac{\pi}{6} = \frac{1}{2}(I_{Lx}+I_{Ly})\tan\frac{\pi}{6} \\ I_{ryr} &= I_{y1}\tan\frac{\pi}{6} = \frac{1}{2}(I_{Lx}+I_{Ly})\tan\frac{\pi}{6} \end{aligned}\right\} \tag{4}$$

$$\left.\begin{aligned} \dot{I}_{x2} &= \dot{I}_{x1} + \dot{I}_{rxr} = \frac{1}{2}(I_{Lx}+I_{Ly})\left[e^{-j30} + \tan\frac{\pi}{6}e^{j60}\right] \\ \dot{I}_{x2} &= \frac{1}{\sqrt{3}}(I_{Lx}+I_{Ly})e^{j0} \end{aligned}\right\} \tag{5}$$

$$\left.\begin{aligned} i_{x2} &= i_{x2m}\sin(\omega t) = \frac{\sqrt{2}}{\sqrt{3}}(I_{Lx}+I_{Ly})\sin(\omega t) \\ i_{y2} &= i_{y2m}\sin\left(\omega t - \frac{2\pi}{3}\right) = \frac{\sqrt{2}}{\sqrt{3}}(I_{Lx}+I_{Ly})\sin\left(\omega t - \frac{2\pi}{3}\right) \\ i_{z2} &= i_{z2m}\sin\left(\omega t + \frac{2\pi}{3}\right) = \frac{\sqrt{2}}{\sqrt{3}}(I_{Lx}+I_{Ly})\sin\left(\omega t + \frac{2\pi}{3}\right) \end{aligned}\right\} \tag{6}$$

$$\left.\begin{aligned} i^*_{rx} &= i_{x2} - i_{Lx} \\ i^*_{ry} &= i_{y2} - i_{Ly} \end{aligned}\right\} \tag{7}$$

4 RPC Control Algorithm

One of the main objectives of the RPC control system is to calculate the final compensation current references that are presented in Eq. (7). The acquisition of both i_{Lx} and i_{Ly} signals is possible by using two current sensors on both load sections of (x) and (y); then, the traction power system currents after compensation, i_{x2} and i_{y2}, should be calculated by the RPC control algorithm. In addition, SMs capacitors voltages should be controlled to keep the same SMs voltages, consequently, maintaining a constant DC-link voltage, which is important to guarantee a good performance.

4.1 Compensation Currents Calculations

The reference compensation currents can be calculated by using the instantaneous load currents waveforms of the two load sections, (x) and (y), where the latter are the most important variables, since they are considered as the input signals of the control system. Electric locomotives normally use power converters to drive the traction motors (DC motors or asynchronous motors). Then, a unity load power factor could be considered in this study and the load sections currents only contain the active currents components I_{Lxa} and I_{Lya}, besides the harmonics currents contents i_{Lxh} and i_{Lyh}. Then, the instantaneous load currents values are as in Eq. (8), where the h symbol refers to the harmonics order of the instantaneous harmonics load currents.

$$\left.\begin{aligned} i_{Lx} &= \sqrt{2}\, I_{Lxa}\ \sin\left(\omega t - \frac{\pi}{6}\right) + \sum_{h=2}^{\infty} i_{Lxh} \\ i_{Ly} &= \sqrt{2}\, I_{Lya}\ \sin\left(\omega t - \frac{\pi}{2}\right) + \sum_{h=2}^{\infty} i_{Lyh} \end{aligned}\right\} \quad (8)$$

Multiplying the instantaneous value of the load section (x) current by $\sin(\omega t - \pi/6)$ and the load section (y) current by $\sin(\omega t - \pi/2)$, then summing the acquired results, it gives a DC current component I_P, as in Eq. (9).

$$I_P = \frac{\sqrt{2}}{2}\left(I_{Lxa} + I_{Lya}\right) \quad (9)$$

Multiplying the DC current component of I_P by $2/\sqrt{3}$ after using a low pass filter (LPF) to obtain the signal without harmonics, it gives the peak value of phase x and phase y currents after compensation, I_{x2m} and I_{y2m}, which have been presented in Eq. (6). Then, it is possible to obtain the instantaneous values of i_{x2} and i_{y2} after multiplying the peak values of I_{x2m} and I_{y2m} by the corresponded sinewaves, $\sin(\omega t)$ for load section x and $\sin(\omega t - 2\pi/3)$ for load section y. Consequently, the final compensation currents references i^*_{rx}, i^*_{ry} are calculated as in Eq. (7). The control strategy of the RPC based on an indirect MMC is shown in Fig. 4.

4.2 Capacitors Voltage Balancing Control

To guarantee a good operation performance of the MMC, voltage balancing control for SMs capacitors must be applied to the system. This control consists of the averaging voltage control, the individual voltage balancing control and the circulating current control as presented in Fig. 5 [12]. The averaging control ensures that the voltage of each capacitor in the leg is close to the average voltage that is provided as a reference. It is implemented by summing the measured capacitors voltages for each leg and dividing the result by the number of SMs per leg. The actual average voltage value in this case is calculated and compared to the reference average voltage value V^*_{ci}. Then, a proportional-integral controller (PI) is used to correct the difference between the actual and the reference values [13, 20]. The output of the first PI controller is considered as a reference for a circulating current controller, which is important to limit the SMs capacitors ripples. The last controller is implemented by summing the arms currents i_{ckm}, i_{cknm} for each MMC leg, where (k) refers to the load section (x) or the load section (y), and (m) defines the upper or the lower arm for each leg. The final output signal of the averaging voltage balancing controller is A_k, which is multiplied by the correspondent sinewaves, as shown in the control diagram of Fig. 4.

The individual voltage balancing control is responsible to set every capacitor voltage to its reference. Therefore, a proportional (P) controller is used to correct the error and to act dynamically in the balancing process. The output of this controller is multiplied by 1 if the current's direction in the arm is to charge the capacitors, or by −1 if its direction is to discharge the capacitors as shown in Fig. 5. This is very useful to maintain a limited value for every SMs capacitor voltage [12, 13]. The final signal B_k will be added to the final control signals as shown in the Fig. 4. A hysteresis controller is used to track the reference currents for each converter to ensure a fast response. The width of the hysteresis controller H should have a suitable value. A large value of H causes low switching frequency and large tracking error. A small value of H leads to high switching frequency but small tracking error [16].

5 Simulation Results

The simulation model has been built by using the *PSIM* software for power electronics simulation, in order to validate the system of RPC based on an indirect MMC and the associated control strategy. This simulation model consists of a 5-level indirect MMC with a total number of 32 SMs, and it is used for 25 kV, 50 Hz electrified traction grids. The main parameters of the simulation model are shown in Table 1.

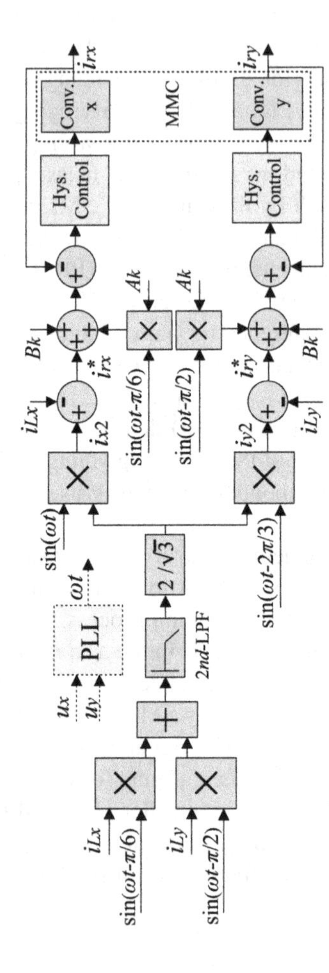

Fig. 4 Control strategy of the RPC based on an indirect MMC

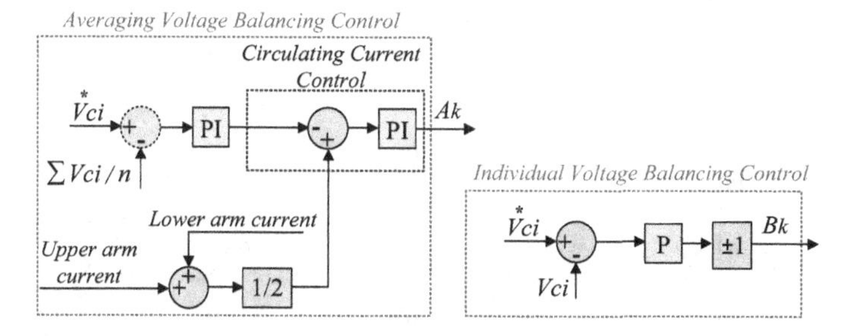

Fig. 5 Capacitors voltage balancing control of the MMC

Table 1 Simulation model parameters

Parameters	Symbols	Values	Units
Public power grid voltage	U_{AB}, U_{BC}, U_{CA}	220	kV
Traction power grid voltage	U_x, U_y	25	kV
SM capacitor	C_{SM}	100	mF
Buffer inductance	L_{km}, L_{knm}	1	mH
SM voltage	V_{SM}	12.5	kV
DC-link voltage	V_{dc}	50	kV
Number of SMs in each leg	n	8	–

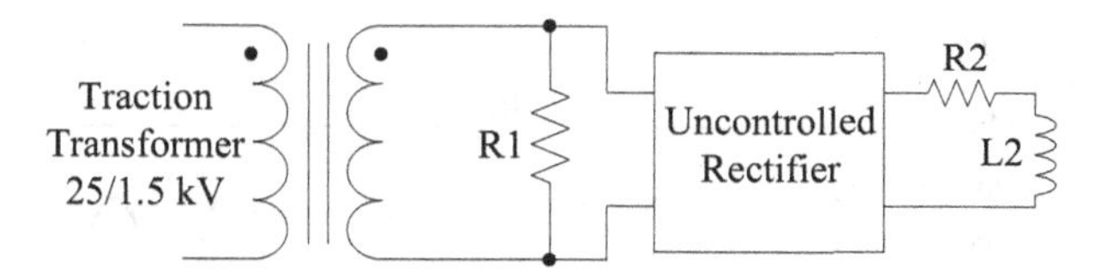

Fig. 6 Locomotive model in simulation

As mentioned before, the load power factor on both load sections was considered to be close to one. Therefore, the locomotives can be modeled as a resistive load connected in parallel with an uncontrolled full bridge rectifier on the secondary side of the locomotive transformer. This full bridge rectifier is considered as a harmonic source, where the full locomotive model is shown in Fig. 6. The simulation results show two different case studies with much possible practical scenarios, when only one load section is loaded, and when both load sections are loaded unequally.

5.1 Only One Load Section Is Loaded

By supposing that the load section (x) is loaded with 4.8 MW and section (y) is without any load, Fig. 7a, b show the waveforms of the public grid currents before

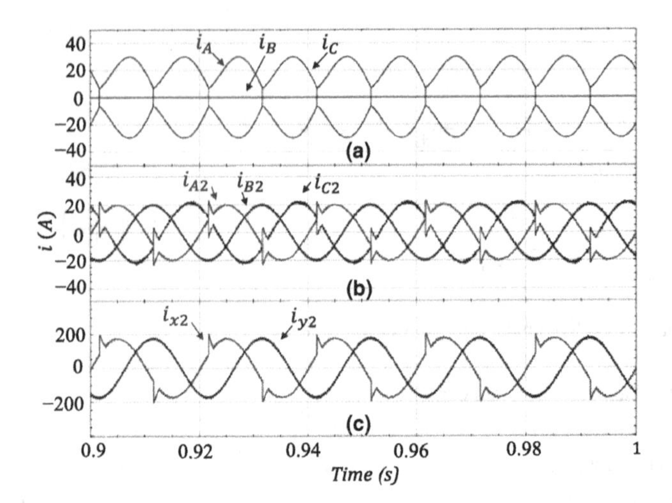

Fig. 7 Simulation results (I_{Ly} =0): **a** Public grid currents before compensation; **b** Public grid currents after compensation; **c** Traction power grid currents after compensation

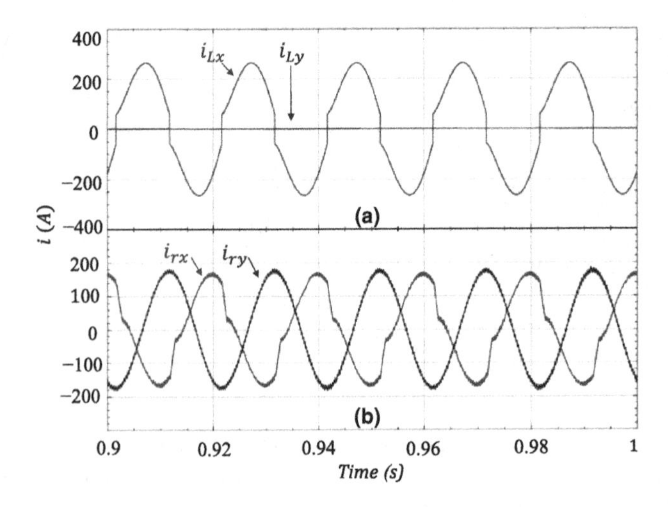

Fig. 8 Simulation results (I_{Ly} =0): **a** Load section currents; **b** Compensation currents

and after applying the compensation strategy. The currents before compensation are totally unbalanced, and after the compensation they are balanced but with some harmonics contents due to a high value of the coupled inductance. Figure 7c shows the currents on the secondary side of the V/V power transformer.

Figure 8a shows the currents of both load sections, and it is clear that the section (x) current (locomotive current) has some harmonics contents. Figure 8b shows the compensation currents injected by the RPC. The current i_{ry}, which corresponds to the section (y) converter, is totally sinusoidal because the load section (y) is unloaded.

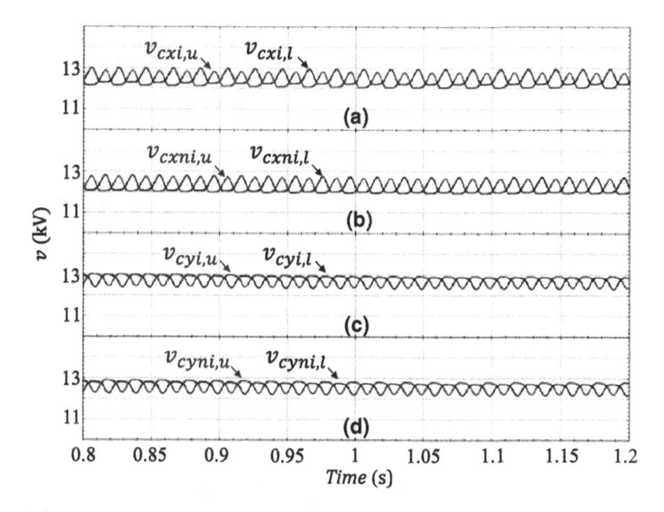

Fig. 9 SMs voltages ($I_{Ly} = 0$): **a** Positive leg of converter (x); **b** Negative leg of converter (x); **c** Positive leg of converter (y); **d** Negative leg of converter (y)

Figure 9 presents the waveforms of the SMs voltages and demonstrates the results of the applied voltage balancing control. Figure 9a shows the waveforms for the upper and the lower arms of the positive leg of section (x) converter. Figure 9b shows the same waveforms but for the negative leg of section (x) converter. Figure 9c shows the waveforms for the upper and the lower arms of the positive leg of section (y) converter. Figure 9d shows SMs voltages waveforms for the upper and the lower arms of the negative leg of section (y) converter. The RPC control keeps the SMs voltages around their reference value of 12.5 kV.

5.2 Both Load Sections Are Loaded Unequally

The second scenario is when both load sections are loaded unequally, the power of section (x) is 4.8 MW and the power of section (y) is 2.4 MW. The public grid side currents before applying the compensation strategy are presented in Fig. 10a and they are unbalanced currents.

Figure 10b shows the same currents after turning on the RPC. The public grid currents are now balanced, as shown in the phasor diagram of Fig. 3a. Figure 10c shows the currents on the secondary side of the V/V power transformer. The load section currents are presented in Fig. 11a, where i_{Lx} has double of the value of i_{Ly}. Both currents are considered as harmonic sources because of using an uncontrolled rectifier in the locomotive model. The compensation currents that injected by the RPC are demonstrated in Fig. 11b, and i_{ry} here is not sinusoidal. Similarly to Fig. 9, when only one load section was loaded, Fig. 12 shows the results of the applied

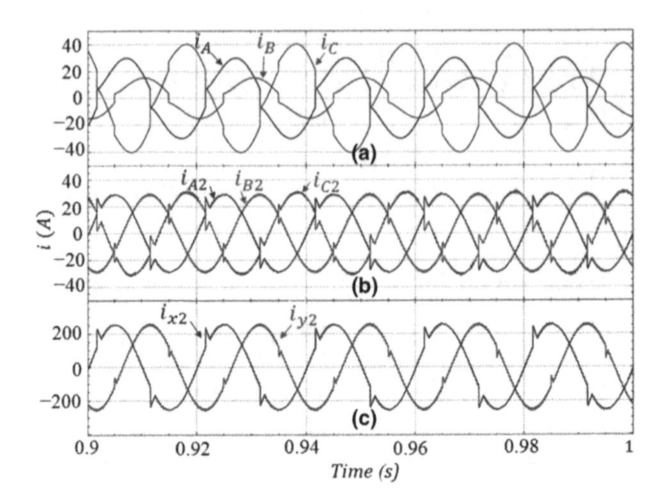

Fig. 10 Simulation results: **a** Public grid currents before compensation; **b** Public grid currents after compensation; **c** Traction power grid currents after compensation

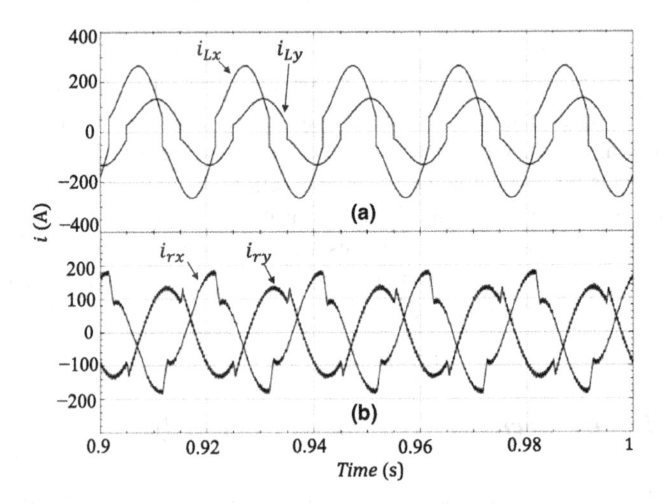

Fig. 11 Simulation results: **a** Load section currents; **b** Compensation currents

voltage balancing control. Besides compensating the NSCs, the main aim of the RPC control is to maintain the SMs voltages at their reference value of 12.5 kV, as shown in Fig. 12.

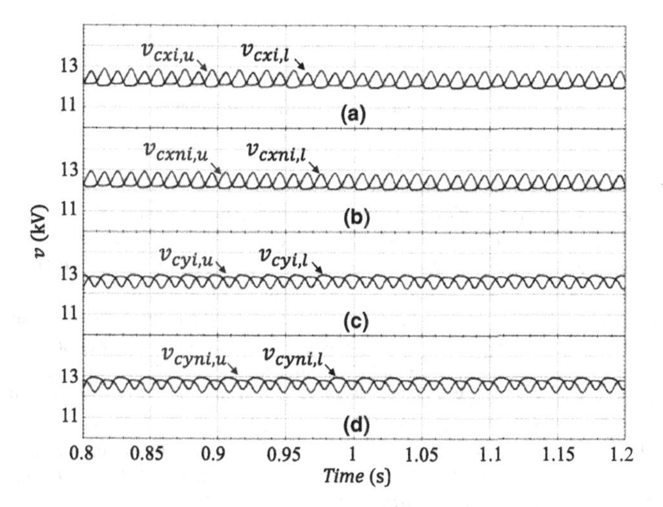

Fig. 12 SMs voltages: **a** Positive leg of converter (x); **b** Negative leg of converter (x); **c** Positive leg of converter (y); **d** Negative leg of converter (y)

6 Conclusion

This chapter discussed a strategy to compensate the negative sequence currents (NSCs) on the public grid by using a rail power conditioner (RPC) based on an indirect modular multilevel converter (MMC) in AC electrified railways. Simulation using *PSIM* has confirmed the operation principle and the control strategy for the proposed system. The results showed balanced public grid currents in different operation scenarios. The RPC reference currents were generated depending on the load sections currents. A control strategy has been explained to calculate the compensation currents references and to perform an equal voltage value for the MMC submodules (SMs) capacitors. The averaging voltage balancing control for each MMC leg, the individual voltage balancing control for each MMC SM, and the circulating current control in each MMC leg, all have been accomplished in order to maintain equal SMs voltages and a constant DC–link voltage.

Acknowledgements This work was supported by COMPETE: POCI-01-0145- FEDER-007043 and FCT—Fundação para a Ciência e Tecnologia within the Project Scope: UID/CEC/00319/2013. Mohamed Tanta was supported by a FCT grant with reference PD/BD/127815/2016.

References

1. I. Perin et al., Application of power electronics in improving power quality and supply efficiency of AC traction networks, in *2015 IEEE 11th International Conference on Power Electronics and Drive Systems*, 2015, pp. 1086–1094
2. A. Steimel, Power-electronic grid supply of AC railway systems, in *2012 13th International Conference on Optimization of Electrical and Electronic Equipment* (*OPTIM*), 2012, pp. 16–25
3. I. Krastev, P. Tricoli, S. Hillmansen, M. Chen, Future of electric railways: advanced electrification systems with static converters for ac railways. IEEE Electrif. Mag. **4**(3), 6–14 (2016)
4. M. Tanta, J.A. Afonso, A.P. Martins, A.S. Carvalho, J.L. Afonso, Rail power conditioner based on indirect AC/DC/AC modular multilevel converter using a three-phase V/V power transformer, in *Proceeding of the World Congress on Engineering*, London, UK, 5–7 July 2017. Lecture Notes in Engineering and Computer Science, pp. 289-294
5. L. Abrahamsson, T. Schütte, S. Östlund, Use of converters for feeding of AC railways for all frequencies. Energy Sustain. Dev. **16**(3), 368–378 (2012)
6. I. Perin, P.F. Nussey, T.V. Tran, U.M. Cella, G.R. Walker, Rail power conditioner technology in Australian Heavy Haul Railway: a case study, in *2015 IEEE PES Asia-Pacific Power and Energy Engineering Conference* (*APPEEC*), 2015, pp. 1–5
7. C. Wu, A. Luo, J. Shen, F.J. Ma, S. Peng, A negative sequence compensation method based on a two-phase three-wire converter for a high-speed railway traction power supply system. IEEE Trans. Power Electron. **27**(2), 706–717 (2012)
8. S. Tamai, Novel power electronics application in traction power supply system in Japan, in *2014 16th International Power Electronics and Motion Control Conference and Exposition*, 2014, pp. 701–706
9. F. Ma, Z. He, Q. Xu, A. Luo, L. Zhou, M. Li, Multilevel power conditioner and its model predictive control for railway traction system. IEEE Trans. Ind. Electron. **63**(11), 7275–7285 (2016)
10. Q. Xu et al., Analysis and comparison of modular railway power conditioner for high-speed railway traction system. IEEE Trans. Power Electron. **32**(8), 6031–6048 (2016)
11. S. Song, J. Liu, S. Ouyang, X. Chen, A modular multilevel converter based railway power conditioner for power balance and harmonic compensation in Scott railway traction system, in *IEEE PEMC-ECCE Asia*, 2016, pp. 2412–2416
12. M. Hagiwara, H. Akagi, Control and experiment of pulsewidth-modulated modular multilevel converters. IEEE Trans. Power Electron. **24**(7), 1737–1746 (2009)
13. P. Asimakopoulos, K. Papastergiou, M. Bongiorno, Design and control of modular multilevel converter in an active front end application. Accelerators and storage rings. Engineering, Chalmers U. Tech, Gothenburg, Sweden, 2013
14. Q. Tu, Z. Xu, H. Huang, J. Zhang, Parameter design principle of the arm inductor in modular multilevel converter based HVDC, in *2010 International Conference on Power System Technology*, 2010, pp. 1–6
15. A.M. Bozorgi, M.S. Chayjani, R.M. Nejad, M. Monfared, Improved grid voltage sensorless control strategy for railway power conditioners. IET Power Electron. **8**(12), 2454–2461 (2015)
16. A. Luo, C. Wu, J. Shen, Z. Shuai, F. Ma, Railway static power conditioners for high-speed train traction power supply systems using three-phase V/V transformers. IEEE Trans. Power Electron. **26**(10), 2844–2856 (2011
17. M. Tanta et al., Simplified rail power conditioner based on a half-bridge indirect AC/DC/AC modular multilevel converter and a V/V power transformer, in *IECON 2017—43rd Annual Conference of the IEEE Industrial Electronics Society*, Beijing, China, 2017, pp. 6431–6436

18. F. Ma, A. Luo, X. Xu, H. Xiao, C. Wu, W. Wang, A Simplified power conditioner based on half-bridge converter for high-speed railway system. IEEE Trans. Ind. Electron. **60**(2), 728–738 (2013)
19. M. Tanta, V. Monteiro, T.J.C. Sousa, A.P. Martins, A.S. Carvalho and J.L. Afonso, Power quality phenomena in electrified railways: Conventional and new trends in power quality improvement toward public power systems, in *2018 International Young Engineers Forum (YEF-ECE)*, Costa da Caparica, Portugal, 2018, pp. 25–30
20. M. Rejas et al., Performance comparison of phase shifted PWM and sorting method for modular multilevel converters, in *2015 17th European Conference on Power Electronics and Applications (EPE'15 ECCE-Europe)*, 2015, pp. 1–10

Jitter and Phase-Noise in High Speed Frequency Synthesizer Using PLL

Ahmed A. Telba

Abstract Jitter happens when data rates increase in high-speed input and output connections for data communications. Characterizing of jitter and measurement is challenge, jitter defined as the misalignment of edges in a sequence of data bits from their ideal positions. Misalignments can result in data errors, and raised bit error rate in digital communication. Tracking these errors over an extended period determines the system stability. Jitter can be due to deterministic and random phenomena, also referred to as systematic and non-systematic respectively. It is worth mentioning that the benefit of jitter is limited to applications using random number generation. There is hardly any other benefit from jitter. Phase noise and jitter are a very important issue when design a phase-locked and delay-locked loops. Different applications may have different emphasis on the jitter specifications. "Cycle-to-cycle" jitter refers to the time difference between two consecutive Cycles of a period signal. A RMS (root mean square) or peak-to-peak value is used to describe a random jitter. According to the noise sources, it can be classified as internal jitters, caused by the building blocks of PLLs and DLLs, and external jitters. Jitters in an Oscillator have been examined for almost half a century and still a hot topic.

Keywords Frequency synthesizer · Jitter noise · Modeling and simulation
Phase-locked loop · Phase noise · Synchronization in digital transmission

1 Introduction

Jitter happens when data rates increase in high-speed input and output connections for data communications. Characterizing jitter is a challenge, as is its measurement. Jitter defined as the misalignment of edges in a sequence of data bits from their ideal positions [1]. Misalignments can result in data errors, and raised bit error rate in

A. A. Telba (✉)
Electrical Engineering Department, King Saud University,
P.O. Box 800, Riyadh 11421, Saudi Arabia
e-mail: atelba@ksu.edu.sa

S.-I. Ao et al. (eds.), *Transactions on Engineering Technologies*,
https://doi.org/10.1007/978-981-13-0746-1_21

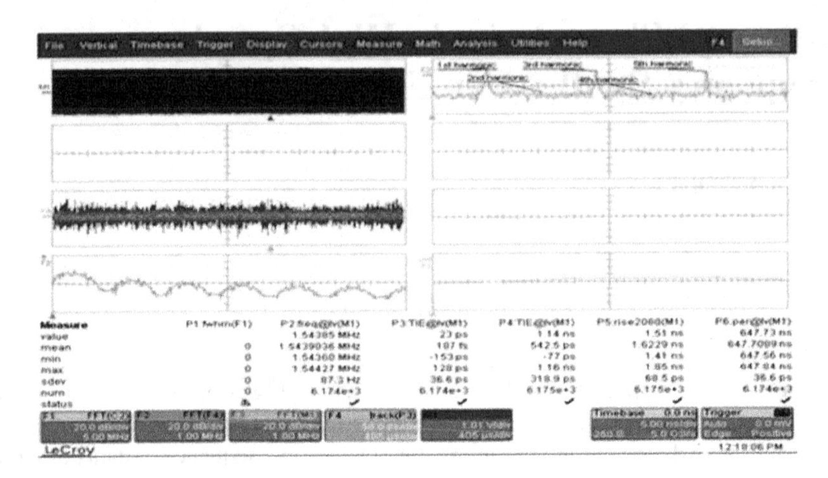

Fig. 1 Jitter in high frequency synthesizer

digital communication. Tracking these errors over an extended period determines the system stability. Jitter can be due to deterministic and random phenomena [2], also referred to as systematic and non-systematic respectively [3].

It is worth mentioning that the benefit of jitter is limited to applications using random number generation. There is hardly any other benefit from jitter. Hence, the disadvantages of jitter highly outweigh its benefits.

Timing jitter is of great concern in high frequency timing circuits. Its presence can degrade the system performance in many high-speed applications [4]. This paper describes the relation between phase noise and jitter in high speed communication as shown in Fig. 1 the real measurements jitter in T1 carrier using Wave Runner LECROY Oscilloscopes 1 GHz, 10 GS/s in this experimental work gives the different measurements such as the minimum and maximum jitter in the frequency range standard deviation of time measurements it gives also the Fourier, 1st, 3rd, and 5th harmonics of the carrier frequency as shown in Fig. 1.

One of techniques used to minimize jitter by using a wide range low jitter clock source using only one crystal oscillator using two phase-locked loops connected in cascade. The first one has a voltage-controlled crystal oscillator to eliminate the input jitter and the second is a wide-band phase-locked loop. Simulating the root-mean-squared jitter of the system is important to analyze system performance [5, 6].

One important advantage of using the proposed system is that it uses only one voltage-controlled crystal oscillator for multiple carrier frequencies, while reducing jitter considerably. The dual phase-locked loop system as proposed designed, built and tested in the laboratory and the results shown in Figs. 1 and 2.

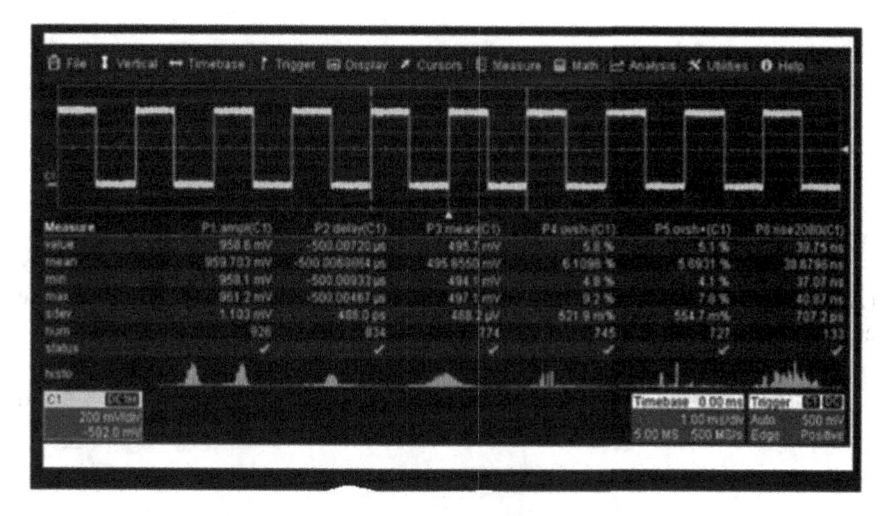

Fig. 2 Real time jitter measurment in high frequency synthesizer

2 How Phase Noise Quantities Relate to Timing Jitter

Timing jitter is the critical performance parameter for clock recovery applications and sampled-data systems [7].

To relate cycle-to-cycle timing jitter to phase noise, we first derive the phase jitter for an oscillator with frequency fc and period Tc over a correlation time T [8],

$$\begin{aligned}\sigma_{\Delta\phi}^2 &= \mathrm{E}\left\{[\phi(t+T)-\phi(t)]^2\right\}\\ &= 2\left[R_\phi(0)-R_\phi(T)\right]\\ &= 2\int_{-\infty}^{\infty} P_\phi(f)\left(1-e^{j2\pi/T}\right)df\\ &= 4\int_{-\infty}^{\infty} S_\phi(f)\sin^2(\pi/T)df \end{aligned} \tag{1}$$

From the integrand in (1), it is apparent that the close-in phase noise near the carrier is significantly attenuated for frequencies much smaller than T − 1. We can gain further insight by assuming a given shape for the phase spectral density. Consider an oscillator with a constant phase noise spectrum for frequencies $f<f1$ and zero everywhere else. Then,

$$\sigma_{\Delta\phi}^{2} = 8\int_{o}^{f_1} L_1 \sin^2(\pi/T)df$$
$$= 4L_1\left(f_1 - \frac{\sin 2\pi f_1 T}{2\pi T}\right) \quad (2)$$

where L_1 is the value of carrier in-band phase noise.

A free-running oscillator phase noise spectrum has a region where $L(\Delta f) \propto (\Delta f)^{-2}$. Hence, we can model the noise as white, frequency modulated (FM) noise as described in [9],

$$L(\Delta f) = \frac{K}{(\Delta f)^2} \quad (3)$$

where

$$K = L(f1)((f1)2 \quad (4)$$

Therefore,

$$\sigma_{\Delta\phi}^{2} = 4\int_{-\infty}^{+\infty} L(\Delta f)\sin^2(\pi\Delta fT)d\Delta f$$
$$= 4\int_{-\infty}^{+\infty}\left(\frac{\sin(\pi\Delta fT)}{(\Delta f)}\right)^2 d\Delta f \quad (5)$$

Using Parseval's relation, the integral can be evaluated as

$$\sigma_{\Delta\phi}^{2} = 4K\pi^2T = L(\Delta f_1)(2\pi\Delta f_1)^2T \quad (6)$$

Relating the variance in phase to timing jitter,

$$\sigma_{\Delta T}^{2} = \frac{\sigma_{\Delta\phi}^{2}}{(2\pi f_c)^2} = L(\Delta f_1)\left(\frac{\Delta f_1}{f_c}\right)^2 T \quad (7)$$

The result has been derived in reference [10].

3 Introduction to Timing Jitter

Timing jitter is an important specification for digital circuits and sampled-data systems. To demonstrate the impact of clock jitter, consider the sampling clock for an ADC. Any error in the sampling instant directly translates into an error in the sampled voltage as displayed in Fig. 1, thus reducing the overall resolution of the converter. For an ADC with Nyquist rate fs full-scale voltage VFS, and resolution of B bits, the largest possible slew rate comes for the signal

How
$$\begin{aligned}\sigma^2_{\Delta\phi} &= 4\int_{-\infty}^{+\infty} \mathrm{L}(\Delta \mathrm{f})\sin^2(\pi\Delta \mathrm{fT})\mathrm{d}\Delta \mathrm{f} \\ &= 4\int_{-\infty}^{+\infty}\left(\frac{\sin(\pi\Delta \mathrm{fT})}{\Delta \mathrm{f}}\right)^2 \mathrm{d}\Delta \mathrm{f}\end{aligned} \tag{8}$$

Using Parseval's relation, the integral can be evaluated as

$$\sigma^2_{\Delta\phi} = 4\mathrm{K}\pi^2\mathrm{T} = \mathrm{L}(\Delta \mathrm{f}_1)(2\pi\Delta \mathrm{f}_1)^2\mathrm{T} \tag{9}$$

Relating the variance in phase to timing jitter,

$$\sigma^2_{\Delta \mathrm{T}} = \frac{\sigma^2_{\Delta\phi}}{(2\pi \mathrm{f_c})^2} = \mathrm{L}(\Delta \mathrm{f}_1)\left(\frac{\Delta \mathrm{f}_1}{\mathrm{f_c}}\right)^2\mathrm{T} \tag{10}$$

the result has been derived in reference [10].

4 Introduction to Supurious Tones

Since all frequency synthesizers generate the carrier frequency by locking to a high-precision low frequency oscillator, the circuits are periodic at the lower reference frequency. This causes spurious tones to appear around the carrier at the output of the frequency synthesizer. For a fixed-frequency synthesizer, the spurs appear at the frequencies

$$f_{spur} = f_c \pm f_{ref},\ f_c \pm 2\,f_{ref},\ f_c \pm 3\,f_{ref}, \ldots\ldots, \tag{11}$$

Spurs can present a problem because any interferer located at a multiple of $\mathrm{f_{ref}}$ away from the desired signal will fall directly in band after mixing. This is especially a problem in cellular systems because the power of the received signal from other users is often several orders of magnitude larger than the desired signal. The location of the other channels is precisely where the spurious tones from the frequency synthesizer lie.

Figure 3 illustrates the suppressed tone in spectrum analyzer output.

Spurious tones can also manifest themselves as systematic timing jitter. Consider an ideal oscillator with a time-varying phase [2], oscillator with a time-varying phase [11],

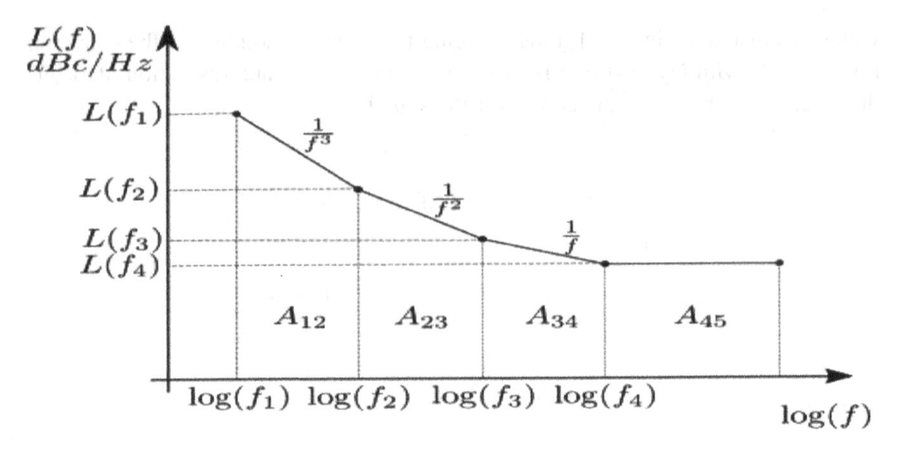

Fig. 3 Spectrum of phase noise

$$\nu_o(t) = A\cos\left\lfloor 2\pi f_c + \hat{\phi}\sin(2\pi\Delta ft)\right\rfloor \tag{12}$$

It was shown previously in (11) and (12) that the power of the spur with respect to the power of the carrier is

$$P_{spur} = (dBc) = 20\log_{10}\left(\frac{\hat{\phi}}{2}\right) \tag{13}$$

Substituting in Eq. (13) we get

$$\nu_o(t) = A\cos\left[2\pi f_c\left(t + \frac{\hat{\phi}}{2\pi f_c}\sin 2\pi\Delta ft\right)\right] \tag{14}$$

The variance of the timing jitter is

$$\sigma^2_{\Delta T} = \frac{\hat{\phi}}{2(2\pi f_c)^2} \tag{15}$$

$$\sigma^2_{\Delta T} = \frac{\hat{\phi}}{2(2\pi f_c)^2} \tag{16}$$

Thus, the spurious tones related to timing jitter by

$$P_{spur}(dBc) = 20\log_{10}(\sqrt{2}\pi f_c\sigma_{\Delta T}) \tag{17}$$

5 Conclusion

Phase noise and jitter are a very important issue when design a phase-locked and delay-locked loops. Different applications may have different emphasis on the jitter specifications. "Cycle-to-cycle" jitter refers to the time difference between two consecutive Cycles of a period signal. A RMS (root mean square) or peak-to-peak value is used to describe a random jitter. Jitter can be considered as time variant of the clock period. When a clock/data signal travels through a non-ideal channel and corrupted with noise, there are some uncertainties about the clock/data edges, which move in time. With large noise, the data eye may close and make data/clock recovery extremely difficult. If the generated clock is jittering, it may not be placed on the center of data eye and make a wrong decision. Such random variation cannot be recovered by simple amplification or clipping. A PLL circuit can be used to efficiently recovery or regenerate the clock/data with low jitter. In the frequency domain, such timing jitter is called phase noise.

Acknowledgements The researcher would like to thank the Research Center College of Engineering King Saud University, Kingdom of Saudi Arabia for the financial support provided for the research project.

References

1. D.H. Wolaver, *Phase-locked Loop Circuit Design* (Prentice Hall, USA, 1991)
2. The system vision environment provides easy-to-use schematic capture, a rich variety of electronic circuit and mechatronic system building-block models, state-of-art simulation technology, http://www.mentor.com/products/sm/systemvision
3. R.E. Best, *Phase Locked Loops: Design, Simulation, and Applications* (McGraw-Hill, New York, 1999)
4. A. Doboli, R. Vemuri, Behavioral modeling for high-level synthesis of analog and mixed-signal systems from VHDL-AMS. IEEE Trans. CAD Integr. Circuits Syst. **11**, 1504–1520 (2003)
5. A. Telba, J.M. Noras, M.A. El Ela, B. Almashary, Jitter minimization in digital transmission using dual phase locked loops, in *Proceedings of the 17th IEEE International Conference on Micro Electronics*, 2005, pp. 270 – 273
6. A. Telba, J.M. Noras, M.A. El Ela, B. Almashary, Simulation technique for noise and timing jitter in phase locked loop, in *Proceedings of the 16th IEEE International Conference on Micro Electronics*, 2004, pp. 501 – 504
7. A. Telba, Phase-noise and jitter in high speed frequency synthesizer, in *Proceedings of The World Congress on Engineering 2017*, 5–7 July, 2017, London, U.K. Lecture Notes in Engineering and Computer Science, pp. 353–356
8. P. Wilson, R. Wilcock, Behavioural modeling and simulation of a switched-current phase locked loop, in *Proceedings of IEEE International Symposium on Circuits and Systems*, vol. 5 (2005), pp. 5174–5177
9. M. Kozak, E.G. Friedman, Design and simulation of Fractional-N PLL frequency synthesizers, in *Proceedings of IEEE International Symposium on Circuits and Systems*, vol. 4, 2004, pp. 780 – 783
10. M. Karray, J.K. Seon, J.J. Charlot, N. Nasmoudi, VHDL-AMS modeling of a new PLL with an inverse sine phase detector (ISPD PLL), in *Proceedings of IEEE International Workshop on Behavioral Modeling and Simulation*, 2002, pp. 80–83

11. G. Breed, Analyzing signals using the eye diagram. High Freq. Electron. Mag. 50–53 (2005)
12. E. Christen, K. Bakalar, VHDL-AMS—a hardware description language for analog and mixed-signal applications. IEEE Trans. Circuits Syst. II Analog Digit Signal Process. **46**(10), 1263–1272
13. P.J. Ashenden, G.D. Peterson, D.A. Teegarden, *The System Designer's Guide to VHDL-AMS: Analog, Mixed-Signal, and Mixed-Technology Modeling* (Morgan Kaufmann, USA, 2003)

Dr. Ahmed A. Telba received his Ph.D. from School of Engineering, Design and Technology, University of Bradford UK Electronics and Telecommunications. Currently he is a postdoctoral research associate in Electronics and Communications, Electrical Engineering Department collage of Engineering, King Saud University Saudi Arabia. Research interests include analogue circuit design, phase locked loop, jitter in digital telecommunication networks, pizo actuator, pizo generation and FPGA.

Detection of Winding *(Shorted-Turn)* Fault in Induction Machine at Incipient Stage Using DC-Centered Periodogram

Ọdun Ayọ Imoru, M. Arun Bhaskar, Adisa A. Jimoh, Yskandar Hamam and Jacob Tsado

Abstract The problem of detecting shorted turns faults in stator windings has been difficult. The risk of the failure or the breaking down of this machine can be circumvented provided there is a proper way to detect the shorted turns faults. From literature, there are many methods of faults detection and diagnosis of the machine, however, DC-centered periodogram has not really been applied to detect and diagnose a fault in the electrical machine. This chapter describes stator winding shorted-turn fault detection of induction machine using DC-centered periodogram. Codes to analyses the DC-centered periodogram for both induction Machine under Healthy and shorted fault conditions were written from the general algorithm of periodogram. It is observed that the abnormality showed from the stator current signals for each condition corresponds to the plots generated by the DC-centered periodogram. The results obtained are also compared with another technique (DWT-Energy) using the same data. The peak values of the shorted turn-(**S**) is greater than the peak of the healthy-(**H**) state in both techniques. Thus, with DC-periodogram method, an electrical machine can be placed under close monitor for fault detection when the peak

Ọ. Imoru (✉) · J. Tsado
Department of Electrical & Electronics, Federal University of Technology, Minna, Nigeria
e-mail: aymorus@gmail.com

J. Tsado
e-mail: tsadojacob@futminna.edu.ng

M. A. Bhaskar
Department of EEE, Velammal Engineering College, Chennai, India
e-mail: m.arunbhaskar@gmail.com

A. A. Jimoh · Y. Hamam
Department of Electrical Engineering, Tshwane University of Technology, Pretoria, South Africa
e-mail: JimohAA@tut.ac.za

Y. Hamam
e-mail: HamamA@tut.ac.za

S.-I. Ao et al. (eds.), *Transactions on Engineering Technologies*,
https://doi.org/10.1007/978-981-13-0746-1_22

value of the PSD of a healthy machine under operation is started deviating from 0 dB/Hz.

Keywords Data capture · Electrical machine · Fault diagnosis · Periodogram Power spectral density (PSD) · Winding faults

1 Introduction

Electrical machine faults are defects that disturb the performance of normal operation on or in the machine. Such faults lead to various manifestations which include, unbalanced air-gap voltages, pulsations in torque and speed, unbalanced line currents, decreased efficiency and average torque, excessive heating, and consequently increased losses. Furthermore, electric machines are often exposed to detrimental or even non-ideal operating environments. These conditions are not limited to insufficient lubrication, overload, inadequate cooling and frequent motor starts/stops. Under such circumstances, electrical machines are subjected to undesirable stresses, which place the motors under risk of developing faults or failures [1, 2]. The chart in Fig. 1 presents the percentage of failures in induction machine components based on the survey carried out by [2, 3] on the distribution of failed components in induction machines. Stator winding faults account for approximately 38% of all faults in induction machines. Similarly, in the survey carried out by EPRI (see Table 1), 36% of all faults in induction machines can be attributed to stator winding faults [4, 5]. Thus, it can be said that the stator (winding) fault contributes a significant proportion of the total number of faults.

There have been many researchers focusing on the new fault diagnosis and condition monitoring techniques for an electrical machine, especially induction machine for the past four decades. Undoubtedly, one of the major parts of the industries that cannot be replaced is induction machine. The machine is considered very important because they are extensively used not only in the industries where it is the core of most of the engineering processes but also in many home appliances. Therefore, it is very crucial that the machine does not break down, particularly for process chains continuity and productions in many industries. The risk of the failure or the breaking down of this machine can be circumvented provided there is a proper diagnostic technique. The technique in question is required to detect coming failure/faults at an early stage. This will prevent production shutdowns, huge financial loss, sudden disruption of the machine and personal injuries if these faults are detected at the incipient stage. From literature, there are many methods for fault detection and diagnosis of the machine, however, DC-centered periodogram has not really been applied to detect and diagnose faults in electrical machine [2, 6–9].

In this chapter, Sect. 2 gives the brief explanation of shorted turns (winding) fault and Sect. 3 discusses the experiment to capture data from both Healthy and Induction machines with shorted-turns fault. Section 4 defines the periodogram and

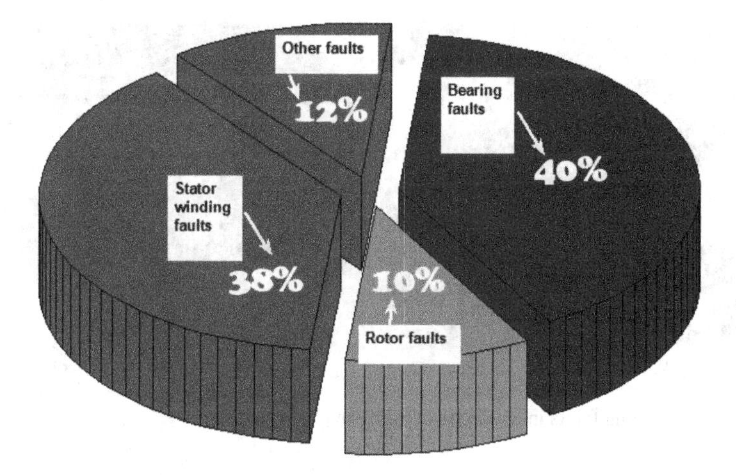

Fig. 1 Percentage chart of faults in an induction machine [2]

Table 1 Comparison IEEE-IAS, Allianz surveys and EPRI [2]

Types of fault	% of faults/failures in induction machine		
	IEEE-IAS	Allianz	EPRI
Bearing related	44	13	41
Stator related	26	66	36
Rotor related	8	13	9
Others	22	8	14

also describes how DC-centered periodogram can be used to detect stator shorted turns fault in an induction machine. The conclusion of the chapter is given in Sect. 5.

2 Winding Fault

Winding faults are related to the turns of wires in both the stator and rotor of electrical machines. The rotor winding faults are not very common. However, stator winding faults contribute a significant proportion (about 38%) of the total number of faults that occur. It can be said that all the faults in the winding commence with the failure of the windings insulation [10, 11].

The stator winding faults are often initiated with inter-turn or shorted-turn which short-circuits a few nearby turns of a phase winding. The shorted turn fault is caused by insulation failure between the turns of the individual windings in either stator of the machine [12]. As the machine continues to operate, the current circulating within the shorted-turns generates heat and temperature increases in the affected area. The rise in temperature leads to further destruction in the insulation of the affected area

(a) Winding shorted turn-to-turn (b) Winding with shorted-coil

Fig. 2 Stator winding faults in an electrical machine [2]

and this can lead to a short circuit between coils of the same phase. This is a more severe fault. However, the machine could still be in operation and increase the severity of the faults into, phase to ground, phase to phase or Open-circuit (in a phase) faults. At this stage, the protective equipment may disconnect the machine from the supply. The general opinion of the users and manufacturers is that there is a longer lead-time between the inception of shorted turns up to failure in the winding. Even if there is no enough knowledge about the time interval from the shorted-turns fault to insulation failure, but it is clear that transition and its rate depend on the severity of the fault. In other words, the number of the shorted-turns has gradual and slow increases to insulation failure. Thus, the earlier the shorted-turn faults are detected the better for the machine.

Figures 2a and b shows winding shorted turn-to-turn and winding with shorted coils faults respectively. Shorted turn forms the genesis and elementary of winding faults in the electrical machine. It can be seen from Fig. 2b that the fault is becoming more severe and this could damage the machine if it continues to operate. As the unit on the machine ages, shorted-turn problems are more likely to be experienced. The stresses involved in each stop-start cycle play an important role in the development of shorted turns faults. The major stresses that caused shouted turns faults are Thermal, Mechanical, Environmental and Electrical [10].

3 Experiment to Capture Data of Induction Machine

Laboratory experiments were carried out on 1.5 kW, 380 V/220 V, 50 Hz, 4-pole induction machine as shown in Fig. 3. The detail parameters of the machine are given in the appendix as shown in Table 3 [9]. Switches are connected to the stator winding on phase *A* on the machine to create a shorted-turns fault in the winding faults on the phase. The data obtained during healthy (normal) and shorted-turns fault conditions are captured by the HIOKI 3197-Power Quality Analyser measur-

Fig. 3 Experimental Set-up for 1.5 kW induction motor for data capturing [2]

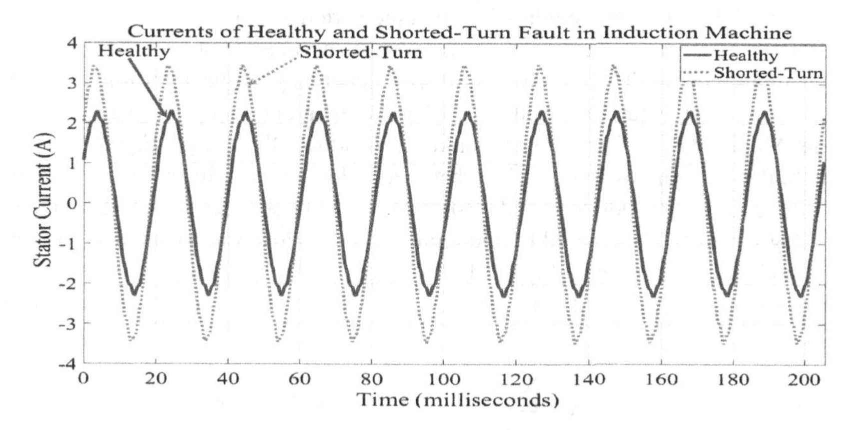

Fig. 4 Comparison of stator currents of healthy and shorted turns fault induction machine [2]

ing device. These data captured are interfaced with the computer for application of DC-centered periodogram. When the machine is in operation, data capture from the HIOKI Power Quality Analyser is recorded and this represents the data for the induction machine healthy state. However, when the switch for the shorted-turn faulty state is on, a shorted-turn fault is created. The data is captured by the HIOKI represent data for induction machine with a stator (winding) shorted-turns fault conditions. Figure 4 shows the comparison of the phase-A current of both healthy and faulty conditions on the machine. A close look at healthy and shorted-turn fault condition is in agreement with similar comparison carried out by [13]. From the Fig. 4, the peak value of current for an induction machine in the healthy condition is 2.32 A and the peak of stator currents for shorted-turns faults condition is 3.48 A. There is an increment of about 50% for shorted-turns. This abnormality is observed and it could grow into more severe winding faults which can destroy the machine if it continues to run. Section 4 discusses the analysis of each signal captured using DC-centered periodogram.

4 Periodogram Algorithm

The periodogram is a nonparametric estimate of the power spectral density (PSD) of an input signal. It is one of the earliest statistical tools utilised for studying periodic tendency and the PSD in time series is periodogram. It is also referred to as the direct method which earlier introduced by Arthur Schuster in 1898. Arthur Schuster directly estimates the PSD by computing the squared-magnitude of the DFT of the signal itself. The periodogram method is one of the primitive, simplest and most classic spectra estimation methods which is still used today, except for the Discrete Fourier Transform (DFT), which is replaced by the FFT. A similar relationship between PSD and DFT can be found in [14, 15]. The spectral estimation method is periodicity. Therefore, the well-known method is called the periodogram.

The periodogram is the Fourier transform of the biased estimate of the auto-correlation sequence. There are various forms depending on the algorithm that will analyse the signal features in a better way. The forms are; Periodogram Using Default Inputs, Modified Periodogram with Hamming Window, DFT Length Equal to Signal Length, Periodogram of Relative Numbers, Periodogram at a Given Set of Normalized Frequencies, Periodogram PSD Estimate of a Multichannel Signal, Reassigned Periodogram and DC-Centered Periodogram [16–18]. However, a general algorithm for periodogram is described Eq. 1 [2]

For a signal, y_n, sampled at f_s samples per unit time, the periodogram is defined as:

$$\hat{P}(f) = \frac{\Delta t}{N}\left|\sum_{n=0}^{N-1} y_n e^{-i2\pi f n}\right|^2, \frac{-1}{2\Delta t} < f < \frac{1}{2\Delta t} \tag{1}$$

where Δt is the sampling interval. For a one-sided periodogram, the values at all frequencies except 0 and the Nyquist, $1/2\Delta$t, are multiplied by 2 so that the total power is conserved. If the frequencies are in radians/sample, the periodogram is defined as:

$$\hat{P}(f) = \frac{1}{2\pi N}\left|\sum_{n=0}^{N-1} y_n e^{-i\omega n}\right|^2, \pi < \omega < \pi. \tag{2}$$

The frequency range in the Eq. 2, has variations depending on the value of the input sampling rate argument. The integral of the true PSD, $P(f)$, over one period, $1/\Delta t$ for cyclical frequency and 2π for normalised frequency, is equal to the variance of the signal in Eq. 3.

$$\sigma^2 = \int_{\frac{-1}{2\Delta t}}^{\frac{1}{2\Delta t}} P(f)df \tag{3}$$

If the normalised frequencies are required for Eq. 3, the limits of integration are replaced appropriately in a similar way to Eq. 2.

4.1 DC-Centered Periodogram Application

The measuring device (HIOKI 3197-Power Quality Analyser) that is used the experiment of Sect. 3 that captures all the signals required before and after the fault conditions. It should be noted that the sampling frequency f_s of the captured signals is very important for the analysis. In this case, the numerical data captured for the samples for 50 Hz (i.e. 20 ms/cycle) consists of 2056 samples/s based on the findings from the measuring device(HIOKI 3197-Power Quality Analyser) manual (10 cycle/s).

Sampling frequency, f_s of the captured signals is very important for the analysis. In this case, the number data captured for samples for 50 Hz (i.e. 20 ms/cycle) is 2056 samples/s based on the findings from the device manual (10 cycle/s). The frequency is measured in cycles/second, more commonly known as "Hertz". For example, the electric power we use in our daily life in South Africa is 50 Hz. This means that if you try to plot the electric current, it will be a sine wave passing through the same point 50 times in 1 s.

The DC-centered periodogram of both healthy and shorted-turns signals is obtained according to the codes in the appendix. The 'centered' option in the codes is used to evaluate and determine the DC-centered periodogram for each condition. Figure 5 shows the plot the results. The peak of the healthy electrical machine at DC (0 Hz) is 0 dB/Hz, however, the peak of the machine with stator shorted-turn fault at DC (0 Hz) shoot above 0 dB/Hz. It is about 4.545 dB/Hz above 0 dB/Hz. This implies that an abnormality noted in Fig. 4 correspond to the 4.545 dB/Hz above 0 dB/Hz in Fig. 5. The DC-centered periodogram is very useful to know when there is a sudden change in the signal of a healthy electrical machine or when it starts deviating from a healthy state.

Figure 6 shows the results of another analysis carried out by [2, 19] to also detect the shorted-turn faults in induction machines using discrete wavelet transform. This result is compared with the one from Fig. 5 to produce Table 2 [2]. In both techniques,

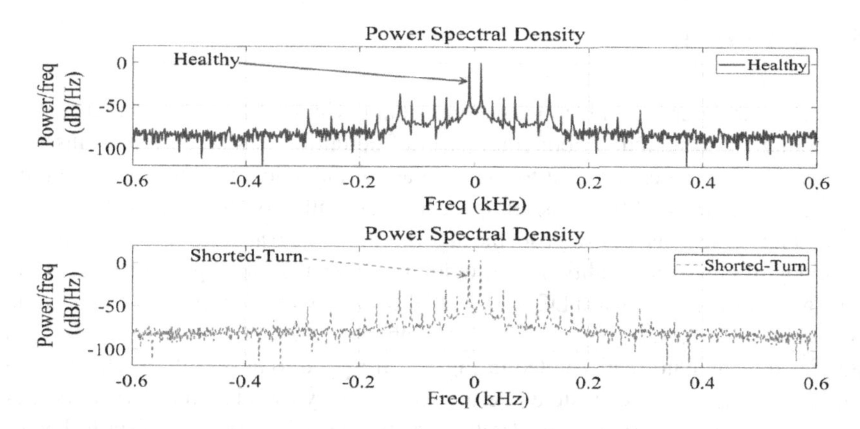

Fig. 5 DC-centered periodogram plot for induction machine under healthy and shorted fault conditions

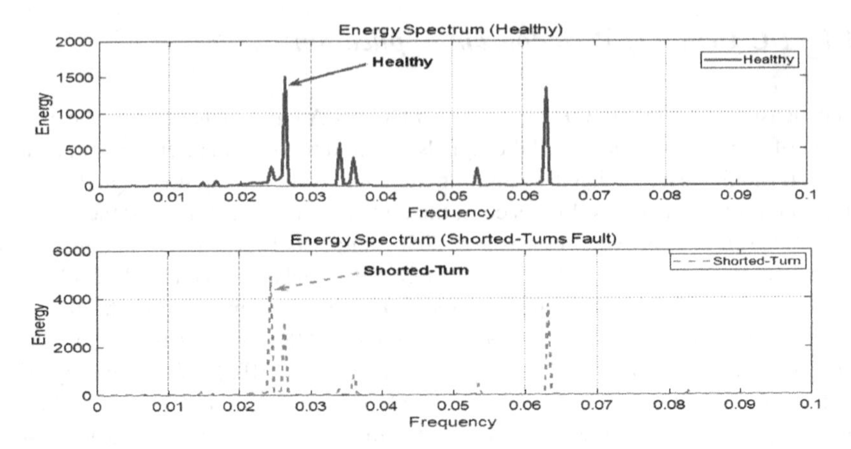

Fig. 6 DWT-Energy plot for induction machine under healthy and shorted fault conditions

Table 2 The comparison for DC-Periodogram PSD and DWT-Energy [2]

State of machine	DC-Periodogram PSD		DWT-Energy	
	Max. energy	Cor. Freq	Max PSD	Cor. Freq
Healthy (H)	0	0.01	1507	0.02626
Shorted-Turn (S)	4.545	0.01	3942	0.02432

the peak values of the shorted turn-(**S**) is greater than the peak of the healthy-(**H**) state. Thus, with the DC-periodogram PSD, the electrical machine can be placed under close monitor when the peak value of a healthy machine under operation is started deviating from 0 dB/Hz.

5 Conclusion

This chapter describes induction machine stator winding shorted-turn fault detection using DC-centered periodogram. A brief definition of shorted-turn faults was discussed. Then an experiment to capture data of induction machine under healthy and shorted fault condition was set up. A brief algorithm of the process of the periodogram was also discussed. From the algorithm, codes to analyses the DC-centered periodogram for each condition were written in Matlab. From Fig. 5, the peak of the healthy electrical machine at DC (0 Hz) is 0 dB/Hz, however, the peak of the machine with stator shorted-turn fault at DC (0 Hz) is about 4.545 dB/Hz above 0 dB/Hz. This implies that an abnormality noted in Fig. 4 correspond to the 4.545 dB/Hz above 0 dB/Hz in Fig. 5. The DC-centered Periodogram is very useful to know when there is a sudden change in the signal of a healthy electrical machine or when it starts deviating from a healthy state. The results obtained are compared with another technique

(DWT-Energy) using the same data and the peak values of the shorted turn-(**S**) is greater than the peak of the healthy-(**H**) state in both methods. Thus, with DC-periodogram method, an electrical machine can be placed under close monitoring for fault detection when the peak value of the PSD of a healthy machine under operation is beginning to deviate from 0 dB/Hz.

Acknowledgements The authors would like to thank Rand Water Professorial Chair (Electrical Engineering) of Tshwane University of Technology, Pretoria for financing the material required to carry out an experiment for the research. The authors would like to thank the National Research Foundation (NRF) for the financial support received for the research work.

Appendix

Induction Machine Parameters

Table 3 Nameplate information

Parameters	Rated values	Remarks
Voltage [V]	380	Y-Connection
Current [A]	3.7	
Power [kW]	1.5	
Power Factor [–]	0.79	
Speed [rev/min]	1500	
Frequency [Hz]	50	
Number of poles [–]	4	

Matlab Codes

```
f_s = 2056 ; % Number of samples/sec
load('I_healthy.mat') % Load Healthy Current
I_Norm=I_1a; %Phase A Current of the healthy Machine
subplot (2, 1, 1)
periodogram(I_Norm,[],length((I_Norm),f_s, 'centered') % DC-centered
periodogram plot for healthy currents
subplot (2, 1, 2)
load('I_Shorted.mat') % Load Shorted turn Current
I_Shorted=I_2a; %Phase A Current of the Machine with shorted turn fault
periodogram(I_Shorted,[],length(I_Shorted),f_s,'centered') % DC-centered
periodogram plot for Machine with shorted turn fault
```

References

1. J. Robinson, C. Whelan, N.K. Haggerty, Trends in advanced motor protection and monitoring. IEEE Trans. Ind. Appl. **40**, 853–860 (2004)
2. O. Imoru, Detection and diagnosis of faults in electrical machines. Doctoral Dissertation, Tshwane University of Technology, South Africa, 2017
3. W.T. Thomson, M. Fenger, Current signature analysis to detect induction motor faults. IEEE Ind. Appl. Mag. **7**, 26–34 (2001)
4. E.D. Mitronikas, A review on the faults of electric machines used in electric ships. Adv. Power Electron. (2013)
5. Z. Pinjia, D. Yi, T.G. Habetler, L. Bin, A survey of condition monitoring and protection methods for medium-voltage induction motors. IEEE Trans. Ind. Appl. **47**, 34–46 (2011)
6. O. Imoru, A. Jimoh, Y. Hamam, Origin and manifestation of electrical machine faults-a review, in *International Conference on Power Engineering and Renewable Energy (ICPERE)*, 2014, pp. 189–194
7. P.S. Bhowmik, S. Pradhan, M. Prakash, Fault diagnostic and monitoring methods of induction motor: a review. Int. J. Appl. Control Electr. Electron. Eng. (IJACEEE) **1**, 1–18 (2013)
8. C. da Costa, M. Kashiwagi, M.H. Mathias, Rotor failure detection of induction motors by wavelet transform and Fourier transform in non-stationary condition case studies. Mech. Syst. Signal Process. **1**, 15–26 (2015)
9. O. Imoru, M.A. Bhaskar, A.A. Jimoh, Y. Hamam, B.T. Abe, J. Tsado, Detection of stator shorted-turns faults in induction machine using dc-centered periodogram? in *Proceedings of The 25th World Congress on Engineering (WCE 2017)*. Lecture Notes in Engineering and Computer Science, 5–7 July 2017, London, U.K., pp. 376–379
10. A. Küçüker, M. Bayrak, Detection of stator winding fault in induction motor using instantaneous power signature analysis. Turk. J. Electr. Eng. Comput. Sci. (The Scientific and Technological Research Council of Turkey) **23**, 1263–1271 (2015)
11. A.H. Bonnett, G.C. Soukup, Cause and analysis of stator and rotor failures in three-phase squirrel-cage induction motors. IEEE Trans. Ind. Appl. **28**, 921–937 (1992)
12. R. Sharifi, M. Ebrahimi, Detection of stator winding faults in induction motors using three-phase current monitoring. ISA Trans. (Elsevier) **50**, 14–20 (2011)
13. O. Imoru, L. Mokate, A.A. Jimoh, Y. Hamam, *Diagnosis of Rotor Inter-turn Fault of Electrical Machine at Speed Using Stray Flux Test Method* (AFRICON, 2015), pp. 1–5
14. S.M. Alessio, *Digital Signal Processing and Spectral Analysis for Scientists: Concepts and Applications* (Springer, 2015)
15. M.H. Hayes, *Statistical Digital Signal Processing and Modeling* (Wiley, 1996)
16. F. Auger, P. Flandrin, Improving the readability of time-frequency and time-scale representations by the reassignment method. IEEE Trans. Signal Process. **43**, 1068–1089 (1995)
17. S.A. Fulop, K. Fitz, Algorithms for computing the time-corrected instantaneous frequency (reassigned) spectrogram, with applications. J. Acoust. Soc. Am. **119**, 360–371 (2006)
18. R. Kasim, A.R. Abdullah, N.A. Selamat, M.F. Baharom, N. Ahmad, Battery parameters identification analysis using periodogram. Appl. Mech. Mater. **785**, 687–691 (2015)
19. O. Imoru, M.A. Bhaskar, A.A. Jimoh, Y. Hamam, Diagnosis of stator shorted-turn faults in induction machines using discrete wavelet transform. Afr. J. Sci. Technol. Innov. Dev. (Taylor & Francis) **9**, 349–355 (2017)

An Intra-vehicular Wireless Sensor Network Based on Android Mobile Devices and Bluetooth Low Energy

José Augusto Afonso, Rita Baldaia da Costa e Silva and João Luiz Afonso

Abstract This chapter presents the development and test of an intra-vehicular wireless sensor network (IVWSN), based on Bluetooth Low Energy (BLE), designed to present to the driver, in real-time, information collected from multiple sensors distributed inside of the car, using a human-machine interface (HMI) implemented on an Android smartphone. The architecture of the implemented BLE network is composed by the smartphone, which has the role of central station, and two BLE modules (peripheral stations) based on the CC2540 system-on-chip (SoC), which collect relevant sensor information from the battery system and the traction system of a plug-in electric car. Results based on an experimental performance evaluation of the wireless network show that the network is able to satisfy the application requirements, as long as the network parameters are properly configured taking into account the peculiarities of the BLE data transfer modes and the observed limitations of the BLE platform used in the implementation of the IVWSN.

Keywords Android · Bluetooth Low Energy · Electric vehicle · Human-machine interface · Intra-vehicular networks · Mobile phone sensing · Performance evaluation · Wireless sensor networks

J. A. Afonso (✉) · R. B. da Costa e Silva
CMEMS-UMinho R&D Center, University of Minho, 4800-058 Guimarães, Portugal
e-mail: jose.afonso@dei.uminho.pt

R. B. da Costa e Silva
e-mail: a58677@dei.uminho.pt

J. L. Afonso
Centro Algoritmi, University of Minho, 4800-058 Guimarães, Portugal
e-mail: jla@dei.uminho.pt

S.-I. Ao et al. (eds.), *Transactions on Engineering Technologies*,
https://doi.org/10.1007/978-981-13-0746-1_23

1 Introduction

Modern vehicles are highly automated, with the main functionalities controlled by multiple microprocessor-based electronic control units (ECU) spread inside the vehicle, which collect data from sensors and communicate with actuators and data sinks. Currently, these devices are mostly connected through cables using a network technology called CAN (Controller Area Network) [1].

The increasing complexity of vehicles and the rise of the number of applications and devices that they encompass increase the quantity of cables required, which introduces challenges such as increased weight, limitations on the placement of sensors and the change of cables when necessary [2]. The replacement of cables by wireless links has the potential to reduce the weight of the vehicle, resulting in lower fuel consumption, increase the mobility and flexibility of the system, and decrease the assembly and maintenance costs.

Given these advantages, the use of wireless technologies was recently proposed to provide the required communication inside the vehicles, forming a new type of network called IVWSN (Intra-Vehicular Wireless Sensor Network) [3].

The transition to a wireless network must be gradual, starting with non-critical systems and sensors in areas not easily accessed with cables, such as inside the car tires. The flexibility and convenience provided by wireless networks also allow the installation of sensors on demand, for example, to measure the temperature or other physical variable on a given place of the car when desired. It also makes possible the integration of wearable sensors [4] to monitor the driver's physiological state and provide alerts when the user is not in condition to drive.

The choice of the wireless sensor node technology for IVWSNs must take into account the following requirements [2]: low cost, low energy consumption, low latency, high reliability and support for messages with different priorities.

As discussed in the next section, Bluetooth Low Energy (BLE) stands as the most promising wireless technology currently available on the market to fulfill these requirements. Another advantage of BLE over some of the competing technologies, such as ZigBee, is the native support on smartphones. These reasons motivated the choice of BLE for the development of the intra-vehicular system described in this chapter. Given the importance of reliability in the context of these networks, this chapter also presents an experimental evaluation of the packet error rate (PER) achieved with BLE under different conditions.

The system described in this chapter was designed, implemented and tested using a plug-in electric vehicle (Fig. 1) named CEPIUM (*Carro Elétrico Plug-In da Universidade do Minho—in Portuguese language*), which was developed by the Group of Energy and Power Electronics (GEPE) of University of Minho.

This car is a Volkswagen Polo where the internal combustion engine parts were replaced by an electric motor, batteries and the electronic circuits required for the conversion into an electric vehicle [5].

This chapter provides a revised and extended version of a conference paper [6]. The main contributions presented in this chapter are: (1) The development of a system

Fig. 1 Electric vehicle used in the development of the IVWSN

for data collection (from real sensor devices inside an electric vehicle), wireless transmission (using BLE) and real-time presentation (on a smartphone installed in the vehicle cockpit); (2) A performance evaluation of BLE in the context of IVWSNs, with the comparison between two data transfer modes: notifications and indications.

2 Related Work

There are several short-range wireless network technologies available currently in the market: IEEE 802.15.4/ZigBee [7], IEEE 802.11/Wi-Fi [8], Bluetooth, among others. Bluetooth Low Energy (BLE), which was introduced in the Bluetooth 4.0 [9] standard, was developed by the Bluetooth Special Interest Group (SIG) as a low-power solution for monitoring and control applications. BLE appeared in response to a need to increase the lifetime of wireless devices powered by batteries, such as fitness and healthcare devices, wireless computer mice and keyboards, among others.

At the physical layer, BLE implements the LE (Low Energy) controller, which is not directly compatible with the BR (Basic Rate)/EDR (Enhanced Data Rate) controllers used by classic Bluetooth. Both versions operate in the 2.4 GHz ISM (Industrial, Scientific and Medical) frequency band. The BLE connection interval used for data transmission can be configured from 7.5 ms to several seconds, depending on the balance between latency and energy consumption requirements of the target application.

In [10], Tsai et al. evaluate the performance of ZigBee inside a vehicle in different places, under varied scenarios, with the engine on and off. The authors analyze the use of the Received Signal Strength Indicator (RSSI) and Link Quality Indicator (LQI) to evaluate the link quality and propose a detection algorithm based on RSSI/LQI/error patterns and an adaptive strategy to increase the link goodput while improving the power consumption.

Ahmed et al. [11] analyze the characteristics and performance of some wireless network technologies in the context of intra-vehicular networks and investigate issues related to the replacement of cabling between sensors/ECUs by wireless links. The authors compares RFID (Passive), Bluetooth, IEEE 802.15.4/ZigBee and UWB (Ultra-Wideband) standards. Based on this analysis and the requirements of IVWSNs, such as support for short payloads with low overhead, low latency and low transceiver complexity, the authors concluded that the most suitable wireless technologies for IVWSNs, among the studied ones, were the IEEE 802.15.4 protocols and an emerging UWB proposal. However, at the time of this study Bluetooth Low Energy was not available yet.

Most available wireless technologies with applicability in the context of IVWSNs share the same spectrum (the 2.4 GHz ISM band) with IEEE 802.11/Wi-Fi and Bluetooth networks. Therefore, it is important to evaluate the performance of potential IVWSNs technologies under interference of these networks. Such evaluation was conducted by Lin et al. [12] for both states of the vehicle: stopped and in movement. Experimental results indicate that Bluetooth Low Energy outperforms ZigBee in the vehicle under Wi-Fi interference. These results can be explained by the use of frequency hopping spread spectrum (FHSS) by the former, which increases the robustness against fading and interference because the transmissions are spread along all the available 2.4 GHz band, whereas ZigBee transmissions are limited to a fixed channel.

The authors in [13] provide an experimental evaluation of the power consumption of BLE using CC2540 modules (the same modules used in this chapter). With one data transmission per second and a sleep current of 10 μA, an average current consumption of just 23.9 μA was obtained. Assuming the use of a common CR2032 coin cell battery with capacity of 230 mAh, it can be concluded that the module would be able to operate continuously for 400 days with the same battery. These results show that BLE is a suitable technology for applications that require low cost and low power consumption.

In [14], Afonso et al. evaluate the performance of BLE using notifications in the context of body area networks (BAN) with multiple sensor nodes generating high data rate traffic. Results show that the BLE protocol is suitable to the task. Moreover, in comparison with an alternative enhanced IEEE 802.15.4-based MAC protocol [15], it is able to consume less energy and supports more sensor nodes due to its higher physical layer data rate.

Unlike [14], this chapter also analyzes the performance of BLE with indications, besides notifications, and in the context of intra-vehicular networks.

3 System Development

The BLE-based intra-vehicular network developed in this work is composed by two types of nodes: a central station (smartphone) placed near the driver at the car cockpit, and multiple peripheral stations (BLE modules) placed all around the car near the sensors, as shown in Fig. 2.

For this prototype two peripheral stations were developed to monitor relevant variables associated to the operation of the electric vehicle: the battery system node, placed below the rear seats, and the traction system node, placed inside the hood. In the future, more peripheral stations can be added to the network as required.

These nodes collect sensor data from two electronic systems developed by GEPE: the battery system and the traction system. The battery system provides information related to the parameters of the battery and the electric charger [16], whereas the traction system provides data related to the parameters of the electric motor controller, the state of charge and the temperature of the motor. All data is collected from the respective electronic systems via UART (Universal Asynchronous Receiver/Transmitter), except for the motor temperature, which is collected using the ADC (Analog-to-Digital Converter) of the BLE module.

The smartphone plays both the roles of central station of the BLE network and human-machine interface (HMI) of the system. It receives the sensor data from the peripheral stations via BLE, converts the integer values contained in the data frames into real values according to the respective units of measurement, and presents the information to the user in real-time.

3.1 *Embedded Software*

The hardware of the peripheral stations is based on the CC2540EM evaluation module, from Texas Instruments (TI). This module integrates an 8051-based microcontroller and a BLE transceiver in the same chip, as well as a connector for an external antenna and auxiliary components. During the development phase, each CC2540EM module was connected to a SmartRF05 EB evaluation board, for easy access to the I/O pins and the download of the developed code to the microcontroller of the CC2540 system-on-chip (SoC). In an early phase the CC2530EM module was also used for the central station, but it was replaced later by an Android smartphone for the development of the HMI. The development of the embedded software in the BLE modules was made using C language and the IAR Embedded Workbench for 8051 IDE (Integrated Development Environment). The SmartRF Packet Sniffer was used for system testing along with a CC2540 USB Dongle.

The first stage of the code development was the establishment of periodic data transfer from the peripheral stations to the central station. It was based on the use of two example projects supplied by TI along with the BLE stack, version BLE-CC2540-1.3.2, named SimpleBLECentral e SimpleBLEPeripheral, as well as an

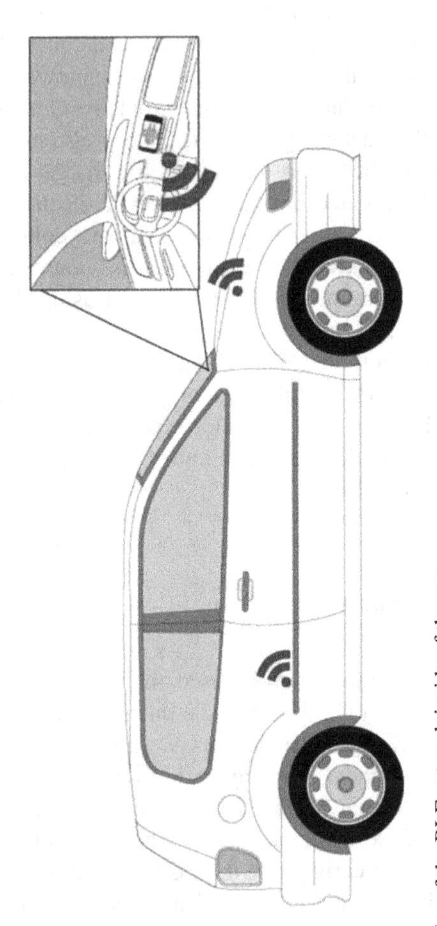

Fig. 2 Placement of the nodes of the BLE network inside of the car

adaptation of the service "Send Data Service" developed in [17]. The data transfer was implemented using both notifications and indications.

For this purpose, it was necessary to implement a profile to provide this service in the peripheral station, which assumed the role of GATT (Generic Attribute Profile) server. The periodic event SBP_PERIODIC_EVT, which was configured to occur every 500 ms, calls the function PeriodicSendDataTask() to send the data from the sensors attached to that node, which are collected with the same period under the event UART_PERIODIC_EVT.

The service responsible for sending the data offers a profile with a single characteristic with maximum length of 20 bytes and a descriptor (CCC—Client Characteristic Configuration). If the network is configured for indications, the peripheral station waits for an acknowledgment (ACK) for each data frame sent.

It was necessary to make changes on the original code to allow the communication using indications. In the peripheral station, it was needed to create a characteristic which assumes the value GATT_PROP_INDICATE and register the CCC descriptor with the value 0x0002. It was also necessary to change the data structure to attHandleValueInd_t and, finally, to use the function GATT_Indication to send the data frames to the central station. In the central station, it was only necessary to add the function ATT_HandleValueCfm to send the acknowledgement for each frame correctly received.

3.2 Communication Protocol

In order to regulate the transfer of data from either the battery system or the traction system and the respective BLE module (peripheral station) via UART, a polling protocol was implemented, with each BLE module as master and the respective electronic system as slave. The master polls the slave with a period defined by the event UART_PERIODIC_EVT and waits for the response. The poll frame contains a single field (1 byte) which indicates the data content requested from the attached electronic system. The data frame sent by the battery system is composed by the system address, frame type, frame length and the samples from its 6 sensors: grid voltage, grid current, battery voltage, battery current, bus voltage and temperature. Likewise, the data frame sent by the traction system is composed by the system address, frame type, frame length and the samples from its 6 sensors: controller state, controller voltage, controller current, state of charge, controller power and controller temperature. An error message frame is sent if the electronic system is not able to send the requested data frame.

3.3 Smartphone Application

The Android [18] application was developed for Android 6.0 using a Motorola Moto G (2nd gen) mobile device and the Android Studio 2.1 IDE. It was designed to provide a simple and flexible human-machine interface, offering access to the data collected from the car sensors in an organized way. For this purpose, the application was created using mainly fragments, simplifying the reuse of components of the user interface.

The developed application uses four activities. The main activity (starts menu) serves as the initial user interface of the application and is responsible for the management of fragments. It provides access to the battery system screen, the traction system screen and the weather information screen. This activity uses a navigation drawer that manages the different fragments in order to guarantee their visibility. This panel can be accessed with a click on the Android hamburger icon or with a swipe from left to right.

The second activity is responsible for the BLE device discovery, connection establishment and activation of the reception of sensor data from the peripheral stations using notifications. The BluetoothLeService file is responsible for managing these features. Besides playing the role of central station, the smartphone also acts as the ATT (Attribute Protocol) client, accessing data from the peripheral stations, which assume the role of ATT servers, maintaining a set of attributes and storing information managed by the GATT protocol. In order to use Bluetooth on an Android smartphone, it is necessary to create the BluetoothAdapter object. For device discovery, the startLeScan() method is called to start searching for active BLE devices, whereas the stopLeScan() is used to stop this search. By default in the Google API, this scan process stops after 4 s, which is sufficient to find the peripheral stations.

The third activity implemented in the smartphone application is responsible for the sensor data presentation. After the notifications are activated, this activity transfers the data frames received from the peripheral stations to the DataHandler class, to extract and process the sensor data contained in the fields of the data frames in real time. The periodic data reception associated to this task is managed by a background service, allowing the user to perform other operations simultaneously.

The last implemented activity provides weather information (current weather and forecast for the next 3 days), based on the current location of the mobile device. This weather information is collected from the online service OpenWeatherMap using its JSON (JavaScript Object Notation) API (Application Program Interface).

Table 1 Sensor values for the battery system

Sensor data	Minimum value	Nominal value	Maximum value	Unit
Grid voltage	210	230	240	V
Grid current	–	16	20	A
Battery voltage	–	300	360	V
Battery current	–	10	13	A
Bus voltage	–	430	450	V
Temperature	–	–	100	°C

4 Results and Discussion

4.1 Overall System

The goal of this first test was to evaluate the correct operation of all parts of the developed system, from the collection of data from the sensors of the electronic system using the developed communication protocol to the presentation of this information on the smartphone. Table 1 presents the expected values for the different sensors of the battery system, whereas Fig. 3 shows an example of the presentation of this sensor values on the screen of the smartphone (which was configured with limits larger than the specified on the table). The same test was performed for the traction system, with similar satisfactory results.

4.2 Setup for the BLE Performance Tests

The following experimental tests evaluate the BLE network reliability by measuring the packet error rate (PER) after the transmission of 1000 data frames from a peripheral station to a central station using CC2540EM modules and either notifications or indications. A packet is only considered lost if the corresponding data frame is not successfully delivered to the central station, either in the first attempt or after retransmissions (in the case of indications).

After a BLE connection between the central station (master) and a peripheral station (slave) is established, time is divided into connection events [9]. Each connection event starts with a packet from the master (poll frame), which is followed by a packet from the slave (data frame). Depending on the configuration, the connection event may then terminate or continue with the transmission of subsequent packets, alternating from the master and the slave. If the slave latency parameter is equal to zero, the interval between active connection events is equal to the connection interval.

The tests were performed using four different transmitter (TX) power levels: −23, −6, 0 and 4 dBm. These tests used the maximum allowed payload length (20 bytes) and a single data frame per notification or indication. The period between

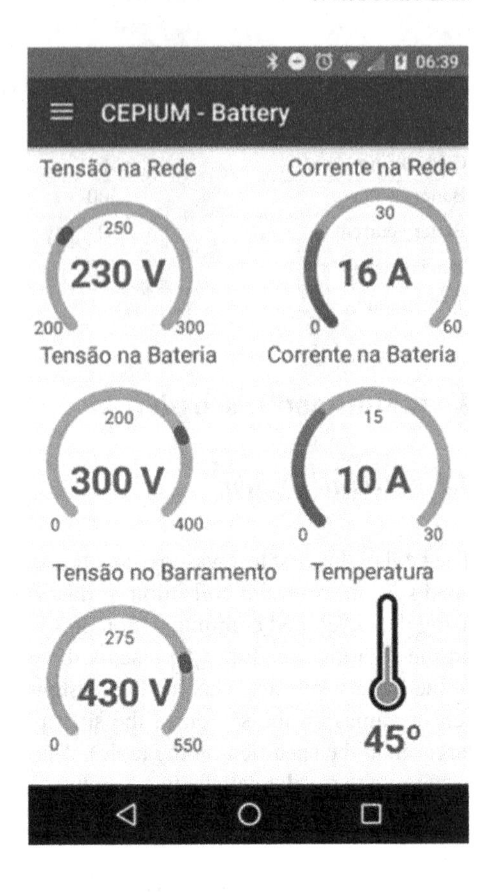

Fig. 3 App screen for presentation of the battery system sensor values

connection events is a BLE parameter called connection interval (T_{ci}) and the period of generation of data frames is a TI stack parameter called the data send period (T_{ds}) in this chapter. All tests with notifications were made with $T_{ds} = T_{ci}$, whereas in the tests with indications the value of T_{ds} was a multiple of T_{ci}.

4.3 Performance Tests Inside of the Car

These tests were performed with the goal to evaluate the communication reliability inside of the vehicle, in order to assess if its metallic parts might have effect on the strength of the signal and if the interference from the electric motor might cause errors. This tests were performed using notifications and the minimum TX power (−23 dBm), with the central station placed at the car cockpit and the peripheral station at different places on the vehicle. For the first test, the peripheral station was placed near the electric motor, with the hood closed, while the motor was accelerated. No packet errors were registered during this test. The same test was repeated with

Table 2 PER with notifications at distance of 1 m

Connection interval (ms)	PER (%)
7.5	78
10	0

the peripheral station near the battery system, the front lights, the rear lights and the four tires, with the same result (PER = 0). Therefore, it was not necessary to make further tests with higher TX power levels.

4.4 Tests at Short Distance

The goal of the following tests is to determine the minimum value for the connection interval and the data send interval from which communications are not affected by errors that are caused exclusively due to limitations of the TI BLE platform implementation, both for notifications and for indications. For this purpose, the tests were made in a controlled setup to exclude any other possible causes of packet errors: low path loss (short distance of 1 m between the stations), no obstructions and no interference from other sources such as Wi-Fi networks.

Table 2 presents the results with notifications using TX power of −23 dBm. The same results were obtained with higher power levels. No packet errors were registered for T_{ci} equal to 10 ms and above. The Bluetooth specifications state that the connection interval may be configured to any multiple of 1.25 ms in the range of 7.5 ms to 4.0 s. However, the tested BLE platform does not perform as expected with $T_{ci} = 7.5$ ms.

The same test was repeated using indications instead of notifications. In this case, besides defining the value of the connection interval, it is also necessary to specify the value of the data send period as a multiple of the former. For each data frame sent it is necessary to wait for the reception of an ACK frame in the next connection event. When the ACK is not received the BLE stack retransmits the data frame at each subsequent connection interval until the ACK is received. If the data send period is exceeded while waiting for the ACK, the peripheral station blocks the transmission of the next data packet, causing its loss. Therefore, for proper operation of indications, it is necessary to make T_{ds} at least twice the value of T_{ci}, even when no retransmissions are required due to the absence of channel errors. This point was confirmed through tests using $T_{ds} = T_{ci}$, with connection intervals ranging from 7.5 to 25 ms, which resulted in PER values ranging from 53 to 55%.

Further tests were made using $T_{ds} = 2\ T_{ci}$ for different values of the connection interval. Table 3 shows the results obtained with −23 dBm. The same results were obtained with −6, 0 and 4 dBm. Although, theoretically, configuring $T_{ds} = 2\ T_{ci}$ should be sufficient to obtain PER = 0, since the occurrence of channel errors (and consequently, the need for retransmissions) was excluded by the way the test was set

Table 3 PER with indications at distance of 1 m when $T_{ds} = 2\,T_{ci}$

Connection interval (ms)	Data send period (ms)	PER (%)
7.5	15	25
10	20	17
15	30	11
20	40	7
25	50	7
50	100	3
60	120	3
100	200	0

up, these results show that the tested BLE platform introduces errors for low values of connection interval. Based on these results, it can be concluded that the minimum connection interval with indications recommended in this case is 100 ms. Moreover, to allocate the required connection events for retransmissions in real world scenarios susceptible to channel errors, it is necessary to make $T_{ds} > 2\,T_{ci}$.

To accommodate at least one retransmission with indications, it is necessary to make the data send period at least three times higher than the connection interval. Given these results, we conclude that for a reliable operation with the used BLE platform and indications, the minimum connection interval should be 100 ms and the minimum data send period should be 300 ms.

These results also show that the maximum possible throughput with indications in scenarios without channel errors is much lower than with notifications, because the minimum recommended data send period is much higher. Moreover, it is only possible to transmit one indication per connection event, whereas the same platform allows the transmission of up to three notifications in the same event, as verified in [14]. Nevertheless, for the purpose of the intra-vehicular network presented in this chapter, the platform may operate with either notifications or indications, since it was configured to send a single data frame each time with a data send period of 500 ms.

4.5 *Tests at Larger Distance*

The final test aims to evaluate the influence of the TX power and the connection interval, using notifications, on the data frame transmission reliability in a scenario where the stations are separated by a larger distance (10 m) and with obstructions. Table 4 presents the results for this test. The minimum value for the connection interval presented in the table was 25 ms because it was not possible to establish connection with lower values, which means that the distance also has influence on the recommended minimum connection interval.

These results show that the TX power has strong influence on the BLE transmission reliability for larger distances and that the minimum TX power is not adequate

Table 4 PER with notifications at 10 m for different TX power values

Data send period (ms)	PER (%) −23 dBm	PER (%) −6 dBm	PER (%) 0 dBm
25	100	30	0
33.75	88	0	0
50	76	0	0
60	50	0	0
80	48	0	0
100	43	0	0

in this case. With −6 dBm it was possible to achieve error-free communication using connection intervals starting from 33.75 ms and with 0 dBm from 25 ms.

5 Conclusion and Future Work

The main objectives of this work were the development of a BLE network to collect sensor data inside a vehicle, the development of an Android app to present the collected information to the driver in real-time and the execution of experimental tests to evaluate the reliability of the BLE network with both notifications and indications.

Based on the literature review and obtained experimental results, it can be concluded that the characteristics and performance of Bluetooth Low Energy make it one of the most promising technologies for the deployment of intra-vehicular wireless sensor networks when compared to the existing alternatives. Nevertheless, eventual limitations of the used BLE platform should be taken into account during the implementation and configuration of the network.

In the future, the developed system will evolve to connect the smartphone application to an IoT (Internet of Things) [19] architecture composed by an online database service, a client application and a residential WSN (Wireless Sensor Network) [20]. Together with the integration of new sensors and actuators into the system, this evolution will allow the monitoring and control of parameters of the vehicle anywhere in the world through the Internet, enabling the provision of several new features, such as the manual/automatic remote control of the battery charging current according to the user's preferences.

Acknowledgements This work is supported by FCT with the reference project UID/EEA/04436/2013, COMPETE 2020 with the code POCI-01-0145-FEDER-006941.

References

1. M. Di Natale, H. Zeng, P. Giusto, A. Ghosal, *Understanding and Using the Controller Area Network Communication Protocol: Theory and Practice* (Springer Science & Business Media, 2012)
2. J. Lin, T. Talty, O. Tonguz, On the potential of Bluetooth Low Energy technology for vehicular applications. IEEE Commun. Mag. **53**(1), 267–275 (2015)
3. L.M. Borges, F.J. Velez, A.S. Lebres, Survey on the characterization and classification of wireless sensor network applications. IEEE Commun. Surv. Tutor. **16**(4), 1860–1890 (2014)
4. A. Pantelopoulos, N.G. Bourbakis, A survey on wearable sensor-based systems for health monitoring and prognosis. IEEE Trans. Syst. Man Cybern. Part C Appl. Rev. **40**(1), 1–12 (2010)
5. D. Pedrosa, V. Monteiro, H. Gonçalves, J.S. Martins, J.L. Afonso, A case study on the conversion of an internal combustion engine vehicle into an electric vehicle, in *IEEE Vehicle Power and Propulsion Conference (VPPC 2014)*, Oct 2014
6. R.B.C. Silva, J.A. Afonso, J.L. Afonso, Development and test of an intra-vehicular network based on Bluetooth Low Energy, in *Lecture Notes in Engineering and Computer Science: Proceedings of The World Congress on Engineering 2017*, 5–7 July 2017, London, U.K., pp. 508–512
7. P. Castro, J.L. Afonso, J.A. Afonso, A low-cost ZigBee-based wireless industrial automation system, in *12th Portuguese Conference on Automatic Control (CONTROLO 2016)*, Sept 2016, pp. 739–749
8. IEEE Std 802.11-2016, IEEE Standard for Information technology—Telecommunications and information exchange between systems Local and metropolitan area networks—Specific requirements—Part 11: Wireless LAN Medium Access Control (MAC) and Physical Layer (PHY) Specifications, 2016
9. Bluetooth SIG, Specification of the Bluetooth System. Master Table of Contents & Compliance Requirements, Version 4.0 [Vol 0], June 2010
10. H.M. Tsai, C. Saraydar, T. Talty, M. Ames, A. Macdonald, O.K. Tonguz, ZigBee-based intra-car wireless sensor network, in *IEEE International Conference on Communications* (2007), pp. 3965–3971
11. M. Ahmed, C.U. Saraydar, T. Elbatt, J. Yin, T. Talty, M. Ames, Intra-vehicular wireless networks, in *IEEE Global Communications Conference (GLOBECOM)* (2007), pp. 1–9
12. J.R. Lin, T. Talty, O.K. Tonguz, An empirical performance study of intra-vehicular wireless sensor networks under WiFi and Bluetooth interference, in *IEEE Global Communications Conference (GLOBECOM)* (2013), pp. 581–586
13. S. Kamath, J. Lindh, Measuring Bluetooth Low Energy power consumption, Application Note AN092, Texas Instruments (2012), pp. 1–24
14. J.A. Afonso, A.J.F. Maio, R. Simoes, Performance evaluation of Bluetooth Low Energy for high data rate body area networks. Wireless Pers. Commun. **90**(1), 121–141 (2016)
15. J.A. Afonso, H.D. Silva, P. Macedo, L.A. Rocha, An enhanced reservation-based MAC protocol for IEEE 802.15.4 networks. Sensors **11**(4), 3852–3873 (2011)
16. V. Monteiro, J.P. Carmo, J.G. Pinto, J.L. Afonso, A flexible infrastructure for dynamic power control of electric vehicle battery chargers. IEEE Trans. Veh. Technol. **65**(6), 4535–4547 (2016)
17. A.F. Maio, J.A. Afonso, Wireless cycling posture monitoring based on smartphones and Bluetooth Low Energy, in *Lecture Notes in Engineering and Computer Science: Proceedings of The World Congress on Engineering 2015*, 1–3 July 2015, London, U.K., pp. 653–657
18. N. Gandhewar, R. Sheikh, Google Android : an emerging software platform for mobile devices. Int. J. Comput. Sci. Eng. **12**, 12–17 (2010)
19. R. Want, B.N. Schilit, S. Jenson, Enabling the internet of things. IEEE Comput. **48**(1), 28–35 (2015)
20. M. Collotta, G. Pau, Bluetooth for internet of things: a fuzzy approach to improve power management in smart homes. Comput. Electr. Eng. **44**, 137–152 (2015)

Extended Performance Research on 5 GHz IEEE 802.11n WPA2 Laboratory Links

J. A. R. Pacheco de Carvalho, H. Veiga, C. F. Ribeiro Pacheco and A. D. Reis

Abstract Wireless communications by microwaves are increasingly important, e.g. Wi-Fi. Performance is a crucial issue, improving reliability and efficiency in communications. Security is also most important. Laboratory performance measurements are presented for Wi-Fi 5 GHz IEEE 802.11n WPA2 links. Our study contributes to performance evaluation of this technology under WPA2 encryption, using available equipments (HP V-M200 access points and Linksys WPC600 N adapters). New results are given, namely for OSI level 4, from TCP and UDP experiments. TCP throughput is measured versus TCP packet length. Jitter and percentage datagram loss are measured versus UDP datagram size. Results are compared for point-to-point, point-to-multipoint and four-node point-to-multipoint links. Comparisons are also made to corresponding data obtained for Open links. Conclusions are drawn about performance of the links.

Keywords Four-node point-to-multipoint · IEEE 802.11n · Point-to-multipoint
Point-to-point · TCP packet size · UDP datagram size · Wi-Fi · WLAN
Wireless network laboratory performance · WPA2

J. A. R. Pacheco de Carvalho (✉) · C. F. Ribeiro Pacheco · A. D. Reis
Departamento de Física, Grupo de Investigação APTEL, Universidade da Beira Interior, 6201-001 Covilhã, Portugal
e-mail: pacheco@ubi.pt

C. F. Ribeiro Pacheco
e-mail: a17597@ubi.pt

A. D. Reis
e-mail: adreis@ubi.pt

H. Veiga
Centro de Informática, Grupo de Investigação APTEL, Universidade da Beira Interior, 6201-001 Covilhã, Portugal
e-mail: hveiga@ubi.pt

S.-I. Ao et al. (eds.), *Transactions on Engineering Technologies*,
https://doi.org/10.1007/978-981-13-0746-1_24

1 Introduction

Contactless communications techniques have been developed using mainly several frequency ranges of electromagnetic waves, propagating in the air. The importance and utilization of wireless fidelity (Wi-Fi) and free space optics (FSO) have been growing. They are relevant examples of wireless communications technologies, using microwaves and laser light, respectively.

Wi-Fi gives versatility, mobility and reasonable prices. It has been increasingly important and used, as it complements traditional wired networks. Both ad hoc and infrastructure modes are possible. In the second case an access point (AP) permits communications of Wi-Fi electronic devices with a wired based local area network (LAN) through a switch/router. Thus, a wireless local area network (WLAN), based on the AP, is formed. At the personal home level a wireless personal area network (WPAN) permits communications of personal devices. Mainly, point-to-point (PTP) and point-to-multipoint (PTMP) 2.4 and 5 GHz microwave links are used, with IEEE 802.11a, 802.11b, 802.11g and 802.11n standards [1]. The increasing use of the 2.4 GHz band has led to growing electromagnetic interference. Therefore, the use of the 5 GHz band is an alternative, in spite of larger absorption and shorter ranges. Wi-Fi communications are not significantly affected by rain or fog, as typical wavelengths are in the range 5.6–12.5 cm. On the contrary, FSO communications are sensitive to rain or fog, as the typical wavelength range of the laser beam is 785–1550 nm.

Wi-Fi has nominal transfer rates up to 11 (802.11b), 54 Mbps (802.11a, g) and 600 Mbps (802.11n). The medium access control of Wi-Fi is carrier sense multiple access with collision avoidance (CSMA/CA). 802.11n offers higher data rates than 802.11a, g. These provide a multi-carrier modulation scheme called orthogonal frequency division multiplexing (OFDM) that allows for binary phase-shift keying (BPSK), quadrature phase-shift keying (QPSK) and quadrature amplitude modulation (QAM) of the 16-QAM and 64-QAM density types. One spatial stream (one antenna) and coding rates up to 3/4 are possible and a 20 MHz channel. 802.11n also uses OFDM, permitting, BPSK, QPSK, 16-QAM and 64-QAM. Up to four spatial streams are possible (four antennas), using multiple-input multiple-output (MIMO). MIMO permits to increase the capacity of a wireless link using multiple transmit and receive antennas to take advantage of multipath propagation. Antenna technology also favours 802.11n by introducing beam forming and diversity. Beam forming can be used both at the emitter and the receiver to achieve spatial selectivity to focus the radio signals along the path. Diversity uses, from a set of available multiple antennas, the best subset to obtain the highest quality and reliability of the wireless link. Coding rates up to 5/6 are possible and a 20/40 MHz channel. The standard guard interval (GI) in OFDM is 800 ns. Additional support for 400 ns GI provides an increase of 11% in data rate. Modulation and coding schemes (MCS) vary from 0 to 31. 600 Mbps are possible using MCS 31, four spatial streams, 64-QAM modulation, 5/6 coding rate, 40 MHz channel and 400 ns GI. 802.11n is suitable for transmitting e.g. high definition video and voice over IP (VoIP). Both the 2.4 and 5 GHz microwave frequency bands are usable.

There are studies on wireless communications, wave propagation [2, 3], practical implementations of WLANs [4], performance analysis of the effective transfer rate for 802.11b PTP links [5], 802.11b performance in crowded indoor environments [6].

Performance increase has been a crucial issue, giving more reliable and efficient communications. Requirements for both traditional and new telematic applications have been given [7].

Wi-Fi security is most important for confidentiality requirements. Microwave radio signals travel through the air and can be easily captured. Security methods have been developed to provide authentication such as, by increasing order of security, wired equivalent privacy (WEP), Wi-Fi protected access (WPA) and Wi-Fi protected access II (WPA2).

Various performance measurements have been published for 2.4 and 5 GHz Wi-Fi Open [8], WEP [9], WPA [10] and WPA2 [11] links, as well as very high speed FSO [12]. Performance evaluation of IEEE 802.11 based wireless mesh networks has been given [13]. Studies exist on modelling TCP throughput [14]. A formula that bounds average TCP throughput is available [15].

It matters to investigate the effects of TCP packet size, UDP datagram size, network topology, security encryption, on link performance and compare equipment performance for several standards. In the present work new Wi-Fi (IEEE 802.11n) results are given from measurements on 5 GHz WPA2 links, namely for OSI level 4, from TCP and UDP experiments. Performance is evaluated and compared in laboratory measurements of WPA2 PTP, three-node point-to-multipoint (PTMP) and four-node point-to-multipoint (4 N-PTMP) links using new available equipments. TCP throughput is measured versus TCP packet length. Jitter and percentage datagram loss are measured versus UDP datagram size. Comparisons are also made to corresponding results obtained for Open links [16]. In comparison to previous work [17], an extended research on performance is carried out.

In prior and actual state of the art, several Wi-Fi links and technologies have been researched. Performance evaluation has been pointed out as a crucially important criterion to assess communications quality. The motivation of the present work is to evaluate and compare performance in laboratory measurements of WPA2 PTP, PTMP and 4 N-PTMP 5 GHz links using new available equipments. This contribution permits to increase the knowledge about performance of Wi-Fi (IEEE 802.11n) links. The problem statement is that performance needs to be evaluated under several TCP and UDP parameterizations, link topologies and security encryption. The proposed solution uses an experimental setup and method, permitting to monitor signal to noise ratios (SNR) and noise levels (N), measure TCP throughput (from TCP connections) versus TCP packet size, and UDP jitter and percentage datagram loss (from UDP communications) versus UDP datagram size.

The rest of the paper is structured as follows: Sect. 2 involves the experimental conditions i.e. the measurement setup and procedure. Results and discussion are given in Sect. 3. Conclusions are drawn in Sect. 4.

2 Experimental Details

The measurements used a HP V-M200 access point [18], with three external dual-band 3 × 3 MIMO antennas, IEEE 802.11a/b/g/n, software version 5.4.1.0-01-16481, a 1000-Base-T/100-Base-TX/10-Base-T layer 2 3Com Gigabit switch 16 and a 100-Base-TX/10-Base-T layer 2 Allied Telesis AT-8000S/16 switch [19]. Three PCs were used having a PCMCIA IEEE.802.11a/b/g/n Linksys WPC600 N wireless adapter with three internal antennas [20], to enable 4 N- PTMP links to the access point. An interference free communication channel (chap. 36) was used to setup the links. This was mainly checked through a portable computer, equipped with a Wi-Fi 802.11a/b/g/n adapter, running Acrylic WiFi software [21]. WPA2 encryption with AES was activated in the AP and the wireless adapters of the PCs, with a key composed of twenty six hexadecimal characters. The experiments were made under far-field conditions. No power levels above 30 mW (15 dBm) were used, as the wireless equipments were close.

A versatile laboratory setup has been planned and implemented for the measurements, as shown in Fig. 1, based in [17]. It provides for up to three wireless links to the AP. At OSI level 4, measurements involved TCP connections and UDP communications using Iperf software [22]. For a TCP client/server connection (TCP New Reno, RFC 6582, was used), TCP throughput was obtained for a given TCP packet size, varying from 0.25 to 64 KB. For a UDP client/server communication with a given bandwidth parameter, UDP jitter and percentage loss of datagrams were determined for a given UDP datagram size, varying from 0.25 to 64 KB. One PC, with IP 192.168.0.2 was the Iperf server. The others, with IPs 192.168.0.6 and 192.168.0.50, were the Iperf client1 and client2, respectively. Jitter, which represents the root mean square of the differences between consecutive transit times, was continuously calculated by the server, as specified by the real time protocol RTP, in RFC 1889 [23]. A control PC, with IP 192.168.0.20, was mainly used to control the settings of the AP. The laboratory setup allowed for three types of experiments to be made: PTP, using the client1 and the control PC as server; PTMP, using the client1 and the 192.168.0.2 server PC; 4 N-PTMP, using simultaneous connections/communications between the two clients and the 192.168.0.2 server PC.

The server and client PCs were HP nx9030 and nx9010 portable computers, respectively. The control PC was an HP nx6110 portable computer. Windows XP Professional SP3 was the operating system. The PCs were configured to allocate maximum resources to the present work. Batch command files have been re-written to enable the new TCP and UDP tests.

The results were obtained in batch mode and written as data files to the client PCs disks. Every PC had a second Ethernet network adapter, to permit remote control from the official IP APTEL (Applied Physics and Telecommunications) Research Group network, via switch.

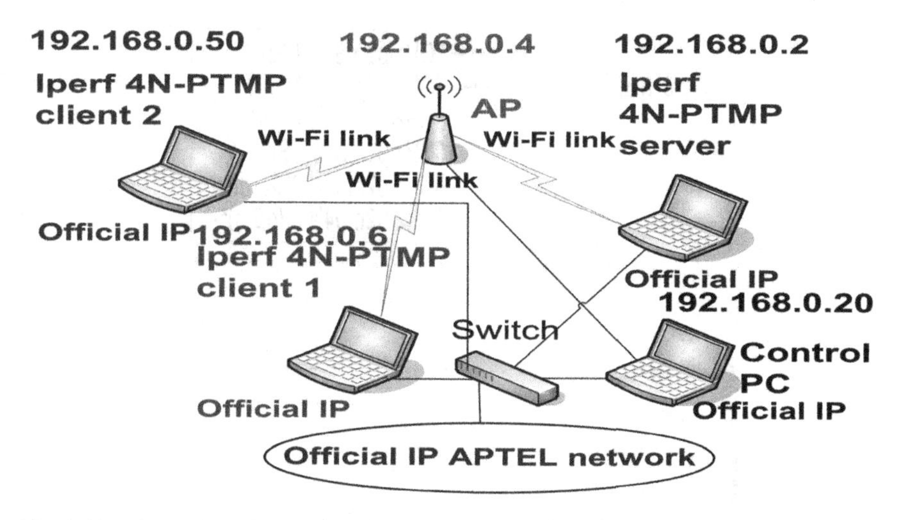

Fig. 1 Experimental laboratory setup scheme

3 Results and Discussion

The wireless network adapters of the PCs were manually configured for best rate. WPA2 encryption with AES was activated in the AP and the wireless adapters of the PCs, as given above. MCS was monitored in the AP along the experiments. It was typically MCS 15 both for transmit and receive. For every TCP packet size in the range 0.25 k-64 KB, and for every corresponding UDP datagram size in the same range, data were acquired for WPA2 PTP, PTMP and 4 N-PTMP links at OSI levels 1 (physical layer) and 4 (transport layer) using the setup of Fig. 1. For every TCP packet size an average TCP throughput was calculated from a set of experiments. This value was included as the bandwidth parameter for every corresponding UDP test, giving average jitter and average percentage datagram loss.

For OSI level 1, signal to noise ratios (SNR, in dB) and noise levels (N, in dBm) were measured in the AP. Signal indicates the strength of the radio signal the AP receives from a client PC, in dBm. Noise means how much background noise, due to radio interference, exists in the signal path between the client PC and the AP, in dBm. The lower, more negative, the value is, the weaker the noise. SNR gives the relative strength of client PC radio signals versus noise in the radio signal path, in dB. SNR is a good indicator for the quality of the radio link between the client PC and the AP. All types of experiments have given similar data. Typical values are shown in Fig. 2. The links exhibited good, high, SNR values.

The main average TCP and UDP results are summarized in Table 1, both for WPA2 and Open PTP, PTMP and 4 N-PTMP links. The statistical analysis, including calculations of confidence intervals, was done as in [24].

In Fig. 3 polynomial fits were made (shown as y versus x), using the Excel worksheet, to the TCP throughput data for PTP, PTMP and 4 N-PTMP links, where

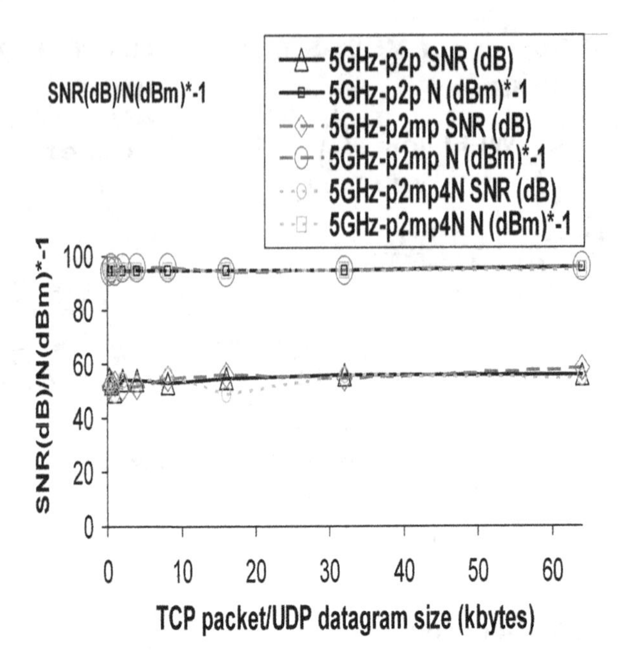

Fig. 2 Typical SNR (dB) and N (dBm)

Table 1 Average IEEE 802.11n 5 GHz WPA2 and open results: PTP; PTMP; 4 N-PTMP

Link type	WPA2			Open		
Parameter/link type	P2P	P2MP	P2MP-4 N	P2P	P2MP	P2MP-4 N
TCP throughput (Mbps)	50.8 ± 1.5	23.4 ± 0.7	15.0 + 0.5	51.2 ± 1.5	25.3 + 0.8	15.8 ± 0.5
UDP-jitter (ms)	3.0 ± 0.6	2.2 ± 0.9	6.0 ± 4.6	2.1 ± 0.1	2.5 ± 0.2	2.5 ± 0.8
UDP-% datagram loss	1.5 ± 0.2	6.4 ± 0.2	13.2 ± 8.6	3.1 ± 0.3	5.1 ± 0.2	5.5 ± 1.2

R^2 is the coefficient of determination. It indicates the goodness of fit. If it is 1.0 it means a perfect fit to data. It was found that, on average, the best TCP throughputs are for PTP links (Table 1). In passing from PTP to PTMP, throughput reduces to 46%; in comparison to P2P, 4 N-PTMP throughput reduces to 30%. Similar trends are visible for Open links. This is due to increase of processing requirements for the AP, so as to maintain links between PCs. In comparison to Open links, throughput data are lower for WPA2 links, where data length increases due to encryption. Generally, this throughput decrease does not significantly exceed the experimental error. Figure 3 suggests a fair increase in TCP throughput with packet size. For small packets the overhead is large, as there are small amounts of data that are sent in comparison to the protocol components. The role of the frame is very heavy in Wi-Fi. For larger packets, overhead decreases; the amount of sent data overcomes the protocol components.

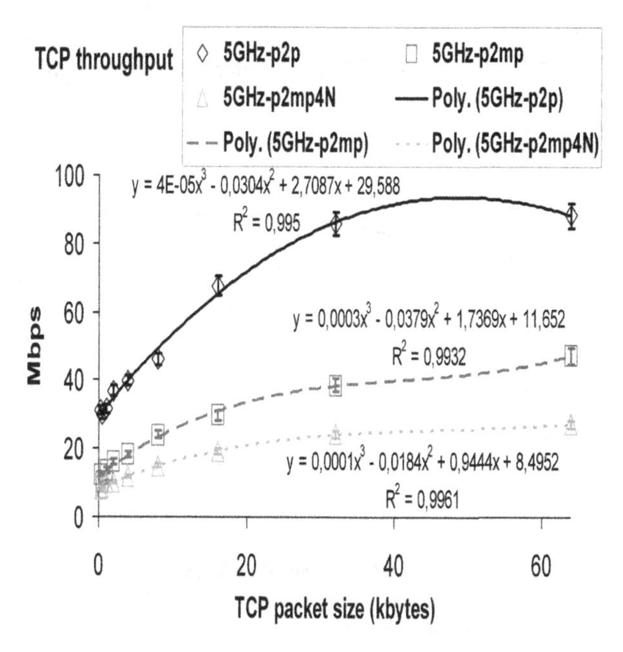

Fig. 3 TCP throughput (y) versus TCP packet size (x)

In Figs. 4, 5, the data points representing jitter and percentage datagram loss were joined by smoothed lines. Log 10 based scales were applied to the vertical axes. It was found that generally, both for Open and WPA2 links, jitter performance degrades, on average, due to link topology by increasing the number of nodes. This arises from higher processing requirements for the AP, to maintain links between PCs. Mainly for PTP and 4 N-PTMP it can be seen that, for small sized datagrams, jitter is small. There are small delays in sending datagrams. Latency is also small. Jitter increases for larger datagram sizes. For PTMP and both for Open [16] and WPA2 links there are oscillations of the jitter curves that are so far unexplained.

Concerning average percentage datagram loss, the best performances were found for PTP links, both for WPA2 and Open links (Table 1). Performance degrades in passing to PTMP and 4 N-PTMP links. This is due to increase of processing requirements for the AP, for maintaining links between PCs. Generally, in comparison to Open data, performance degradation is larger for WPA2. This is specially visible for 4 N-PTMP links. Figure 5 shows larger percentage datagram losses for small sized datagrams, when the amounts of data to send are small in comparison to the protocol components. There is considerable processing of frame headers and buffer management. For larger datagrams, percentage datagram loss is lower. However, large UDP segments originate fragmentation at the IP datagram level, leading to higher losses.

TCP throughput, jitter and percentage datagram loss were generally found to show performance degradations due to link topology, in passing from PTP to PTMP and to

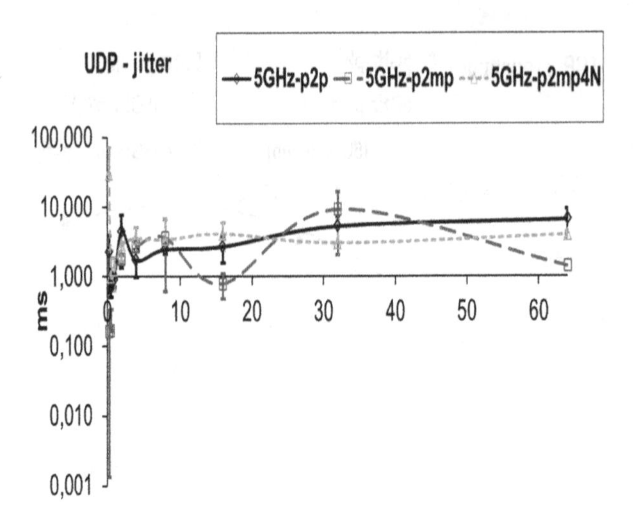

Fig. 4 UDP—jitter versus UDP datagram size

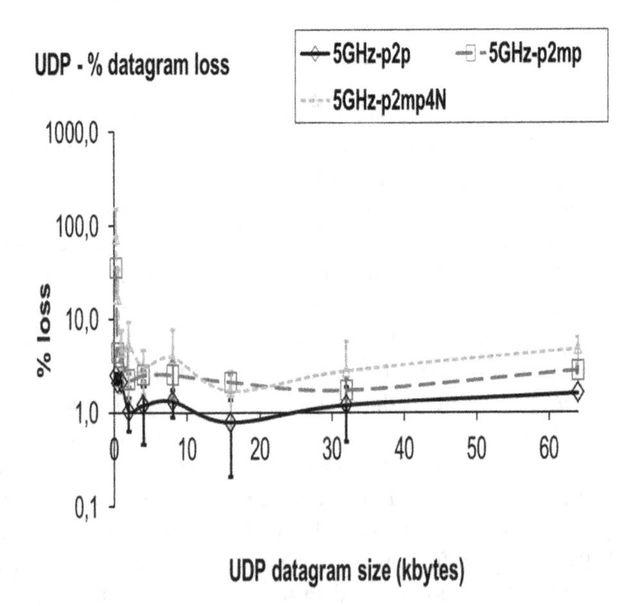

Fig. 5 UDP—percentage datagram loss versus UDP datagram size

4 N-PTMP i.e. by increasing the number of nodes, where processing requirements for the AP are higher so as to maintain links between PCs. As CSMA/CA is the medium access control, the available bandwidth and the air time are divided by the nodes

using the medium. WPA2, where there is increase in data length due to encryption, also played its role, but it did not show a very significant influence in our study.

TCP and UDP performance aspects versus TCP packet size and UDP datagram size were found as given above.

Additional experiments were made for Open 802.11a links at 54 Mbps. TCP and UDP measurements were carried out as for 802.11n, for variable TCP packet size and UDP datagram size in the range 0.25–64 KB. The best average performances for TCP throughput, jitter and percentage datagram loss were found by descending order for PTP, PTMP and 4 N-PTMP links. In comparison to 802.11a, performances were found better for 802.11n.

4 Conclusions

In the present work a versatile laboratory setup arrangement was devised and implemented, permitting systematic performance measurements with new available wireless equipments (V-M200 access points from HP and WPC600 N adapters from Linksys) for Wi-Fi (IEEE 802.11n) in 5 Ghz WPA2 PTP, PTMP and 4 N-PTMP links.

Through OSI layer 4, TCP and UDP performances were measured versus TCP packet size and UDP datagram size, respectively. TCP throughput, jitter and percentage datagram loss were measured and compared for WPA2 PTP, PTMP and 4 N-PTMP links. Comparisons were also made to corresponding results obtained for Open links. TCP throughput was found to increase with packet size. Generally jitter, for small sized datagrams, is found small. It increases for larger datagrams. Concerning percentage datagram loss, it was found high for small sized datagrams. For larger datagrams it diminishes. However, large UDP segments originate fragmentation at the IP datagram level, leading to higher losses. In comparison to PTP links, TCP throughput, jitter and percentage datagram loss were generally found to show significant performance degradations due to topology by increasing the number of nodes, where the AP experiments higher processing requirements for maintaining links between PCs. WPA2, where there is increase in data length due to encryption, also played its role but it did not show a very significant influence in our study.

Further performance studies are planned using several standards, equipments, topologies, security settings and noise conditions, not only in laboratory but also in outdoor environments involving, mainly, medium range links.

Acknowledgements Supports from University of Beira Interior and FCT (Fundação para a Ciência e a Tecnologia)/PEst-OE/FIS/UI0524/2014 (ProjectoEstratégico-UI524-2014) are acknowledged.

References

1. IEEE, 802.11a, 802.11b, 802.11g, 802.11n, 802.11i standards, http://standards.ieee.org/getieee802
2. J.W. Mark, W. Zhuang, *Wireless Communications and Networking* (Prentice-Hall Inc, Upper Saddle River, NJ, 2003)
3. T.S. Rappaport, *Wireless Communications Principles and Practice*, 2nd edn. (Prentice-Hall Inc., Upper Saddle River, NJ, 2002)
4. W.R. Bruce III, R. Gilster, *Wireless LANs End to End* (Hungry Minds Inc, NY, 2002)
5. M. Schwartz, *Mobile Wireless Communications* (Cambridge University Press, 2005)
6. N.I. Sarkar, K.W. Sowerby, High performance measurements in the crowded office environment: a case study, in *Proceedings of the ICCT'06-International Conference on Communication Technology*, Guilin, China, 27–30 Nov 2006, pp. 1–4
7. E. Monteiro, F. Boavida, *Engenharia de Redes Informáticas*, 10th edn. (FCA-Editora de Informática Lda, Lisbon, 2011)
8. J.A.R. Pacheco de Carvalho, H. Veiga, C.F. Ribeiro Pacheco, A.D. Reis, Extended performance research on Wi-Fi IEEE 802.11a,b,g laboratory open point-to-multipoint and point-to-point links, in *Transactions on Engineering Technologies,* ed. by S.-I. Ao, G.-C. Yang, L. Gelman, (Springer, Singapore 2016), pp. 475–484
9. J.A.R. Pacheco de Carvalho, H. Veiga, N. Marques, C.F. Ribeiro Pacheco, A.D. Reis, Wi-Fi WEP point-to-point links—performance studies of IEEE 802.11a, b, g laboratory links, in *Electronic Engineering and Computing Technology, Series: Lecture Notes in Electrical Engineering*, vol. 90, ed. by S.-I. Ao, L. Gelman (Springer, Netherlands 2011), pp. 105–114
10. J.A.R. Pacheco de Carvalho, H. Veiga, C.F. Ribeiro Pacheco, A.D. Reis, Extended performance studies of Wi-Fi IEEE 802.11a, b, g laboratory WPA point-to-multipoint and point-to-point links. in *Transactions on Engineering Technologies: Special Volume of the World Congress on Engineering 2013*, ed. by G.-C. Yang, S.-I. Ao, L. Gelman, (Springer, Gordrecht, 2014), pp. 455–465
11. J.A.R. Pacheco de Carvalho, H. Veiga, C.F. Ribeiro Pacheco, A.D. Reis, Performance evaluation of IEEE 802.11a, g laboratory WPA2 point-to-multipoint links, in *Lecture Notes in Engineering and Computer Science: Proceedings of the World Congress of Engineering 2014*, WCE 2014, 2–4 July 2014, London, UK, pp. 699-704
12. J.A.R. Pacheco de Carvalho, N. Marques, H. Veiga, C.F.F. Ribeiro Pacheco, A.D. Reis, Performance measurements of a 1550 nm Gbps FSO Link at Covilhã City, Portugal, in *Proceeding of the Applied Electronics 2010—15th International Conference*, 8–9 Sept 2010, University of West Bohemia, Pilsen, Czech Republic, pp. 235–239
13. D. Bansal, S. Sofat, P. Chawla, P. Kumar, Deployment and evaluation of IEEE 802.11 based wireless mesh networks in campus environments, in *Lecture Notes in Engineering and Computer Science: Proceedings of the World Congress on Engineering 2011,* WCE 2011, 6–8 July, 2011, London, UK, pp. 1722–1727
14. J. Padhye, V. Firoiu, D. Towsley, J. Kurose, Modeling TCP throughput: a simple model and its empirical validation, in *Proceedings of the SIGCOMM Symposium Communications, Architecture and Protocols,* Aug 1998, pp. 304–314
15. M. Mathis, J. Semke, J. Mahdavi, The macroscopic behavior of the TCP congestion avoidance algorithm. ACM SIGCOMM Comput. Commun. Rev. **27**(3), 67–82 (1997)
16. J.A.R. Pacheco de Carvalho, H. Veiga, C.F. Ribeiro Pacheco, A.D. Reis, Performance Measurements of Open 5 GHz IEEE 802.11n Laboratory Links, in *Lecture Notes in Engineering and Computer Science: Proceedings of the World Congress on Engineering 2016, WCE 2016*, 29 June–1 July 2016, London, UK, pp. 607–611
17. J.A.R. Pacheco de Carvalho, C.F. Ribeiro Pacheco, A.D. Reis, H. Veiga, Performance Evaluation of 5 GHz IEEE 802.11n WPA2 Laboratory Links, in *Lecture Notes in Engineering and Computer Science: Proceedings of the World Congress of Engineering 2017*, *WCE 2017*, 5–7 July, 2017, London, UK, pp. 524–528

18. HP, HP V-M200 802.11n access point management and configuration guide (2010), http://www.hp.com. Accessed 3 Jan 2017
19. Allied Telesis, AT-8000S/16 level 2 switch technical data (2009), http://www.alliedtelesis.com. Accessed 10 Dec 2016
20. Linksys, WPC600 N notebook adapter user guide (2008), http://www.linksys.com. Accessed 10 Jan 2012
21. Acrilyc WiFi, Acrylic WiFi software (2016), http://www.acrylicwifi.com. Accessed 16 Nov 2017
22. Iperf, *Iperf software*; 2017; http://iperf.fr; accessed 20 Feb 2017
23. Network Working Group, RFC 1889-RTP: a transport protocol for real time applications (1996), http://www.rfc-archive.org. Accessed 3 Jan 2014
24. P.R. Bevington, *Data Reduction and Error Analysis for the Physical Sciences* (Mc Graw-Hill Book Company, 1969)

Membrane and Resins Permeation for Lactic Acid Feed Conversion Analysis

Edidiong Okon, Habiba Shehu, Ifeyinwa Orakwe and Edward Gobina

Abstract The process intensification of cellulose acetate membrane impregnation with resin catalysts and carrier gas transport with membrane was carried out. The different catalysts used were amberlyst 36, amberlyst 16, dowex 50w8x and amberlyst 15. The carrier gases used for the analysis of the esterification product were tested with a silica membrane before being employed for gas chromatography analysis. The different carrier gases tested were helium (He), nitrogen (N_2), argon (Ar) and carbon dioxide (CO_2). The experiments were carried out at the gauge pressure range of 0.10–1.00 bar at the temperature range of 25–100 °C. The carrier gas transport results with the membrane fitted well into the Minitab 2016 mathematical model confirming the suitability of Helium gas as a suitable carrier gas for the analysis of lactic acid feed with GC-MS. The esterification reaction of lactic acid and ethanol catalysed with the cellulose acetate membrane coupled with the different cation-exchange resins gave a conversion rate of up to 100%.

Keywords Cation-exchange · Conversion · Esterification · Gas transport Lactic acid feed · Membrane

E. Okon · H. Shehu · I. Orakwe · E. Gobina (✉)
School of Engineering, Centre for Process Integration and Membrane Technology (CPIMT), The Robert Gordon University Aberdeen, Aberdeen AB10 7GJ, UK
e-mail: e.gobina@rgu.ac.uk

E. Okon
e-mail: e.p.okon@rgu.ac.uk

H. Shehu
e-mail: h.shehu@rgu.ac.uk

I. Orakwe
e-mail: i.r.orakwe@rgu.ac.uk

S.-I. Ao et al. (eds.), *Transactions on Engineering Technologies*,
https://doi.org/10.1007/978-981-13-0746-1_25

1 Introduction

Biomass products are organic materials in which the solar energy is stored through chemical bonds such as trees, crops and animal manure. In contrast to other renewable sources of energy, biomass is the only renewable feedstock that can be converted into energy and chemical products. The usage of biomass as a source of energy has a lot of advantages including: low sulphur content, production of less ash than coal combustion and prevention of more carbon dioxide emissions in the atmosphere [1]. The high dependence on oil for energy and production of several chemicals and products has been linked to climate change caused by high concentration of greenhouse gas in the atmosphere, has attracted significant attention recently on the possible solution for alternative renewable resources to produce biofuels and chemicals. The generation of products through biotechnological processes makes it possible to discover and explore several chemical routes to obtain products with very low environmental impact and high yields [2]. Solvents represent an important class of chemicals. They are used in many industries for their abilities to dissolve, remove or dilute other compounds without causing environmental problems. Ethyl lactate can be obtained from the esterification of biomass feedstocks such as lactic acid and ethanol in the presence or absence of a catalyst. This solvent has numerous applications in different industries including food, pharmaceutical and agricultural processes. The selective removal of water from the reaction products can improve reaction yields [3]. Ethyl lactate and esters of fatty acids are biosolvents which can be used alone or with a mixture. Their mixtures give effective solvency for a wide range of applications [4]. Ethyl lactate (EL) is one of the solvents in high demand in the chemical industry and it is a sustainable alternative organic solvent with advantages that include biodegradability, non-carcinogenic, non-corrosive and non-ozone depleting and is miscible with water and hydrocarbons. Because of its outstanding advantages, EL has been described by the U.S Environmental Protection Agency (EPA) as a "green solvent" [5]. EL is therefore a suitable replacement for several hazardous organic solvents such as toluene, benzene and hexane. It is used in the different industries such as food, pharmaceutical, paint, adhesive, agriculture and petroleum refinery [6].

The use of heterogeneous catalysts has several inherent advantages over homogenous catalyst including purity of products, avoidance of corrosive environment and removal of catalysts from the reaction mixture by decantation or filtration process. Several varieties of solid catalysts can be used for esterification reactions with cation-exchange resins being the most commonly used solid acid catalysts in organic reactions [7]. Membranes are fabricated from many materials such as inorganics including alumina or silica or organics including polyethersulfone, polyamides, or cellulose acetate. Membranes are commercially available in different module formats, including tubular, hollow fibre, flat sheet, spiral wound [8]. Membrane-based separation technologies has been successfully employed over the years in several industrial applications [9] including food, biotechnology, pharmaceutical and in the treatment of industrial effluents [10] and has also replaced a lot of conventional technologies because of the following advantages including: reliability, simple to operate, absence of moving parts and ability to tolerate fluctuations in flow rate and feed composition [11]. Based on the number of applications in the industry as well as in laboratory research, ceramic and polymeric membranes are the two main classes of separation membranes [12]. The concept behind incorporating a membrane into a reaction process is to separate a reactant or product from a stream and hence drive the reaction further providing an enhancement to process efficiencies [13]. Generally, membrane gas separation process is driven by a pressure difference across the membrane [14]. The transport through porous membranes can be explained using various transport mechanisms include surface diffusion, Knudsen diffusion, capillary condensation, viscous flow, and molecular sieving mechanisms [15]. In Knudsen diffusion mechanism, gas molecules diffuse through the pores of the membrane and then get transported by colliding more frequently with the pore walls [16]. Viscous flow mechanism takes place if the pore radius of the membrane is greater than the mean free path of the permeating gas molecule [17]. Currently, multifunctional reactors have attracted a lot of attention in both industrial and academic sectors. Most researchers have focused on the use of membrane reactor and reactive distillation for most esterification reactions which normally suffer from chemical equilibrium limitations [18]. In contrast to other traditional reactors, membrane reactor is more compact, less capital intensive, giving higher conversion due to high selectivity for the thermodynamically and kinetically controlled esterification reactions. Therefore, the combination of the esterification of alcohol and carboxylic acid with inorganic and polymeric membrane separation can lead to process intensification by pervaporation or vapour permeation arrangements. However, process intensification requires novel equipment with less process step and low energy consumption as well as new process methodology. Besides the methodology and equipment, other requirements include reusable and recyclable catalysts, less by-product, process safety, reduced plant volume, safety operation conditions, non-hazardous and renewable raw material [19]. Pervaporation membrane reactor is a type of membrane reactor that combines both reaction and separation in one single unit. In process intensification by pervaporation, one or more products (normally water) in a reaction liquid mixture interacting on one side of a membrane, permeate preferentially across the membrane and the permeated stream is removed as a vapour from the other side of the membrane reactor, to enhance the equilibrium shift to the forward reaction [18].

2 Materials and Methods

(a) Flat sheet cellulose acetate membrane

A flat sheet cellulose acetate membrane with the thickness of 0.035 mm, effective membrane area of 0.0155 m^2 with the dimension of 150 mm × 150 mm was supplied by Good fellow, Cambridge Limited, England, UK. The carboxyl methyl cellulose, boric acid, lactic acid (80% purity), ethanol (99% purity) and ethyl lactate

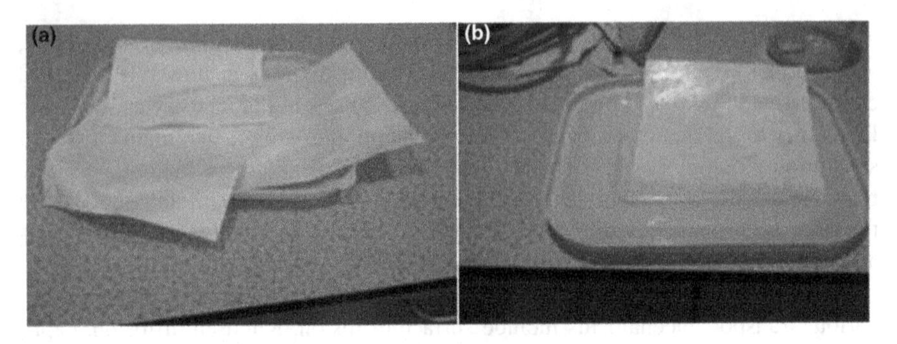

Fig. 1 Pictorial view of the cellulose acetate immersion (**a**) and drying, **b** at room temperature

(98% purity) were all purchased from Sigma Aldrich, UK. The deionized water that was used for the analysis was supplied by the Center for Process Integration and Membrane Technology (CPIMT), RGU, UK. The resistivity of the instrument was programmed at 18.9 mΩcm. The cation-exchange resin catalysts were amberlyst 16, amberlyst 15, dowex 50w8x and amberlyst 36 (Sigma Aldrich, UK).

(b) **Process Intensification of Resin Catalysts Impregnated Cellulose Acetate Membrane**

Prior to the analysis, the cellulose acetate membrane was prepared based on a similar work by Nigiz et al. [20]. Two layers were prepared i.e the catalytic and the separation layer. The catalytic layer was prepared as thus: The carboxyl methyl cellulose (CMC) with weight of 0.5wt% was measured into a 100 mL beaker and 50 mL of deionized water was used to dissolve the solid CMC. The solution was then allowed to stir for 10 h to attain a homogeneous mixture. The separation layer was prepared separately by weighing 2wt% acid into a 100 mL beaker and adding a 50 mL of the deionized water to the beaker containing boric acid and allowed to stir for 10 h to dissolve the boric acid (in a tablet form). After 10 h, the solution containing the CMC was poured into the beaker containing the boric acid solution. The two solutions (catalytic and separation solutions) were mixed together while stirring for 3 h to obtain a homogenous mixture. The cellulose acetate membrane was then immersed into the mixture. The significance of coating the support surface was to obtain a thin layer on the surface of the support to be effective in the removal of water from the esterification system when in contact with the reactants. Figure 1a, b shows the pictorial view of the separation and the catalytic solutions that were prepared. Prior to the cellulose acetate membrane immersion process. A similar method to that of Nigiz et al. [20] was adopted for the cellulose acetate membrane preparation with some modification in the solvent composition. The compositions of the CMC and boric acid that were used for the preparation of the catalytic and separation layers are presented in Table 1.

After 3 h, the cellulose acetate membrane was immersed into the solution consisting of the CMC, boric acid and deionized water and was allowed in the solution for 3 min to allow a uniform coating. After 3 min, the membrane was taken out of the solution and allowed to air dry for 3 days at room temperature before the ester-

Table 1 Composition of solvents used for the cellulose acetate membrane preparation

Substance	Amount (mL) and g
Carboxyl methyl cellulose (CMC)	5 g
Boric acid	2 g
Deionized water	50 mL
Beaker	100 mL

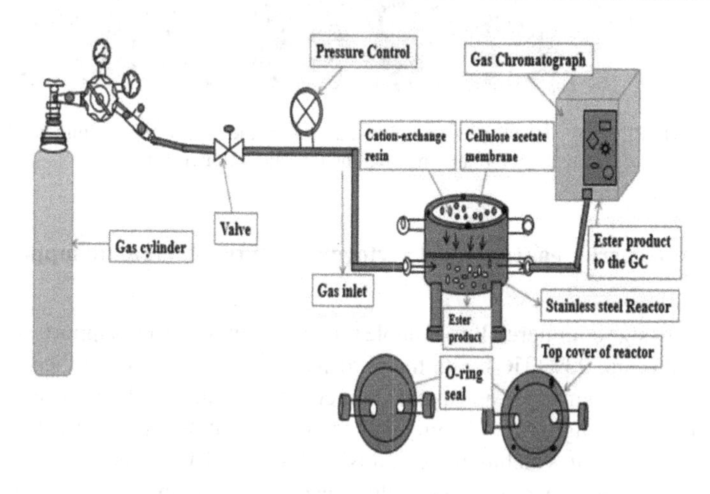

Fig. 2 Schematic diagram for cellulose acetate membrane process intensification setup

ification reaction. The importance of drying the cellulose acetate membrane in air was for the homogenous solution of the catalyst to penetrate the porous surface of the membrane sample to obtain a uniform coating on the surface and to avoid the acetate membrane burning in the oven. A similar method by Collazos et al. [21] was adopted for the catalysts preparation. Figure 1a, b shows the pictorial view of the membrane immersion and drying process.

The intensification process by pervaporation was performed for the removal of water from the reaction mixture as shown in Fig. 2. The reactor was made up of two stainless steel compartments. The cellulose acetate membrane weighing 1.5 g was placed in between the lower compartment of the cell and the core holder containing 0.5 g of each cation-exchange catalyst. After the addition of the cation-exchange resin catalysts, the core holder and the rubber gasket were placed on the reactor and the reactant consisting of the lactic acid and ethanol solutions which had been prepared separately were then added to the reactor through the openings of the core holder. The experimental set up was based on a similar work carried out by Khadijah et al. [22].

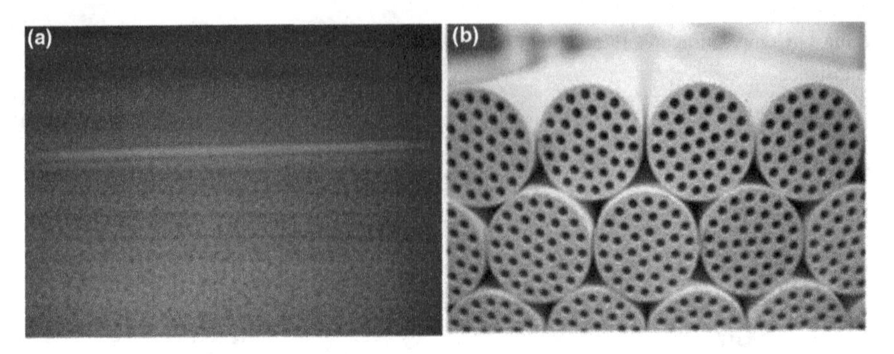

Fig. 3 **a** and **b**: Pictorial view of the 15 nm alumina ceramic membrane fresh commercial (3a) and the bottom view of the stack membrane showing the internal channels (3b)

(c) **Study of carrier gas transport performance through ceramic support membrane.**

A 15 nm pore size commercially available tubular ceramic porous support, consisting of 77% Al_2O_3 and 23% TiO_2 with the porosity of 45% was used for the study. The support possesses an inner and outer diameter of 7 mm and 10 mm respectively with a permeable length of 34.2 mm and a total length of 36.6 mm. The support was supplied by ceramiques techniques et industrielles (CTI SA), France. Figure 3a and b shows the pictorial view of the 15 nm alumina ceramic membrane fresh commercial membrane (3a) and the view of the stack membrane showing the internal channels (3b) [23].

(d) **Single Gas Transport through Tubular Support**

The transport of the single gases through ceramic support was carried out using five different gases which could serve as carrier gases for esterification reaction including argon (Ar), helium (He), carbon dioxide (CO_2) and nitrogen (N_2) with the purity of at least 99.999 (% v/v). The carrier gases were supplied by BOC, UK. The permeability experiment was carried out at the gauge pressure of between 0.10–1.00 bar and at room temperature. The different carrier gases were feed into the reactor through the feed gas opening where the gases interact with the support and exit through the permeate. The flow rate of the gases was measured using a flow meter. Figure 4 shows the carrier gas permeation setup.

3 Results and Discussion

(a) **Selectivity Comparison at different Pressure Gauge**

Tables 2 and 3 shows the comparison between the theoretical selectivity and the experimental selectivity of the He, Ar, N_2 and CO_2 that were used for the analysis.

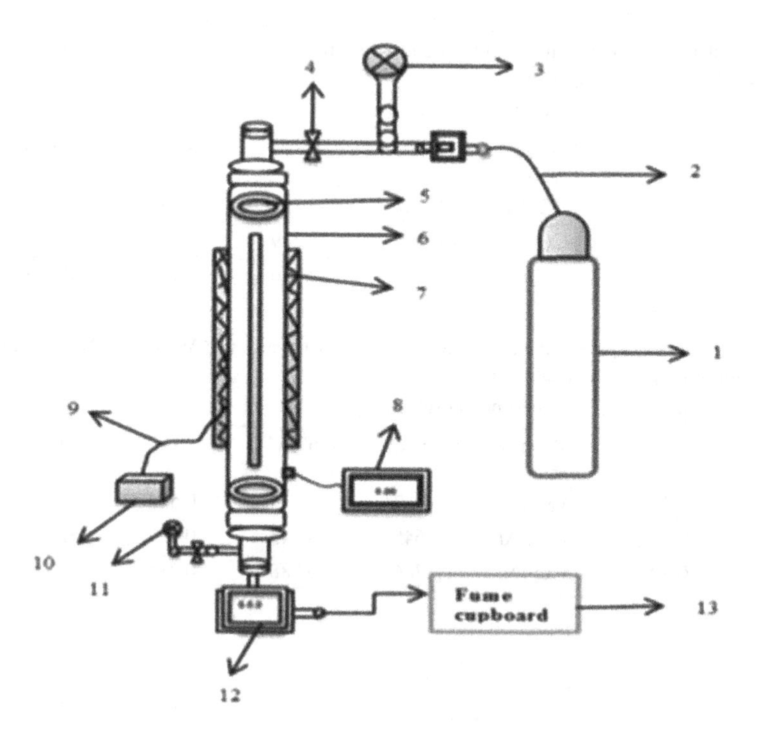

Fig. 4 Carrier gas permeation test setup which consists of; carrier gas cylinder (1), gas feed inlet (2), permeate pressure gauge (3), control valve (4), O-ring graphite seal (5), reactor (6), heating tape (7), temperature regulator (8), thermocouple (9), thermocouple box (10), retentate pressure gauge (11), flow meter (12) and fume cupboard (13)

The selectivity of the gases was compared at the gauge pressure of 0.4 bar (Table 2) and 0.9 bar (Table 3) at the temperature of 60 °C. From the result obtained, it was found that CO_2 permselectivities over He, Ar and N_2 with the membrane at 0.40 bar and the temperature range of 25–100 °C was higher than the theoretical selectivity value indicating a good separation of the carrier gases at 0.40 bar in contrast to the selectivity values at 0.90 bar. However, it was found that the experimental selectivity of argon gas with the membrane at 0.40 bar shows a drastic increase in contrast to the selectivity of argon at 0.90 bar which indicate a good separation of argon at this gauge pressure [24].

The permeance ratio of CO_2 over Ar, He and N_2 gas was also plotted against the feed gauge pressure (bar) at 298 k as shown in Fig. 5. From the result obtained, it was found that the theoretical selectivity value of argon was higher than that of the experimental values suggesting Knudsen flow as the dominant mechanism of transport. Experimental error bars were determined on the graph to further confirm the accuracy of the results. Also, in Fig. 6, it was found that the permeance of the carrier gas decreases with respect to the gauge pressure with helium and argon gas

Table 2 Calculated experimental and theoretical selectivity values of the different gas over CO_2 at 0.40 bar for 3rd dip-coated membrane

Theoretical selectivity		Experimental selectivity				
At 0.40 bar		Permeance ratio	25 °C	60 °C	80 °C	100 °C
He/CO_2	0.3	CO_2/He	0.502	0.755	0.831	0.613
Ar/CO_2	0.95	CO_2/Ar	0.606	1.219	1.318	0.899
N_2/CO_2	0.79	CO_2/N_2	0.915	0.966	0.983	0.833

Table 3 Calculated experimental and theoretical selectivity values of Ar, He, and N_2 over CO_2 at 0.90 bar for 3rd dip-coated membrane

Theoretical selectivity		Experimental selectivity				
At 0.40 bar		Permeance ratio	25 °C	60 °C	80 °C	100 °C
He/CO_2	0.3	CO_2/He	0.467	0.543	0.546	0.417
Ar/CO_2	0.95	CO_2/Ar	0.645	0.945	0.957	0.726
N_2/CO_2	0.79	CO_2/N_2	0.766	0.788	0.786	0.649

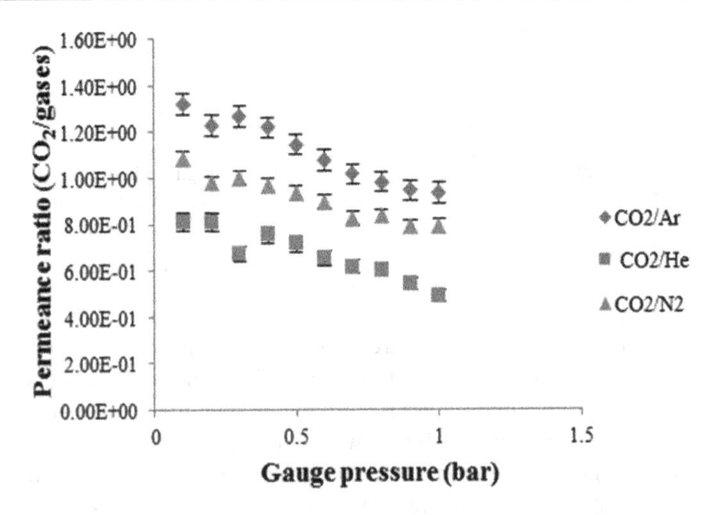

Fig. 5 Permeance ratio of CO_2 over Ar, He and N_2 carrier gases at 298 K

exhibiting the highest permeation rate. The order of the gas permeance with respect to the gauge pressure was given as He > Ar > N_2 > CO_2.

(b) **Mathematical Model**

In order to validate the carrier gas experimental results that were obtained and to further predict results for the parameter, Minitab 2016 mathematical model was used to design the graphical plots for the experimental results. Figure 7a–c shows the mathematical model plots of permeance with respect to gauge pressure for the

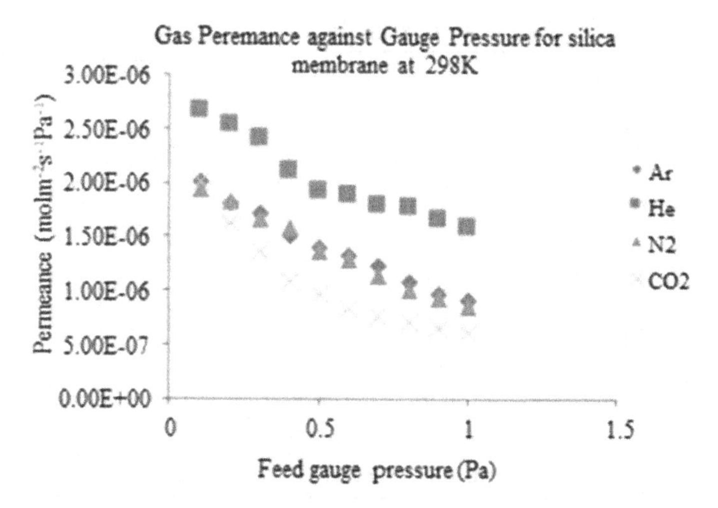

Fig. 6 Permeance against feed gauge pressure (Pa) for support at 298 K

Table 4 Mathematical model and gas permeation parameters

Model parameter	Permeation parameters
Sample numbers	Different gauge pressure (0.1, 0.2, 0.3, 0.4, 0.5, 0.6, 0.7, 0.8, 0.9, 1.0 bar)
PCs	Experimental result for gases (permeance $molm^{-2}s^{-2}Pa^{-1}$)
Variable	Gases (He, Ar, N_2 and CO_2)
Scores	Calculated values from the experimental results
Eigenvalues	Calculated values from the experimental results for the gas

support membrane at 298 K. The mathematical model was designed to describe three plots including a "scree plot model" which considers the gas permeance and flow rate through the support and silica membranes at room and high temperatures with respect of gauge pressure, the model also put into consideration the Eigen analysis of the correlation matrix i.e how many principal components (PCs) are necessary to describe the experimental data "loading plot model" which is used to determine the experimental flow rate and feed gauge pressure based on the trend of the gas flow and a "score plot model" which takes into consideration the permeance, feed gauge pressure and flow rate. Table 4 depict the Eigen analysis for the correlation matrix and the principal components analysis for the four variables (He, Ar, N_2 and CO_2) from the Minitab 2016. Table 4 explained the different permeation parameters.

From Fig. 7a, b, it can be see that helium gas permeate faster through the porous ceramic support for both experimental (as compared in Fig. 6) and model results. The experimental results fitted well into the model as helium gas with a lower molecular weight demonstrate the highest permeation rate for the support at 298 K, although there were some alteration in the trend for CO_2, N_2 and Ar gases in the model results.

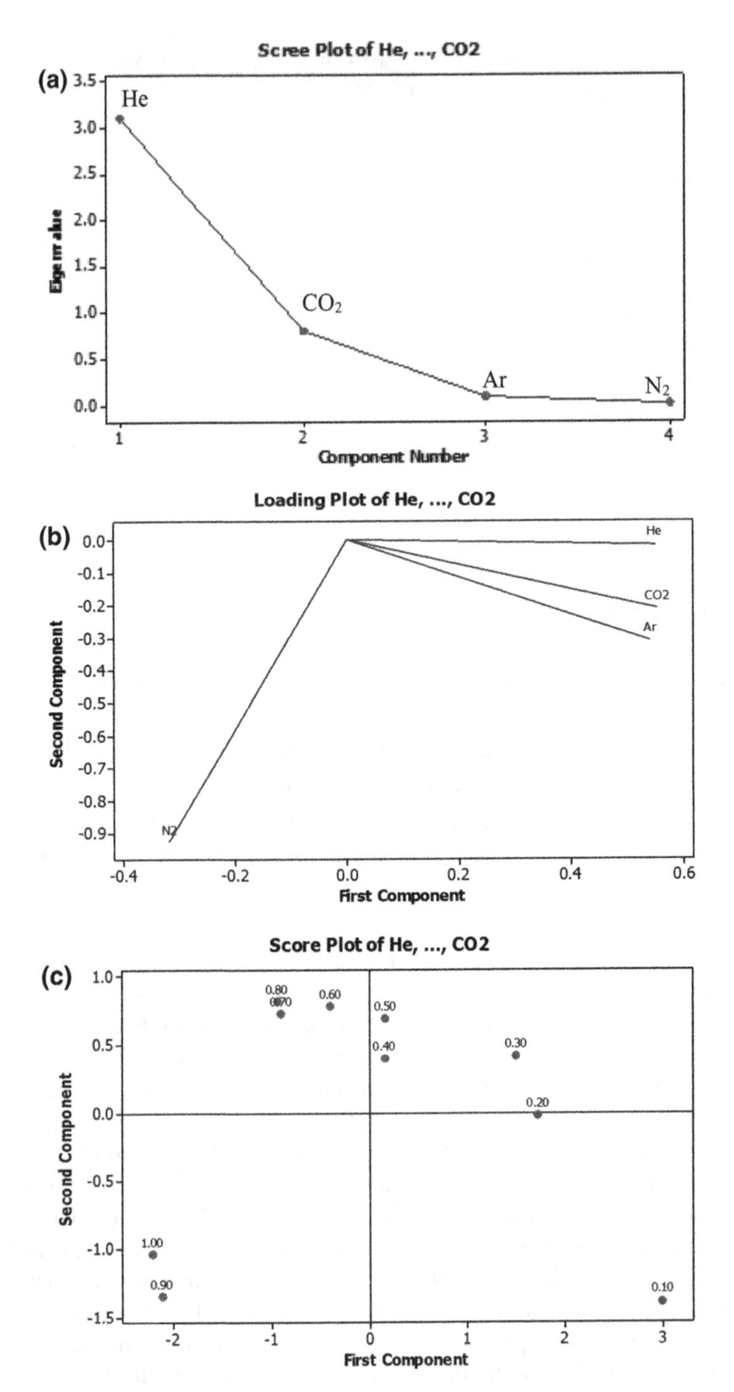

Fig. 7 **a** Minitab scree plot of eigenvalue of gas permeance against gauge pressure for support at 298 K. **b** Minitab loading plot of gas permeance values against gauge pressure for support at 298 K. **c** Minitab score plot of gas permeance against gauge pressure for support at 298 K

Table 5 Eigen analysis for the correlation matrix and the principal components analysis for the four variables

Eigenanalysis of the correlation matrix				
Eigenvalue	3.818	0.117	0.062	0.004
Proportion	0.954	0.029	0.015	0.001
Cumulative	0.954	0.984	0.999	1.000
Variable	PC1	PC2	PC3	PC4
He	0.491	0.803	−0.296	0.164
Ar	0.503	−0.503	−0.149	0.687
N_2	0.500	0.040	0.852	−0.152
CO_2	0.506	−0.318	−0.406	−0.692

However, it was also found that the eigenvalue obtained from the principal component analysis (permeance of the gases) decreases with respect to gauge pressure as seen in Fig. 7a for the respective gases. Figure 7b shows the loading plot for the variables (gases). This plot explains the contribution of each variables to the first two components (He and Ar with the highest permeation rate). It can be seen that He, CO_2 and Ar gases demonstrate a positive loading as the gauge pressure increases except for N_2 gas with negative loading suggesting that this could be due to the non-metallic character of N_2 as the non-metallic character of group V elements decreases down then group in the periodic table. The dissimilarities in the trend of the carrier gas permeation for the experimental and mathematical model were suggested to be due to the fact that the model does not account for the different mechanisms of gas transport and also due to some form of systematic experimental errors.

Figure 7c shows the score plot model which explain the grouping of the sample number (gauge pressure). It was found that the samples number were found to be in three different groups with respect to the principal components. It was found that the gauge pressure of 0.1 bar, gauge pressure 0.2, 0.3, 0.4, 0.5 bar and gauge pressure 0.6, 0.7, 0.8, 0.9, 1.0 bar appear to be in different groups. The three groups also fall on the positive axis indicating that the gauge pressure increase with decrease in permeance (PCs) for the support at 298 K. Table 5 shows the Eigen analysis for the correlation matrix and the principal components analysis for the four variable (He, Ar, N_2 and CO_2). In Table 5, the first line of the Eigenvalue shows how the variance is distributed between the four PCs with PC1 (He) having a variance of 3.8180, PC2 (Ar) = 0.1166, PC3 (CO_2) = 0.0618 and PC4 (N_2) = 0.0036. It can be seen that PC1 has the highest variance followed by PC2, PC3 and PC4. The second line of the table describe the proportion of the data variation by each PCs whereas the third line shows the cumulative proportion. This suggest that both PC1 and PC2 account for 98.4% of the variable for the data, which further confirm that accuracy of the model. The bottom half of the table depict the coefficients of the principle components. This were also compared with the silica coated membrane.

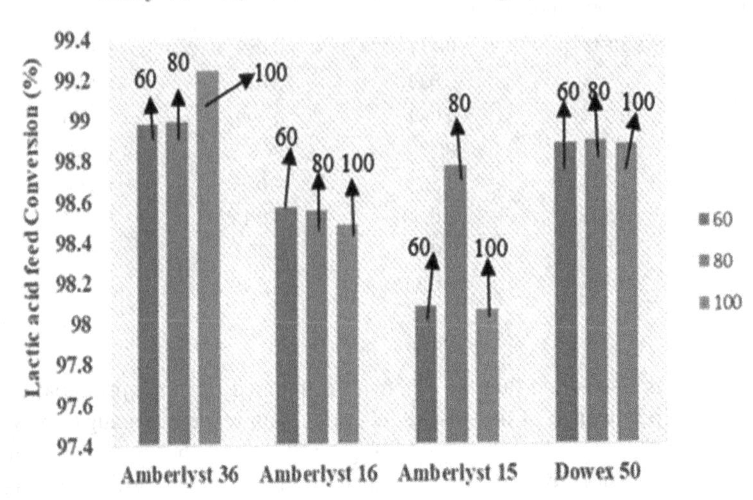

Fig. 8 Conversion of lactic acid feed catalysed with amberlyst 36, amberlyst 16, amberlyst 15 and dowex 50W8x at different temperatures (°C) and at 1.0 μg/L

(c) **Lactic Acid Feed Conversion**

Figure 8 depict the graph of the effect of temperature on lactic acid feed conversion catalysed with the different cation-exchange resin at the injection concentration of 1.0 μg/L. From Fig. 8, it can be seen that the lactic acid feed catalysed by amberlyst 36 showed a higher conversion of 99.2% at 100 °C and 98.9% for both 60 and 80 °C. It can also be seen from Fig. 8 that dowex 50W8x also showed a good conversion rate of 98.8% at 1.0 μg/L. Also, the lactic acid feed catalysed with amberlyst 36 and dowex 50W8x gave a good conversion rate at 60, 80 and 100 °C in contrast to amberlyst 15 and amberlyst 16 at the same temperatures which confirms the effectiveness of the catalyst.

From the result obtained in Fig. 9, it can be seen that the conversion rate of the lactic acid feed increases with respect to temperature for the lactic acid feed catalysed with amberlyst 36 and dowex50W8x at the concentration of 0.50 μg/L. It was found that although the feed gave a lower conversion of 97.8% for both amberlyst 15 and 16 cation exchange resin at 60 °C. However, dowex50W8x showed a good conversion of upto 99.9%. It was found that 60 °C, amberlyst 36, 15 and dowex50W8x showed a good conversion of 99%, 98.7% and 98.6% respectively. At 100 °C, it can be seen that amberlyst 15 demonstrate a conversion of 99.8% indicating that the esterification reaction with ambrelyst 15 was more favourable at 100 °C.

Also, from Fig. 10, it can be seen that the lactic acid feed catalysed with amberlyst 36 gave a conversion of 99.9 and 98.9% at 80 °C. It was also found that the three temperatures favored the conversion of the feed catalysed with dowex 50W8x with

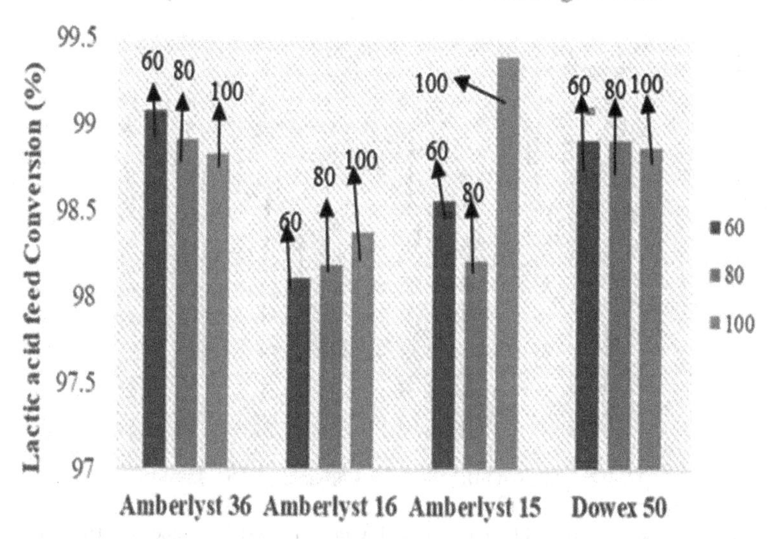

Fig. 9 Conversion of lactic acid feed catalysed with amberlyst 36, amberlyst 16, amberlyst 15 and dowex 50W8x at different temperatures (°C) and at 0.5 μg/L

the highest conversion rate of 98.85% at 60 °C. It was found that the conversion rate of the lactic acid feed catalysed with amberlyst 15 and amberlyst 16 were a bit low for the two catalysts at the three temperatures. It was found that amberlyst 36 and dowex 50W8x cation exchange resin exhibited a good conversion and can also withstand the effect of temperature at the concentration of 2.0 μg/L. From this result, it was also confirmed that temperature plays a major role in the batch process esterification reaction, as it might favoured some catalyst to produce a good conversion of the lactic acid feed.

4 Conclusion

The tubular support membrane was characterized by gas permeation method to determine the selectivity of the carrier gas for coupling with the GC-MS for the analysis of lactic acid feed at different temperatures. The process intensification of the cellulose acetate membrane with amberlyst 36 cation-exchange resin exhibited a higher conversion (99.9%) at the flow rate of 2.0 μg/L in contrast to other cation-exchange resin when coupled with the membrane. The conversion of the feed was found to increase with respect to the gas flow rate. The permeance of the carrier gases showed a decrease with respect to the gauge pressure at 298 K. The results further confirmed

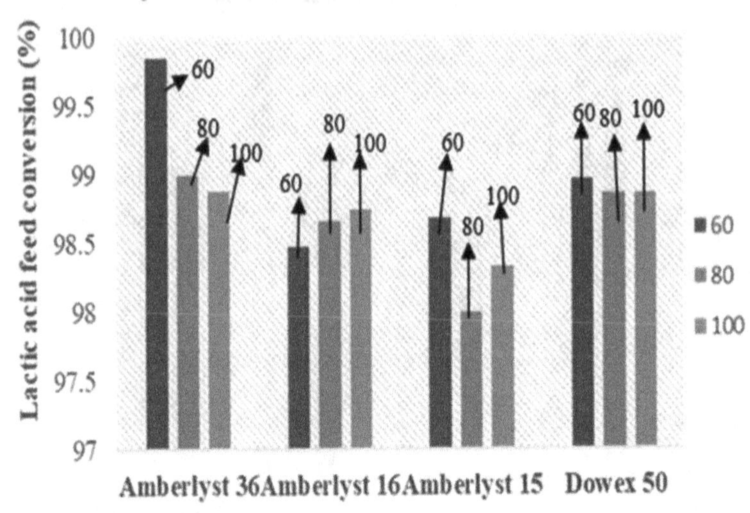

Fig. 10 Conversion of lactic acid feed catalysed with amberlyst 36, amberlyst 16, amberlyst 15 and dowex 50W8x at different temperatures (°C) and at 2.0 μg/L

the effectiveness of cellulose acetate membrane coupled with resin catalysts in the selective removal of the water from the esterification reaction feed. The gas flow through the membrane was dominated by different mechanisms including viscous and Knudsen flow. The membrane exhibited a straight-line dependence of flux with gauge pressure with R^2 value of up to 0.97. The experimental selectivity of Ar/CO_2 gas showed a higher value of 1.318 at gauge pressure of 0.90 bar in contrast to the theoretical selectivity of 0.95. The order of the CO_2 permselectivity over other gases with respect to the gauge pressure was $CO_2/Ar > CO_2/H_2 > CO_2/N_2 > CO_2/He$. The standard deviation with 5% error bar confirms the accuracy of the experimental results. The mathematical model further confirms helium as the suitable catalysts.

Acknowledgements The Authors acknowledge the Centre for Process Integration and Membrane Technology (CPIMT) at RGU for providing the research infrastructure. Additionally, the sponsorship provided by the petroleum technology development fund (PTDF) is gratefully acknowledged.

References

1. M. Buonomenna, J. Bae, Membrane processes and renewable energies. Renew. Sustain. Energy Rev. **43**, 1343–1398 (2015)
2. B.H. Lunelli, E.R. De Morais, M.R.W. Maciel, R. Filho, Process intensification for ethyl lactate production using reactive distillation. Chem. Eng. Trans. **24**, 823–828 (2011)
3. M. Zhu, Z. Feng, X. Hua, H. Hu, S. Xia, N. Hu et al., Application of a mordenite membrane to the esterification of acetic acid and alcohol using sulfuric acid catalyst. Microporous Mesoporous Mater. **233**, 171–176 (2016)
4. T.H.T. Vu, H.T. Au, T.H.T. Nguyen, T.T.T. Nguyen, M.H. Do, N.Q. Bui et al., Esterification of lactic acid by catalytic extractive reaction: an efficient way to produce a biosolvent composition. Catal. Lett. **143**(9), 950–956 (2013)
5. F.U. Nigiz, N.D. Hilmioglu, Green solvent synthesis from biomass based source by biocatalytic membrane reactor. Int. J Energy Res. (2015)
6. D.T. Vu, C.T. Lira, N.S. Asthana, A.K. Kolah, D.J. Miller, Vapor-liquid equi-libria in the systems ethyl lactate ethanol and ethyl lactate water. J. Chem. Eng. Data **51**(4), 1220–1225 (2006)
7. P. Jagadeesh Babu, K. Sandesh, M. Saidutta, Kinetics of esterification of acetic acid with methanol in the presence of ion exchange resin catalysts. Ind. Eng. Chem. Res. **50**(12), 7155–7160 (2011)
8. S. Datta, Y. Lin, S. Snyder, Current and emerging separations technologies in biorefining, in *Advances in Biorefineries. Biomass and Waste Supply Chain Exploitation*, ed. by K. Waldron, 2014, pp. 112–151
9. A. Labropoulos, C. Athanasekou, N. Kakizis, A. Sapalidis, G. Pilatos, G. Romanos et al., Experimental investigation of the transport mechanism of several gases during the CVD post-treatment of nanoporous membranes. Chem. Eng. J. **255**, 377–393 (2014)
10. J.I. Calvo, A. Bottino, G. Capannelli, A. Hernández, Pore size distribution of ceramic UF membranes by liquid–liquid displacement porosimetry. J. Membr. Sci. **310**(1), 531–538 (2008)
11. G. Clarizia, Strong and weak points of membrane systems applied to gas separation. Chem. Eng. Trans. **17**, 1675–1680 (2009)
12. T. Van Gestel, H.P. Buchkremer, Processing of Nanoporous and Dense Thin Film Ceramic Membranes, in *The Nano-Micro Interface: Bridging the Micro and Nano Worlds*, 2015, pp. 431–458
13. S. Smart, S. Liu, J.M. Serra, J.C. Diniz da Costa, A. Iulianelli, Basile A. 8—Porous ceramic membranes for membrane reactors, in *Handbook of Membrane Reactors*, ed. by A. Basile (Woodhead Publishing, 2013), pp. 298–336
14. A.F. Ismail, K.C. Khulbe, T. Matsuura, *Gas Separation Membranes* (Springer, 2015)
15. J. Coronas, J. Santamarıa, Catalytic reactors based on porous ceramic membranes. Catal. Today **51**(3), 377–389 (1999)
16. H. Lee, H. Yamauchi, H. Suda, K. Haraya, Influence of adsorption on the gas permeation performances in the mesoporous alumina ceramic membrane. Sep. Purif. Technol. **49**(1), 49–55 (2006)
17. A. Marković, D. Stoltenberg, D. Enke, E. Schlünder, A. Seidel-Morgenstern, Gas permeation through porous glass membranes: Part II: transition regime between Knudsen and configurational diffusion. J. Membr. Sci. **336**(1), 32–41 (2009)
18. S. Assabumrungrat, J. Phongpatthanapanich, P. Praserthdam, T. Tagawa, S. Goto, Theoretical study on the synthesis of methyl acetate from methanol and acetic acid in pervaporation membrane reactors: effect of continuous-flow modes. Chem. Eng. J. **95**(1), 57–65 (2003)
19. F.U. Nigiz, N.D. Hilmioglu, Green solvent synthesis from biomass based source by biocatalytic membrane reactor. Int. J. Energy Res. (2015)
20. F.U. Nigiz, N.D. Hilmioglu, Simultaneous separation performance of a catalytic membrane reactor for ethyl lactate production by using boric acid coated carboxymethyl cellulose membrane. React. Kinet. Mech. Catal. 1–19 (2016)

21. H.F. Collazos, J. Fontalvo, M.Á. Gómez-García, Design directions for ethyl lactate synthesis in a pervaporation membrane reactor. Desalination Water Treat. **51**(10–12), 2394–2401 (2013)
22. S.K. Hubadillah, Z. Harun, M.H.D. Othman, A. Ismail, P. Gani, Effect of kaolin particle size and loading on the characteristics of kaolin ceramic support prepared via phase inversion technique. J. Asian Ceramic Soc. **4**(2), 164–177 (2016)
23. V. Gitis, G. Rothenberg, *Ceramic Membranes: new Opportunities and Practical Applications* (Wiley, 2016)
24. O. Edidiong, S. Habiba, E. Gobina, Effect of Resins and membrane permeation for improved selectivity, in *Lecture Notes in Engineering and Computer Science: Proceedings of The World Congress on Engineering 2017, WCE 2017, 5–7 July, 2017, London, UK*, pp. 1059–1065

DNA Sequences Classification Using Data Analysis

Bacem Saada and Jing Zhang

Abstract The technological progress of recent years has led biologists to extract, examine and store huge quantities of genetic information. Through this work, we propose a comparative study based on the percentages of similarity of height genera's species from Bacillales order and their taxonomic classification and detect if the two classification methods perform the same results or not.

Keywords Ascending Hierarchical Classification · Bioinformatics Data analysis · DNA sequence alignment · High Throughput Sequencing Multidimensional Scaling

1 Introduction

In the field of DNA sequences and genomes alignment, due to the large amount of information available in databases and the complexity of the algorithms used in data processing, researchers proposed algorithms that reduce the execution time of classical DNA sequences alignment algorithms. This improvement, however, has no impact on our work, which consists on reducing the amount of existing information in data banks. We will also present the hierarchical classification of species. We will base our work on the results of this approach and attempt to propose solutions to its limitations.

B. Saada (✉) · J. Zhang
Harbin Engineering University, College of Computer Science and Technology, Harbin 150001, China
e-mail: basssoum@gmail.com

J. Zhang
e-mail: zhangjing@hrbeu.edu.cn

S.-I. Ao et al. (eds.), *Transactions on Engineering Technologies*,
https://doi.org/10.1007/978-981-13-0746-1_26

2 Species Classification Using Data Analysis Approaches

The similarity percentages of the algorithms for alignment and comparison of DNA sequences are used to classify the species studied in order of similarity. The similarity percentages is also used to verify the conformity of these algorithms To the classification methods implemented by biologists.

In what follows, we will examine the classification of species based on Data Analysis.

Among the most used methods of data analysis, we could mention:

- Factorial Correspondence Analysis: Factorial Correspondence Analysis was initially designed to process data tables. Unlike the principal component analysis, in factorial correspondence analysis, the array to be analyzed is symmetrical with respect to the indices i and j. Two lines are considered close if they are associated in the same way with all the columns. Factorial correspondence analysis aims to gather, in two dimensions, most of the initial information by focusing on the correspondence between the variables.
- Ascending Hierarchical Classification: Through this approach, we seek to obtain, from an I-set, a system of non-empty classes such that every individual i belongs to only one class. If set I is divided into a finite number of classes, each of which is also divided into a finite number of classes the hierarchical representation is called a hierarchy of nested classes. The best-known example of hierarchical classification is that provided by biology where living beings are divided into two kingdoms: the animal kingdom and the vegetable kingdom; each of these two kingdoms is itself subdivided [1].
- Classification Approaches: Much of the data acquired in genetics or population biology should be analyzed using mathematical and informatics tools to identify the observations and conclusions that best describe data representation and the relationships between them. Since biology is defined as the science that studies living beings, including their hierarchical classifications and similarity relationships or the common ancestors between different species, we will present the approaches to data analysis that we will apply to reach a good analysis of the percentages of similarities between the DNA sequences. We will also present the reasons behind the use of these approaches.

We will also the multidimensional scaling and hierarchical ascending classification approaches that are used in the hierarchical classification of DNA sequences.

2.1 *Multidimensional Scaling*

2.1.1 Presentation of the Approach

Multidimensional scaling is a set of statistical techniques used to explore the similarities and dissimilarities of data. In general, the purpose of this analysis is to detect underlying dimensions that allow the researcher to explain the similarities or differences observed between the studied objects. With this approach, it is possible to analyze similarity matrices and dissimilarity matrices. Multidimensional Scaling makes it possible to reorganize the objects so as to construct a configuration that best represents the observed distances. It uses a minimization of a function that evaluates different configurations in order to maximize the quality of fit or minimize the lack of fit.

2.1.2 Presentation of the results

One of the advantages of this approach is that it can be used both for the measurement of similarity and dissimilarity. In addition, it can be used on numbers, probabilities or percentages. This approach could therefore be applied, to:

- Voting results,
- Results of an advertising survey,
- Distance between geographical places,
- Degree of similarity between geometric shapes,
- Similarity percentages between biological species.

The results obtained make it possible to group individuals into a set of classes and to draw up dendrograms of dissimilarity (or similarity) between these individuals.

This approach is used by biologists since it helps:

- Calculate the similarity rate between the different DNA sequences through the percentages of similarity between them.
- Visualize the results via a two-dimensional diagram which makes it possible to correctly interpret the similarities between the species.
- Construct dendrogram that are used to study similarities between species.
- Classify species by degree of dissimilarity or similarity.

2.2 Ascending Hierarchical Classification

2.2.1 Presentation of the Approach

The purpose of the Ascending Hierarchical Classification is to obtain an automatic classification of the entities according to the used statistical data.

This approach is based on a matrix of similarity weight between the different entities to subsequently attempt to predict or explain the collection of several entities in a single class [2]. The techniques available have much in common with the techniques used in the methods of the discriminant analysis of similarities between objects.

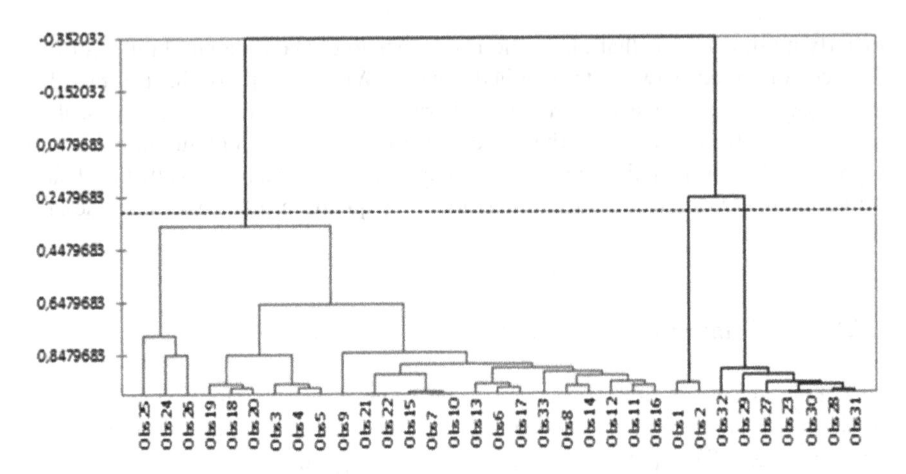

Fig. 1 Classes configuration

2.2.2 Presentation of the results

The interpretation of the results could be achieved through various means. For example, we can construct a histogram of level indices to allow the user to see how level indices vary, and to indicate at what level we can cut the classification tree to have a suitable partition (stable classes). We can also construct a hierarchical classification tree that is defined by a set of classes. The downward reading of the tree, in the opposite direction of its construction, makes it possible to examine the partitions comprising few classes (Fig. 1).

This approach is used as it allows to:

- Group a set of individuals into a set of classes
- Represent the data as a two-dimensional point cloud to facilitate the analysis and interpretation of the obtained results.
- Determine the species that best characterize the classification.
- Compare the taxonomic classification of species on the genera level and compare it to the classification obtained by applying this approach.

2.3 Application of Classification Methods for the Similarity Percentages of Smith and Waterman Algorithm

In this part, we will present a study of the Smith and Waterman [3] algorithm based on the multidimensional scaling approach and the hierarchical classification method. We will present the species used in our study as well as the data and the software used to perform the statistical calculations. We analyze the obtained results and

Table 1 Representative table of all species used

Genus	*Alicyclobacillus*	*Anoxybacillus*	*Bacillus*	*Geebacillus*
Number of species	2	3	12	3
Genus	*Lactobacillus*	*Lysinibacillus*	*Peanibacillus*	*Sporosarcina*
Number of species	3	2	7	1

present a comparative study of the classification of species, resulting from these two approaches, with the taxonomic classification of biological species.

2.3.1 Data Presentation

In order to analyze, through statistical methods, the hierarchical classification of species made by taxonomists, we used DNA sequences from several genera but of the same order. In this way, it is possible to analyze at once:

- **Gender discrimination**: find out if species of the same genus are grouped together.
- **The homogeneity between classes**: to find out if only one class represents each genre.
- **Tolerance to similarity errors**: to find out if the alignment between an unidentified sequence and a sequence of which the species is known makes it possible to correctly identify the exact species of this new sequence.
- **The relations between species of the same order**: to find out if the choice of species of the same order will have, or have not, an influence on the classification of the species.

The table below presents all the species used in our analysis (Table. 1):

We opted for taking the strain called "Type-Strain" of each species [4], to have an accurate analysis of the DNA sequences. This DNA sequence is the representative strain of each species in the online databases. Indeed, this choice makes it possible to have no errors in the genetic information of the sequence and that there would be no confusion between this sequence and another sequence that is poorly identified or whose genetic information is false.

2.3.2 Software Environment

To carry out this study, we use XLSTAT software which is a tool for data analysis and statistics whose particularity is to be integrated with Excel. Access to the different modules is possible through menus and toolbars from the Excel interface. XLSTAT uses Microsoft Excel as an interface to retrieve data and display results. However, all mathematical calculations are done outside of Excel.

Our statistical analysis [5] was performed by this tool which allows to visualize the results in several forms and therefore facilitates their interpretation (point clouds, classification, two-dimensional graph, matrix of distance between the entities, etc.).

The computer used to perform this analysis has the following characteristics:

- Intel i3 2200 MHz processor,
- Ram 4 GB,
- 500 GB hard drive,
- Windows 7 Ultimate 64-bit version.

2.3.3 Multidimensional Scaling

In this section, we will present the analysis made through the multidimensional scaling approach by presenting the type of data presented and the interpretation of the obtained results.

Data Preparation:

We have developed an algorithm based on the Smith and Waterman algorithm which allows drawing up a similarity matrix that contains the percentages of similarity between the different DNA sequences. This percentage can be replaced by a variable 0 or 1 which presents the Euclidean distance between the variables.

As a formula for calculating the correspondence between the sequences, we used the Kruskal Stress function, as follows

$$\sqrt{\frac{\sum\sum\left(f(x_{ij}) - d_{ij}\right)^2}{\sum\sum d_{ij}^2}}$$

We have also specified that the graphical representation will be in 2D and that the initial configuration is random, regarding e computation of the values of correspondence between the variables (Fig. 2).

Presentations of the Results:

The results obtained are varied. From these results, we can mention:

- Configuration: The coordinates of the objects in the representation space are displayed. If the space is two-dimensional, a graphical representation of the configuration is provided.
- Table of measured distances in the representation space: this table corresponds to the distances between the objects in the two-dimensional representation space.
- Table of disparities between the different entities: this table represents the degree of dissimilarity between the different points.
- Residual Distances: These distances are the difference between the dissimilarities calculated from the values of the initial matrix and the distances measured in the representation space.

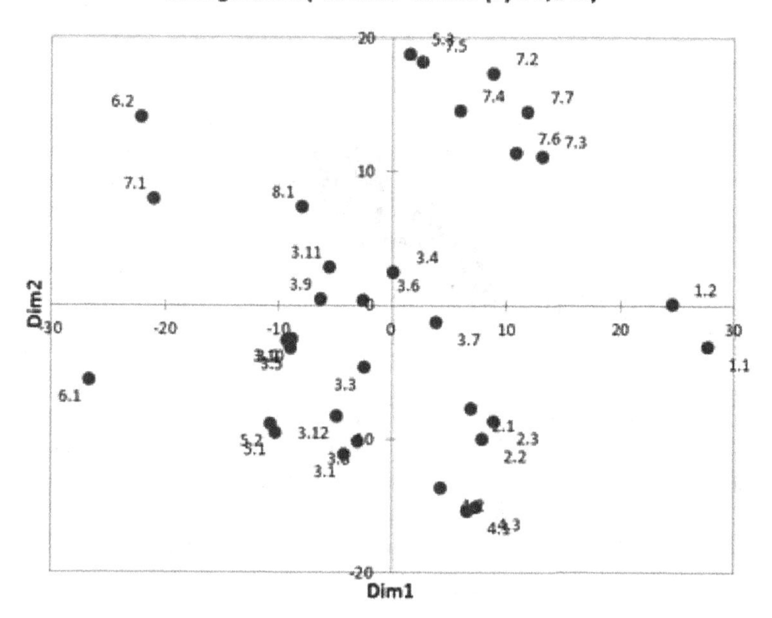

Fig. 2 Configuration diagram

- Comparison Table: This table allows to compare dissimilarities, disparities and distances, as well as the ranks of these three measures for all two-by- two object combinations.
- Shepard Diagram: This graph compares disparities and distances to dissimilarities (Fig. 3).

Interpretations

We will base our interpretations on the graph 2.6 of configuration. It allows to visualize the distribution of the points, to deduce the distances between them and to, therefore, calculate the similarity and dissimilarity between these points.

By analyzing the graph in the figure above, we can deduce that some species are well grouped by genus (such as species of the genus *Alicyclobacillus* [1.1 and 1.2]) while there are others that are far enough from each other (such as the genus *Lysinibacillus* [6.1 and 6.2]) (Fig. 4).

Through this graph, we notice that some species of the same genus are not close enough in the percentages of similarity. This calls into question the similarity that exists between these species and even leads to the conclusion that a review of the classification made by taxonomists is strongly recommended. We will also notice that sequences of different genera are quite close.

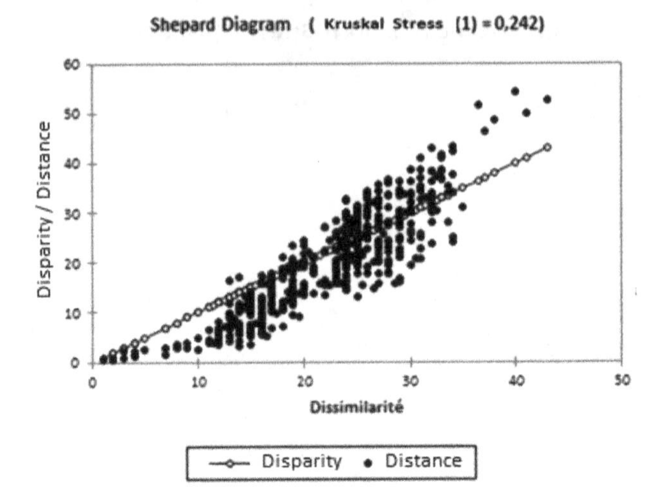

Fig. 3 Shepard diagram

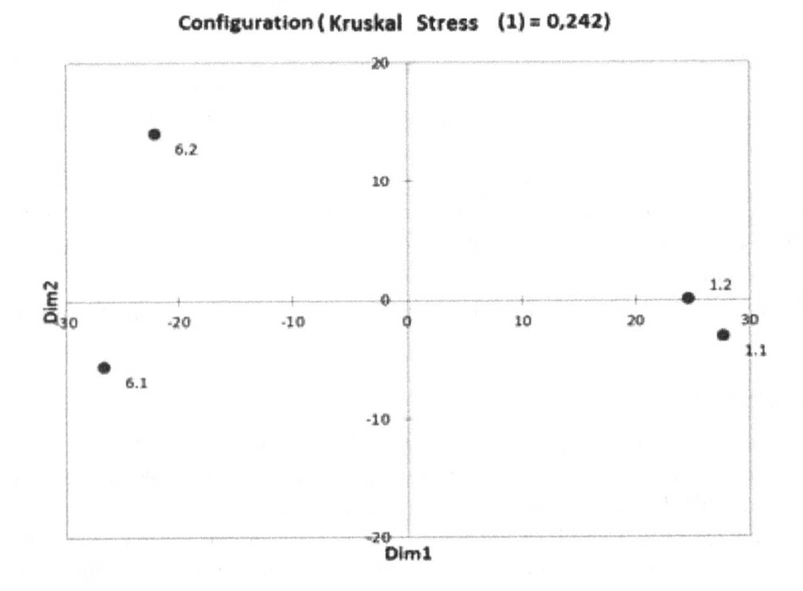

Fig. 4 Representation of species 1.1, 1.2, 6.1 and 6.2

The second example, below, shows that the species *Lactobacillus paraplatarum* (5.3) is closer to the species *Paenibacillus polymyxa* (7.5) than to the other sequences of the genus *Lactobacillus* (5.1 and 5.2).

We also noticed that for the genus *Paenibacillus*, which is represented in our work by 7 different species, has a species that is quite far from other species (*Paenibacillus azotofixans* 7.1).

Presentations of the Results:

The results obtained are varied. From these results we can mention:

Configuration: The coordinates of the objects in the representation space are displayed. If the space is two-dimensional, a graphical representation of the configuration is provided.

The major problem is that in case we would like to associate an unidentified DNA sequence with a given genus or species, the result may be wrong.

To determine if the error rate is important, we will consider two factors:

- **Probability of error**: It represents the probability that the association with a species or a genus is erroneous.
- **Frequency of appearance in the databases**: in the databanks, each species is represented by a definite number of sequences. If the probability of error is large and the occurrence frequency is large then the percentage of identification error believes. Will also be considerable.

In this part, we will present three cases illustrating the influence of the probability of error and the frequency of occurrence of these errors in the databases in the species classification results.The frequency of appearance of the DNA sequences will be calculated.

Example 1 ***Lactobacillus paraplatarum*** **(5.3) and** ***Paenibacillus polymyxa*** **(7.5)**
From Fig. 5, we notice that these two species are almost confounded. The most likely hypothesis is that these two species are but one species. However, a bad identification or an error occurred when one of them was put online in the database and its strain was declared as a new species (Fig. 6).

Moreover, by consulting the database, we note that *Lactobacillus paraplatarum* is represented by only two strains whereas *Paenibacillus polymyxa* is represented by 214 strains.

Example 2 ***Paenibacillus azotofixans*** **(7.1) and** ***Lysinibacillus sphaericus*** **(6.2)**
From the point cloud shown in Fig. 2, these two species are quite close to each other. The species *Lysinibacillus sphaericus* is closer to the species *Paenibacillus azotofixans* than to the other species of the genus *Lysinibacillus* 6.1.

Let's consider the example where an unidentified strain is a *Lysinibacillus sphaericus* and we want to identify the genus of this strain.

The probability of misidentification of the species of this strain is 12 since two sequences are only close enough for the results of identifications to be incorrect.

Regarding the frequency of occurrence indatabases, *Lysinibacillus sphaericus* appears 202 times in the database while *Paenibacillus azotofixans* appears 21 times only.

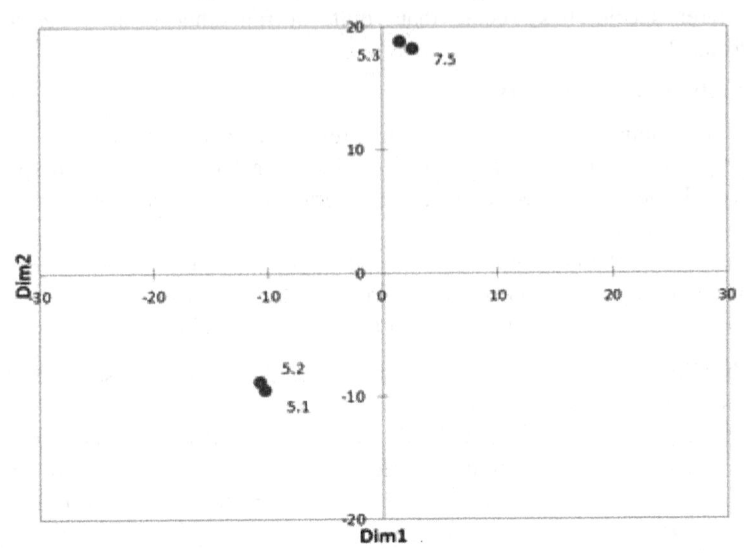

Fig. 5 Representation of species 5.1, 5.2, 5.3 and 7.5

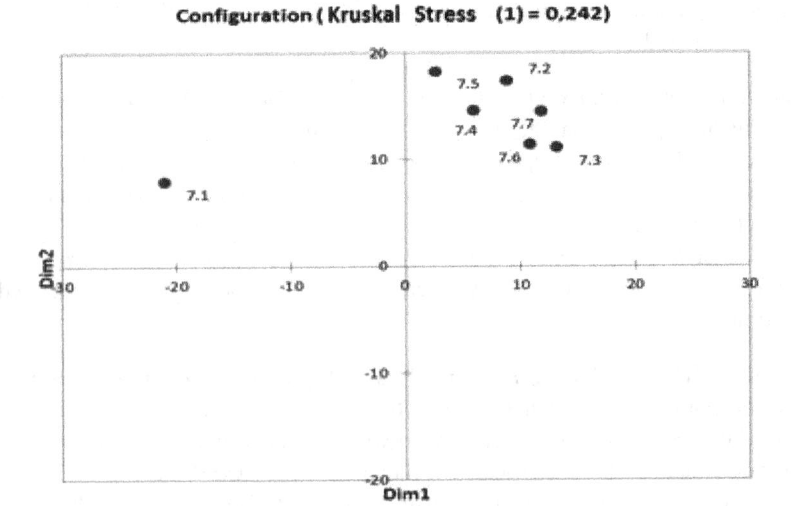

Fig. 6 Representation of species of the genus *Paenibacillus*

The Euclidean distance between the two species is 0.062 which is also quite important. Even thoughthe error rate is not too high itremains important as the probability of misidentification is significant.

Example 3 ***Paenibacillus azotofixans*** **(7.1) and other** ***Paenibacillus*** **species**
If an identification operation of a genus is made with the different species of this genus, the risk of error may also arise.

The total sequence of *Paenibacillus* is 2548. Among them, 21 strains represent the species *Paenibacillus azotofixans* and since this species is closer to the species *Lysinibacillus sphaericus*, the analysis can deduce that this strain belongs to the genus *Lysinibacillus* and not to the genus *Paenibacillus*. Similarly, the error rate remains low since the *Paenibacillus azotofixans* represents only 0.8% of all sequences of the genus *Paenibacillus*.

Synthesis:
Through these three examples, we can deduce that the two metrics that we have presented above allows to calculate this error rate. As a result, we this error rate by the following formula:

$$Terrj = \frac{\sum(\mathbf{Pi} * \mathbf{fi} * \mathbf{max}(\mathbf{0}, \sigma -' \Upsilon))}{fj}$$

With:

Terr j Error rate when identifying the species j.
pi probability of having a strain of the species i.
fi frequency of appearance of strains of i in the data banks used.
fj frequency of appearance of the strains of j in the databanks used.
σ threshold which represents the maximum Euclidean distance which means, for the researcher, that his two species are quite close.
Y Euclidean distance between species extracted from the species similarity matrix.

This error rate is proportional to the probability of error and frequency of occurrence of the species in the databanks in question. It is also inversely proportional to the Euclidean distance between species. To have a very low error rate, close to zero, the frequency of appearance of the correct strain must be much higher than that of the species that can distort the final result. The same is true for the distance between the representative points of the species in the cloud of points.

2.3.4 Ascending Hierarchical Classification

In this section, we will present the analysis based on the Ascending Hierarchical Classification approach while listing the type of data to be used, the different representations of the obtained results and the interpretation of these results.

Data Preparation:

Similar to the previous approach, we will present the matrix of similarity between the DNA sequences, which contains the percentages of similarities between the different sequences to the software.

The algorithm will automatically choose the number of resulting classes.

Presentation of the results:

It is true that each of the 5 genera species are grouped in correct classes (*Alicyclobacillus, Anoxybacillus, Bacillus, Geobacillus and Sporosarcina*), but the other three genera are poorly grouped.

We deduce that this approach also shows that the error rate exists in the classification of species and that it may be necessary to re-check the conditions and methods of taxonomic classification made by biologists.

3 Conclusion

As a conclusion, we can say that the hierarchical classification made by taxonomists does not verify mathematical rules or statistical approaches and does not respect the percentages of similarities resulting from the execution of optimal algorithms for alignment and comparison DNA sequences. This classification method leads to errors in the identification of an unclassified strain. In addition, the problem of storing DNA strains also exists in this case since the entire length of the DNA sequences will be stored in the databases.

The other lines of research were based on the parallelization of the Smith and Waterman algorithm.

Currently, there is no research that deals with reducing the size of DNA sequence alignment to decrease the storage capacity needed to store these DNA sequences.

Acknowledgments This work was funded by the International Exchange Program of Harbin Engineering University for Innovation-oriented Talents Cultivation.

References

1. M. Greenacre, *Correspondence Analysis in Practice* (CRC press, 2017)
2. S. Tufféry, *Data mining and statistics for decision making*, vol. 2 (Wiley, Chichester, 2011)
3. T.F. Smith, M.S. Waterman, Identification of common molecular subsequences. J. Mol. Biol. **147**(1), 195–197 (1981)
4. Genbank, http://www.ncbi.nlm.nih.gov/genbank/
5. B. Saada, J. Zhang, Classification of DNA Sequences based on data analysis, Lecture In *Notes in Engineering and Computer Science: Proceedings of The World Congress on Engineering 2017*, 5–7 July 2017, London, UK, pp. 1055–1058

Managing Inventory on Blood Supply Chain

Fitra Lestari, Ulfah, Ngestu Nugraha and Budi Azwar

Abstract There is unbalance the amount of blood demand and the availability of blood for each component at *Blood Transfusion Unit* in Indonesia. As the result, this component run into inventory shortage so management need to maintain the strategy of blood supply chain for the patients. Purpose of this is to manage inventory on the blood component of *Packed Red Cells* which it to be the highest blood component requirement for patient in this case study. Managing inventory is done through several stages including forecasting method, safety stock, and re-order point. Finding of this study was obtained that *exponential smoothing* ($\alpha = 0.95$) to be the best forecasting method. Then, to manage inventory, this agency need to prepared 34 blood bags for safety stock and 76 blood bags for re-order point. This results able to give recommendation to the *Blood Transfusion Unit* at Indonesia regarding with the number of blood component provided and how much re-order to be made at the time of reaching the lead time. Further study is suggested to conduct simulation method in order to evaluate policy in managing blood inventory and prepare scenario for optimizing inventory level.

Keywords Blood · Blood transfusion unit · Inventory · Production Re-order point · Safety stock · Supply chain

F. Lestari (✉) · N. Nugraha
Industrial Engineering Department, Sultan Syarif Kasim State Islamic University, Pekanbaru, Indonesia
e-mail: fitra_lestari@yahoo.com

N. Nugraha
e-mail: ngestu.nugraha@yahoo.com

Ulfah
AlMadinah Health Consulting in Indonesia, Pekanbaru, Indonesia
e-mail: ul_pha7@yahoo.com

B. Azwar
Islamic Economic Department, Sultan Syarif Kasim State Islamic University, Pekanbaru, Indonesia
e-mail: budiazwar79@yahoo.com

S.-I. Ao et al. (eds.), *Transactions on Engineering Technologies*,
https://doi.org/10.1007/978-981-13-0746-1_27

1 Introduction

Blood Transfusion Unit (BTU) serves to ensure the availability of blood, quality, safety, and blood donor information. To improve this services, the agency seeks to provide integrated information related with the blood supply activities for the patients. Blood deserved to be donated for the patient was blood that passed the screening test and fitting with the patient's blood [1]. In addition, BTU collaborates with various entities in conducting activities of blood donation involving donors, blood processing centers, hospitals and patients. For more details, the interrelation of activities among entities at BTU is to receive blood supplies from donors which it is transferred in form of blood bags. Then, blood is processed in BTU in order to determine the blood components and its quality. It aims to provide blood for patients infected with the virus or disease. Furthermore, the blood component that has been provided will be distributed to the patients. Obviously, the activity of blood distribution from donors to patients in form a flow of product which it can be studied based on supply chain strategy. Lestari et al. [2] revealed that the supply chain strategy is the integration of activities including several entities to provide the best service to the end consumer.

Moreover, this research conducted a case study at *Blood Transfusion Unit* in Indonesia. They provided seven blood components that were distributed to the patient involving *whole blood, packed red cell, thrombocyte concentrate, fresh frozen plasma, washed erytrocyte, cryoprecipitate* and *apheresis* [3]. Data of blood demand at BTU in years 2016 showed that there was unbalance the amount of blood demand and the availability of blood for each component which this can be seen in Table 1. The results in this table found that the highest blood component requirement for patient is *Packed Red Cell* (PRC). Furthermore, this component run into inventory shortage; as the result, management needed to maintain the strategy of blood supply chain for the patients. Thus, the main problem in this research is the unbalance of the amount of blood supply with the amount of blood demand. Consequently, it showed that the number of donors over time decreases while the demand for blood products tends to increase.

Table 1 Demand and supply of blood component

Blood component	Average monthly demand (bag)	Average monthly stock (bag)	Inventory
Whole blood (WB)	1203	1280	Surplus
Packed red cell (PRC)	1888	1512	Shortage
Washed erythrocyte (WE)	26	49	Surplus
Fresh frozen plasma (FFP)	222	175	Shortage
Thrombocyte concentrate (TC)	667	873	Surplus
Cryoprecipitate (C)	24	4	Shortage
Apheresis (A)	25	18	Shortage

There were several studies related with production planning and inventory control [4, 5]. They revealed that managing inventory able to support production activities and avoid unbalance between demand and supply on inventory. Most of study adopted this method for manufacturing case. Thus, activity of business process at *Blood Transfusion Unit* able to employed this method. This study aims for managing inventory on the blood component of *packed red cells* (PRC) in *Blood Transfusion Unit* in Indonesia. Furthermore, the data of blood component demand was taken from January to December 2016. This data was used to predict demand of blood component in year 2017 by looking at data history of blood component.

2 Inventory (S)

Managing inventory is the activity to determine the level and composition of the stock of parts, raw materials and finished products to achieve the effectively and efficiently of production and sales [6]. Moreover, managing inventory requires to implement in order to maintaining and controlling the production scheduling which it can be done by conducting forecasting demand as well as a technique of providing goods based on the amount and time as planned [7]. Furthermore, Vlckova and Patak [8] explained that forecasting activity is early step in business function that seeks to estimate demand and use of products. Thus, the production line was able to produce in the right quantity. Indeed, forecasting technique is implemented to predict future demand based on some predictor variables using historical time series data. Then, results of forecasting technique are used to perform production scheduling and control inventory.

In addition, to anticipate the unstable production activities, it need to be done arrangement of safety stock. Rădăşanu [9] stated that the safety stock is an extra supply that must be held for protection or safety in avoiding depletion of inventory. Furthermore, safety stock purpose to determine how much stock needed during the grace period to meet the demand [10]. Safety stock also was used to consider re-order point in production scheduling. Re-order point model occurs when the amount of inventory contained in the stock decreases steadily. Thus it is necessary to determine how much the minimum level of inventory should be considered so that there is no shortage of inventory. Thus, the expected amount can be calculated during the grace period. There were several models in conducting the re-order point to support production scheduling. The model was classified based on the pattern of demand rate and lead time [11]. Obviously, a Re-order point commonly referred to as a limit or a re-order point includes requests that are desired or needed during the grace period.

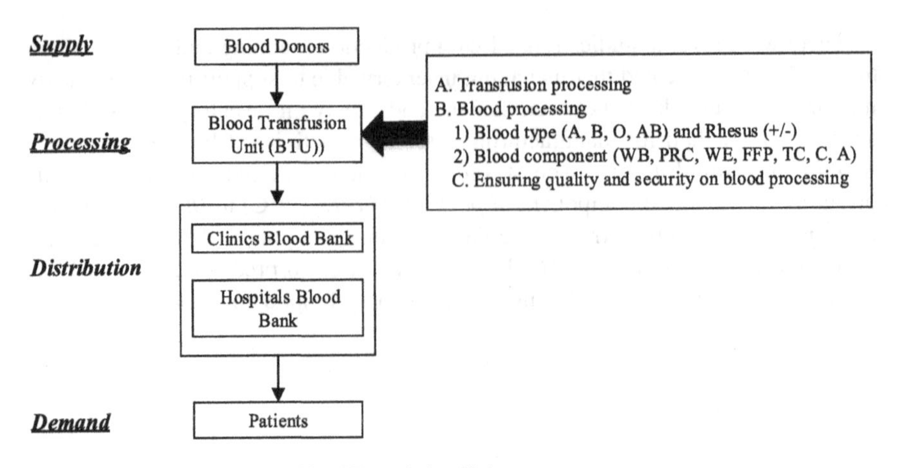

Fig. 1 Blood supply chain on *Blood Transfusion Unit*

3 Blood Supply Chain

Blood are capable of carrying oxygen (O_2) and nutrients to the tissues through flowing in the arteries, capillaries and veins and carrying carbon dioxide (CO_2) and other waste products. Moreover, Patton and Thibodeau [12] in their study described that blood consists of 4 parts including red blood cells (erythrocytes), white blood cells (leukocytes), blood platelets (thrombocytes), and blood fluid (blood plasma). Furthermore, blood can be channeled from one person to another's circulatory system due to the need for medical conditions such as trauma, surgery, shock and malfunction of red blood cell organs.

The case study of blood components demand at BTU in Indonesia was the integration of several entities into a supply chain strategy. Activity of supply chain represented of blood suppliers which they were be classified into three blood donors including *permanent blood donors, blood donors volunteers, blood donors substitute*. Furthermore, blood that had been transfused by donors at the *Blood Transfusion Unit* was used to obtain four blood types and seven blood components based on the consumer's need. Moreover, it served to obtain blood in term of safe and quality. Patients through the doctor's recommendation ordered blood through blood Bank such as Hospital and Clinic. Figure 1 shows the business activity on *Blood Transfusion Unit* based on blood supply chain strategy.

4 Finding

Managing inventory in this case study is done through several stages including forecasting demand, safety stock, and re-order point. This mechanism is used as a reference in managing blood component of the *Packed Red Cell* (*PRC*).

Table 2 forecasting demand of *Packed Red Cell* using *exponential smoothing* with $\alpha = 0.95$

Month	Demand (bag)	Forecasting (bag)
January	1888	2001
February	1799	1894
March	1956	1804
April	1915	1949
Mei	1877	1917
June	1655	1879
July	1996	1667
Augustus	2186	1980
September	1921	2176
October	2041	1934
November	2214	2036
December	2559	2206
Total	24,007	23,443

- Forecasting

Calculation of forecasting demand on blood component of the *Packed Red Cell (PRC) is* initial stages in the managing inventory. This study used four methods in forecasting demand involving *moving average*, *weighted moving average*, *exponential smoothing* and *exponential smoothing with trend*. To select the appropriate forecasting method, it is determined with selecting the smallest error accuracy of each forecasting model. Then, it done verification forecasting within the control limit by using tracking signal. Moreover, positive tracking signal showed that actual value is greater than forecast, while negative value indicated actual value is less than predicted. Rangkuti [11] revealed that the best tracking signal whenever it had a low running sum of forecast error (RSFE) value or approximates the number zero. The results of this study was obtained that *exponential smoothing* with $\alpha = 0.95$ to the best forecasting method. Thus, results of forecasting demand from the selected method will be estimated data of the blood component of *Packed Red Cell* (PRC) for year 2017. Table 2 is the result of forecasting blood component demand for the *Packed Red Cell (PRC)*. Furthermore, data from result of forecasting taken place into graphic. Figure 2 is trend data of demand and forecasting on *Packed Red Cell (PRC)*. Obviously, the data of blood component of *Packed Red Cell* in form of time series analysis into cyclic pattern.

- Safety Stock and Re-order Point

Calculation of safety stock and re-order point in this study was done based on several regulations related in blood donor. The policy of Ministry of Health in Indonesian revealed that blood donation is limited on 3 months for each donor [13]. Furthermore, World Health Organization (2014) also concluded that 3 month to be constraint for blood donor. Thus, this study decided lead time for blood donor during 3 months

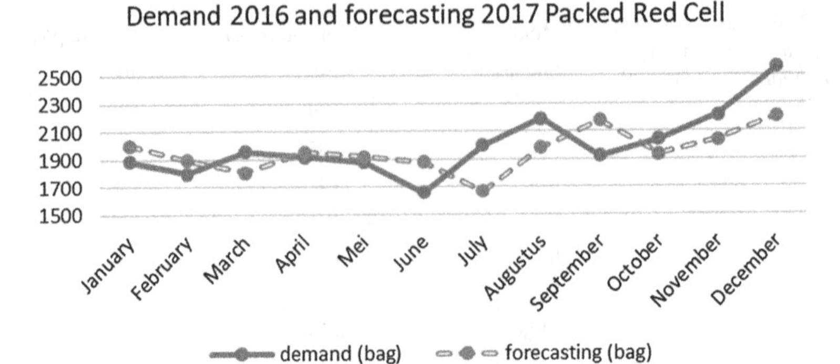

Fig. 2 Trend data of demand and forecasting on *Packed Red Cell* (*PRC*)

which it was assumed as constant lead time. Moreover, requirement of blood component of *Packed Red Cell* (PRD) for each month during one year changes greatly according to the patient's request. Thus, according to Rangkuti [11] who research related with strategy for managing inventory concluded that the above problem able to solve using a safety stock model that adopted variable demand rate with constant lead time.

This study required safety stock model to avoid the occurrence of lack of blood components of *Packed Red Cell* (PRC) in activities of business process in *Blood Transfusion Unit*. Then, it was continued to calculate re-oder point method to determine the minimum number of blood component for re-ordering. To determine the amount of safety stock, this agency established a risk of running out of inventory for this blood component of the by 5%. In other words, the blood component demand of *Packed Red Cell* (PRC) able to meet customer requirement was 95%. Thus, the standard deviation (Z) was used approximately 1.65. Moreover, lead time was used by policy by Ministry Of Heath Indonesia involving 4 month (0.021). Finally, Result of safety stock for *Packed Red Cell* (PRC) found 34 bags. Then, the calculation of re-order point for blood component on *Packed Red Cell* (PRC) found 76 bags.

- Managing Inventory

The current conditions showed that blood component of *Packed Red Cell* (*PRC*) tend to change due to uncertainty of demand which it was caused by a few people to do blood donation. Thus, *Blood Transfusion Unit* must consider to establish the safety stock of inventory in order to maintain occurring the lack of supplies of blood component. Result of this study showed that the number of safety stock for *Packed Red Cell* (*PRC*) was 34 bags and the number of re-order point was bags 76. This implication described that the inventory must has arrived at *Blood Transfusion Unit* on this number of 76 bags. Then, this agency prepared for making the reservations again. Figure 3 is strategy for managing inventory on blood supply chain to *Packed Red Cell* (*PRC*).

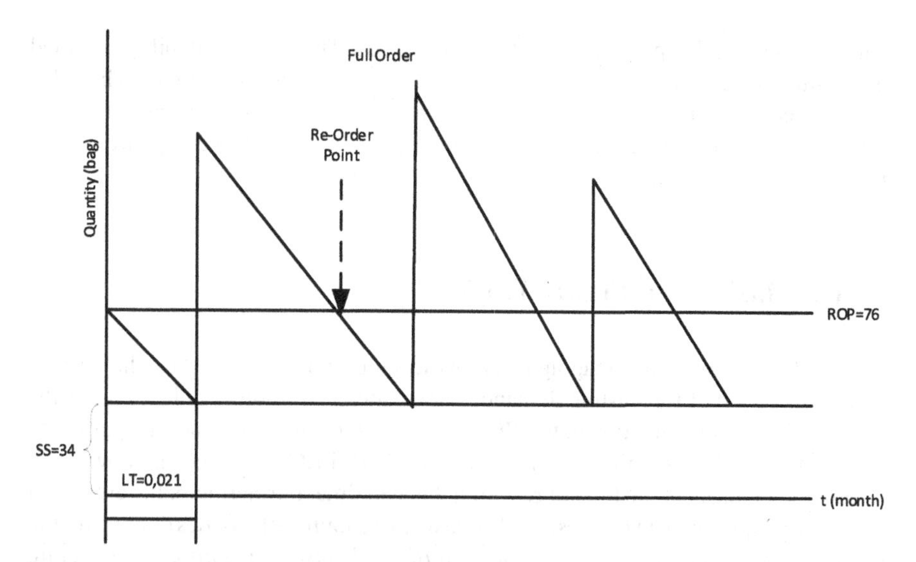

Fig. 3 Managing Inventory for *Packed Red Cell* (*PRC*)

5 Discussion

Blood Transfusion Unit must consider for managing inventory for blood supply chain. This research has done managing inventory through forecasting demand and re-order point which it also considered the schedule of order blood component based on the safety stock approach. This serve to arrange the production scheduling of blood components of based on when orders must be received and how many safety stocks should be provided. Especially for blood component of *Packed Red Cell* (PRC), there was threat for *Blood Transfusion Unit* whenever they do not notice this inventory. This blood component cannot be found if there are patients who desperately need blood. This is fatal for the patient because this blood component required for medical condition in form of accident and give birth which they need to do blood transfusion soon. In addition, shortage of inventory must be avoided by *Blood Transfusion Unit* and they also should consider viability of blood for each component. Therefore, the managing inventory for blood components is very important to maintain the inventory in this agency safely and available. Okhovat et al. [14] emphasized that irregular production schedulers able to disturb production and consumer dissatisfaction.

Moreover, there were four main entities of the blood supply chain involving donors, blood processing, hospitals and patients. They represented in supply chain relationship in form of supplier-manufacture-distributor-customer. The production process undertaken by *Blood Transfusion Unit* as the filtering the blood component to the patient. The results showed that the supply chain strategy in this business is still dependent on the supplier. Kähkönen [15] explained that the supplier dominant will consider their forecasting to deliver products to the buyer. Furthermore, every entity in the blood supply chain had different roles. Thus, the availability of blood was influenced by efficiently and effectively relationship between entities. Thus, this encouraged for balancing between supply and demand of blood inventory. Kleab [16] asserted that the balance of demand and supply can provide a competitive value for the entities involved.

6 Conclusion and Future Work

This study has provided managing inventory to support the blood supply chain based on consideration of forecasting demand, safety stock and re-order point. The results able to give recommendation to the *Blood Transfusion Unit* at Indonesia regard with the number of blood component of *Packed Red Cell* (PRC) provided and how much re-order to be made at the time of reaching the lead time. Furthermore, the entities in the blood supply chain in this case study consists of supplier of blood is donors. for the process of blood production, it is located in *Blood Transfusion Unit* as manufacture that serves as a place of blood donor, blood supply and storage. Then, the blood component is distributed to the Hospital and clinic as distribution which it is used by the patient as the customer for health purposes. Furthermore, this study limited on minimizing shortage technique and optimizing the safety stock policy. Further study is suggested to conduct simulation method in order to evaluate performance policy in managing blood inventory and prepare scenario for optimizing inventory level.

Acknowledgements The authors thank to *Blood Transfusion Unit* in Indonesia and *Sultan Syarif Kasim State Islamic University* who supported this research.

References

1. WHO, *Towards 100% Voluntary Blood Donation: A Global Framework for Action*, Geneva, Switzerland (2010)
2. F. Lestari, K. Ismail, A.B.A. Hamid, W. Sutopo, Designing supply chain analysis tool using SCOR model (Case study in palm oil refinery), in *2013 IEEE International Conference on Industrial Engineering and Engineering Management (IEEM)*, Bangkok, Thailand (2013), pp. 919–923
3. F. Lestari, U. Anwar, N. Nugraha, B. Azwar, Forecasting demand in blood supply chain (Case study on blood transfusion unit), in *Lecture Notes in Engineering and Computer Science: Proceedings of The World Congress on Engineering 2017*, 5–7 July 2017, London, U.K. (2017), pp. 764–767
4. P. Wanke, A conceptual framework for inventory management: focusing on low-consumption items. Prod. Inventory Manag. J. **49**(1), 6–23 (2014)
5. P. Vrat, Basic concepts in inventory management, in *Materials Management: An Integrated Systems Approach* (Springer, India, 2014), pp. 21–36
6. K.B. Prempeh, The impact of efficient inventory management on profitability: evidence from selected manufacturing firms in Ghana. MPRA 1–6 (2015)

7. N. Liu, S. Ren, T. Choi, C. Hui, S. Ng, Sales forecasting for fashion retailing service industry: a review. Math. Probl. Eng. **2013**, 1–10 (2013)
8. V. Vlckova, M. Patak, Role of demand planning in business process management, in *6th International Scientific Conference*, 13–14 May 2010, Vilnius, Lithuania (2010), pp. 1119–1126
9. A.C. Rădăşanu, Inventory management, service level and safety. J. Public Admin. Finance Law Inventory **9**, 145–153 (2016)
10. C. Mekel, P.T. Bank, C. Asia, L. Lahindah, Stock out analysis: an empirical study on forecasting, re-order point and safety stock level at PT. Combiphar, Indonesia. Rev. Integr. Bus. Econ. Res. **3**(1), 52–64 (2014)
11. F. Rangkuti, *Inventory Management*, vol. 2 (RajaGrafindo Persada, Jakarta, 2007)
12. K.T. Patton, G.A. Thibodeau, *Anatomy and Physiology*, 8th edn. (Elsevier Health Sciences, 2014)
13. Ministry of Health Indonesia, *Standard of Blood Transfusion Unit*: Regulation No. 91, 2015 (2015)
14. M.A. Okhovat, M. Khairol, A. Mohd, T. Nehzati, Development of world class manufacturing framework by using six-sigma, total productive maintenance and lean. Sci. Res. Essays **7**(50), 4230–4241 (2012)
15. A.-K. Kähkönen, The influence of power position on the depth of collaboration. Supply Chain Manag.: Int. J. **19**(1), 17–30 (2014)
16. K. Kleab, Important of supply chain management. Int. J. Sci. Res. Publ. **7**(9), 397–400 (2017)

Pore-Scale Modeling of Non-Newtonian Fluid Flow Through Micro-CT Images of Rocks

Moussa Tembely, Ali M. AlSumaiti, Khurshed Rahimov and Mohamed S. Jouini

Abstract Most of the pore-scale models are concerned with Newtonian fluid flow due to its simplicity and the challenge posed by non-Newtonian fluid. In this paper, we report a non-Newtonian numerical simulation of the flow properties at pore-scale by direct modeling of the 3D micro-CT images using a Finite Volume Method (FVM). To describe the fluid rheology, a concentration-dependent power-law viscosity model, in line with the experimental measurements of the fluid rheology, is proposed. The model is first applied to a single-phase flow of Newtonian fluids in 2 benchmark rocks samples, a sandstone and a carbonate. The implemented FVM technique shows a good agreement with the Lattice Boltzmann Method (LBM). Subsequently, adopting a non-Newtonian fluid, the numerical simulation is used to perform a sensitivity study on different fluid rheological properties and operating conditions. The normalized effective mobility variation due to the change in polymer concentration leads to a master curve while the flow rate displays a contrast between carbonate and sandstone rocks.

Keywords Digital rock physics · Finite volume method · Lattice boltzmann method · Non-newtonian fluid · Polymer flooding · Pore-scale modeling Porous media

M. Tembely (✉) · A. M. AlSumaiti · K. Rahimov · M. S. Jouini
Petroleum Institute, Khalifa University of Science and Technology,
P.O. Box 2533, Abu Dhabi, UAE
e-mail: mtembely@pi.ac.ae

A. M. AlSumaiti
e-mail: aalsumaiti@pi.ac.ae

K. Rahimov
e-mail: khrahimov@pi.ac.ae

M. S. Jouini
e-mail: mjouini@pi.ac.ae

S.-I. Ao et al. (eds.), *Transactions on Engineering Technologies*,
https://doi.org/10.1007/978-981-13-0746-1_28

1 Introduction

Many applications ranging from hydrology, environment, water management, to oil and gas industry would benefit from an accurate model of fluid flowing inside porous media. In order to optimize reservoir management, the fluid flow processes in porous media should be investigated through a multiscale approach ranging from the field to the core level, down to the pore-scale. Recent development in image acquisition techniques based on X-ray micro-Computed Tomography (micro-CT), Focused Ion Beam (FIB) and Scanning Electron Microscopy (SEM) allow investigating core samples to model pore networks at very high resolution and simulate rock properties at pore-scale. This approach known as Digital Rock Physics (DRP) is seen as a complementary tool to conventional laboratory measurements [1–3]. Nevertheless due to physical limitations of detectors size in scanning devices it is not possible to image macro, micro and nano pores at the same scale. Indeed, the best resolution to scan a core plug sample of 1.5 inches diameter using X-ray micro-CT with a detector size of 2000×2000 pixels is 20 μm. At this scale, usually, macro pores and main textural heterogeneities can be visualized. In order, to image smaller pores, subsets of few millimetres are extracted physically and scanned at higher resolution around 1 to 2 μm. At this scale, several rocks like carbonates are still revealing unresolved porous phase including micro and nano pores which might have significant impact on the effective permeability and porosity of the sample. FIB-SEM imaging technique is implemented in this case to investigate the remaining unresolved nano pores. Then, a segmentation processing is applied in order to extract pore network and petrophysical properties can be simulated numerically on the derived model.

It's worth noting that most of the DRP simulations focus on Newtonian fluids and overlook the rheology of the fluid [2, 13]. However, in petroleum engineering many fluids such as heavy oil and polymer solutions used for enhanced oil recovery (EOR) are non-Newtonian [15, 16]. One of the most important petro-physical properties for reservoir rock is the permeability, which prediction through an accurate and efficient numerical tool is highly desirable [2, 6–8]. Despite a large body of work, modeling of non-Newtonian fluid rheology is still challenging and remain an active field of research [5, 11, 16, 17]. In order to use DRP as a future tool to generating accurate, fast, and cost effective special core analysis (SCAL) data to support reservoir characterization and simulation; pore-scale modeling of non-Newtonian fluid is essential. This work proposes to shed a light on that aspect of the DRP.

In the present paper, we use a finite volume method (FVM) coupled with an adaptive meshing technique to perform the pore-scale simulation from the micro-CT images of a sandstone and a carbonate from the literature with both a Newtonian and non-Newtonian fluids. Besides, we performed simulations based on the LBM method for comparison and validation of the FVM technique. Finally, we implement a power-law (or Ostwald deWaele) non-Newtonian fluid to test the pore-scale model sensitivity to both the rheological and operating parameters. This paper, focused on modeling at pore-scale, can be considered as complementary to the works in [5, 9, 10] on simulating non-Newtonian fluids at Darcy or core scale.

This contribution, an extension of the work in [14], is organized as follows: in Sect. 2, the fluid flow transport equations in porous media are presented and in Sect. 3 the pore-scale modeling and validation are provided. Numerical results and sensitivity studies are performed in Sect. 4. Finally, conclusions are drawn in Sect. 5.

2 Fluid Flow Equations in Porous Media

In the next section, the governing equations of the fluid flow transport within the porous media will be presented; and the fluid is assumed to be an incompressible non-Newtonian liquid.

2.1 *Mass and Momentum Conservations*

The continuity and momentum equations to be numerically solved in the finite volume method (FVM) formulation expresses as follows:

$$\nabla \cdot \mathbf{V} = 0 \tag{1}$$

$$\rho \mathbf{V} \nabla \mathbf{V} = -\nabla p + \rho \mathbf{g} + \nabla \cdot \tau \tag{2}$$

where $\mathbf{V}$ is the fluid velocity vector, and $\mathbf{g}$ denotes the gravity, while the fluid is assumed incompressible of density ρ, viscosity μ. The stress tensor τ, assuming the viscosity to depend on both the polymer concentration (C) and shear rate ($\dot{\gamma}$), can be written as:

$$\tau = \mu(C, \dot{\gamma})[\nabla \mathbf{V} + \nabla \mathbf{V}^{\mathrm{T}}] \tag{3}$$

2.2 *Fluid Rheology: A Concentration Power-Law Viscosity Model*

We propose for the viscosity a modified power-law accounting for the effect of the polymer concentration as follows:

$$\mu = \chi \exp[\alpha C] \dot{\gamma}^{n-1} \tag{4}$$

where C is the polymer concentration, α a constant, χ is the consistency factor and n the flow behavior index. The proposed equation (see Fig. 1) is in line with experimental measurements of polymer solutions used for EOR [15]. The typical values used in Eq. (4) are given in Table 1. It is worth mentioning that even though our

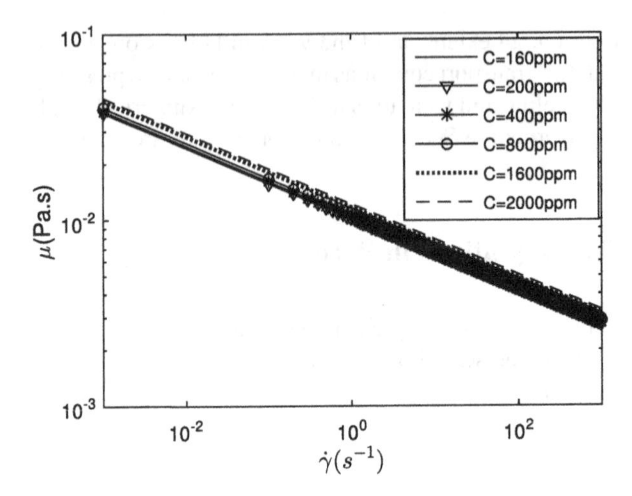

Fig. 1 A concentration dependent power-law viscosity fluid function of the shear rate at different concentrations

Table 1 Typical values of the viscosity model parameters

Parameters	χ	α	n
Values (SI)	10^{-2}	100	0.81

model may approximate well some polymers rheology for EOR, generally polymer solutions exhibit a more complex behavior such as viscoelastic effects.

Due to the non-Newtonian nature of the fluid, we will adopt the effective mobility (M_{eff}), instead of the absolute permeability (k), to characterize the concentration-dependent-power-law fluid flowing inside the porous media as:

$$M_{eff} = -\frac{\mu_{eff} LU}{k} = -\frac{1}{\Delta P}\frac{LQ}{A} \tag{5}$$

where Q is the flow rate, U is the Darcy velocity, μ_{eff} is the effective viscosity, ΔP the pressure gradient imposed on the sample, L and A are the sample length and surface area, respectively. Due to the challenge to perform experiments at the scale of the micro-CT 3D images resolution and the characterization of fluid flow within the pore network, the present work will be based on numerical simulations to gain insight into the physics of modeling complex fluid through a rock. It's worth noting that for a Newtonian fluid, $n = 1$, the effective mobility reduce to $M_{eff} = k/\mu$. For comparison with experimental data, we will use the normalized mobility (m_{eff}) by scaling using the effective mobility of water such that $m_{eff} = M_{eff}/M_{eff\,W}$.

3 Numerical Methodology and Validation

One of the challenge in simulating flow in porous rocks is that experimentally, laboratory measurements cannot be performed at the scale in which pore phase is resolved. However, pore-scale modeling within the emerging digital rock physics(DRP) paradigm has gained attention recently going from understanding the physics within the rock to a predictive tools to support laboratory measurements [3, 12]. The novelty of the present work is the development of a solver capable of computing at pore-scale non-Newtonian fluid flow through porous rocks and gains insight into the flow behavior at scales unattainable experimentally. The pore-scale solver which can be used for different rock types and fluids can help in predicting flow behavior at reservoir conditions as well as to better understanding polymer flooding.

X-ray micro-CT scanners generate after acquisition and reconstruction 3D grey level images where voxels values depend on material density. High values corresponding to bright voxels are related to solid phase whereas dark ones denote pores. Image segmentation techniques allow identifying pore structure in digital images. The workflow from the micro-CT image to the computation of petrophysical properties such as the permeability is summarized in Fig. 2. The rock is first scanned at high resolution then segmented to discriminate between pore and solid phases; then the connectivity is determined in order to assess the relevance of the numerical simulation. Finally the 3D digital rock model is generated as an input of the numerical model.

In order to run the simulation, the segmented micro-CT image is meshed using SnappyHexMesh/C++ code to perform an Adaptive meshing technique, through refinement and adjustment to fit onto the provided geometries of the rock; the addition

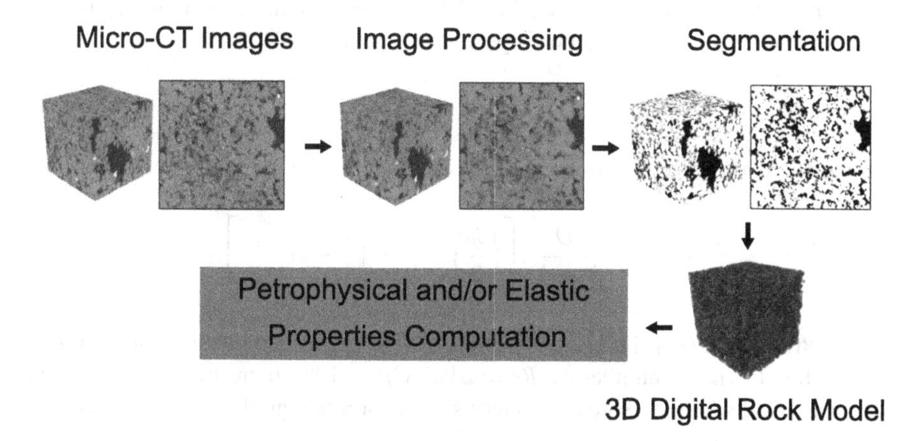

Fig. 2 Workflow of the numerical simulation at pore-scale of the Micro-CT image of the rocks

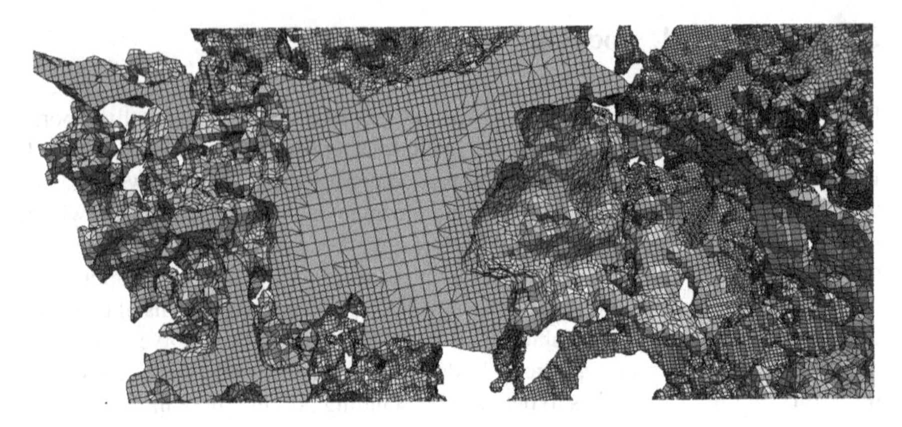

Fig. 3 Adaptive Mesh at the pore-scale from the digital image of the rock

of the boundary layers cells near the solid surface is also performed for better accuracy (Fig. 3).

The SIMPLE algorithm is used to calculate the pressure and velocity fields using a Generalized Geometric-Algebraic Multi-Grid (GAMG) solver in conjunction with a Gauss Seidel smoother. The convergence criteria set for the pressure and velocity fields is of the order of 10^{-6}. The simulations are run in parallel using a domain decomposition method. The validation of the approach will be undertaken in the next section based on both Newtonian and non-Newtonian fluids.

3.1 2D Hagen-Poiseuille Flow

To validate the model a Hagen-Poiseuille flow using both a Newtonian and non-Newtonian fluids between two parallel plates is considered. In the non-Newtonian case, the fluid viscosity is taken as a power-law model:

$$\mu(\dot{\gamma}) = \kappa \dot{\gamma}^{n-1}, \tag{6}$$

for which an analytical solution can be written function of the flow rate Q as:

$$V_x(y) = \frac{Q}{B(n, h)} \left[\left(\frac{h}{2} \right)^{1/n+1} - |\frac{h}{2} - y|^{1/n+1} \right] \tag{7}$$

where $B(n, h) = 2(n+1)(h/2)^{2+1/n}/(2n+1)$. We performed the simulation at relatively low Reynolds number of, $Re = \rho V_0 D_0/\mu = 100$, to ensure a laminar flow. The analytical solution expressed in terms of the pressure gradient, ΔP, yields:

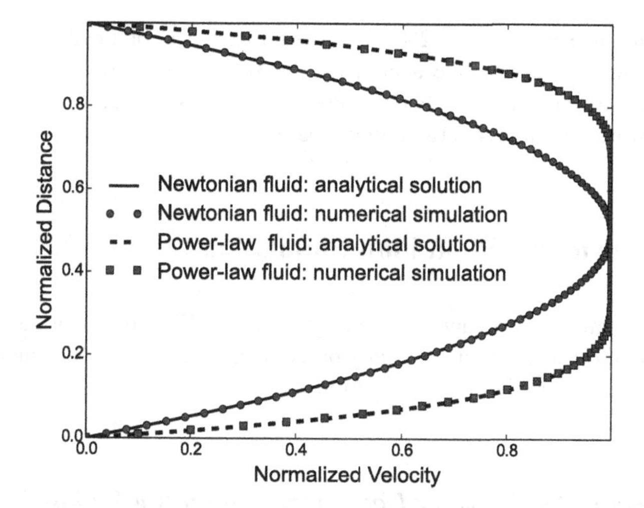

Fig. 4 Results of the Numerical Simulation by the Finite Volume Method (FVM) of the Sandstone. Streamlines Shown at the Top of the Pore Structure for Clarity

$$V_x(y) = \frac{n}{n+1}\left(\frac{\Delta P}{\kappa L}\right)^{1/n}\left[\left(\frac{h}{2}\right)^{1/n+1} - |\frac{h}{2} - y|^{1/n+1}\right] \tag{8}$$

The comparison between analytical and numerical solutions is provided in Fig. (4). A very good agreement of less than 0.5% the relative error is found (Fig. 4).

3.2 *Flow Through Porous Media*

For validation purpose of our model, we apply the FVM model to 2 rocks samples from the literature, the Fontainebleau sandstone and Grosmont carbonate [2]. Similarly, we performed the simulations under the same conditions using the widely used LBM (Palabos library) from the literature. We provide in Table 2 the simulation results of the absolute permeability in (milliDarcy) along with the relative errors. The sample size and resolution are 288 × 288 × 300 voxels at 7.5 μm and 400 × 400 × 400 voxels at 2.02 μm of the sandstone and carbonate, respectively.

Table 2 Numerical simulations results of the absolute permeability on the z-axis in milliDarcy

Sample	FVM (mD)	LBM (mD)	Relative errors (%)
Sandstone	1614	1610	0.2
Carbonate	217	214	1.4

The difference is less than 2% suggesting that the implemented finite volume method is capable to simulate accurately the simulation at pore-scale. Unlike in the LBM the extension of the Navier-Stokes equations to Non-Newtonian fluid is straightforward with no numerical tuning parameters.

4 Non-newtonian Fluid Flow Simulation

Based on the micro-CT images of the rocks, namely Fontainebleau Sandstone and Grosmont carbonate, a sensitivity study based on a concentration-dependent-power-law fluid will be carried out in this section.

4.1 *Effect of the Polymer Concentration on the Mobility*

We simulated the fluid flow at pore-scale using the liquid rheological model given in Eq. (4). In order to investigate the concentration effect, we performed the simulation (see Fig. 5) at different concentrations of the polymer solutions. The results of the

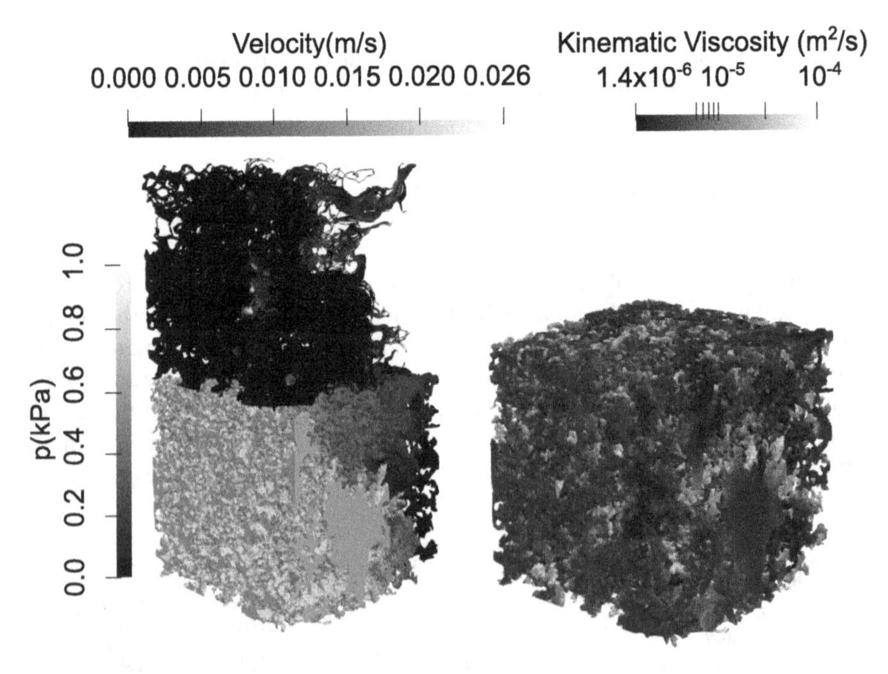

Fig. 5 Results of the Numerical Simulation of the Non-Newtonian Fluid at Pore-Scale (left) Pressure Field and Streamlines and (right) the Kinematic Viscosity Field for the Carbonate

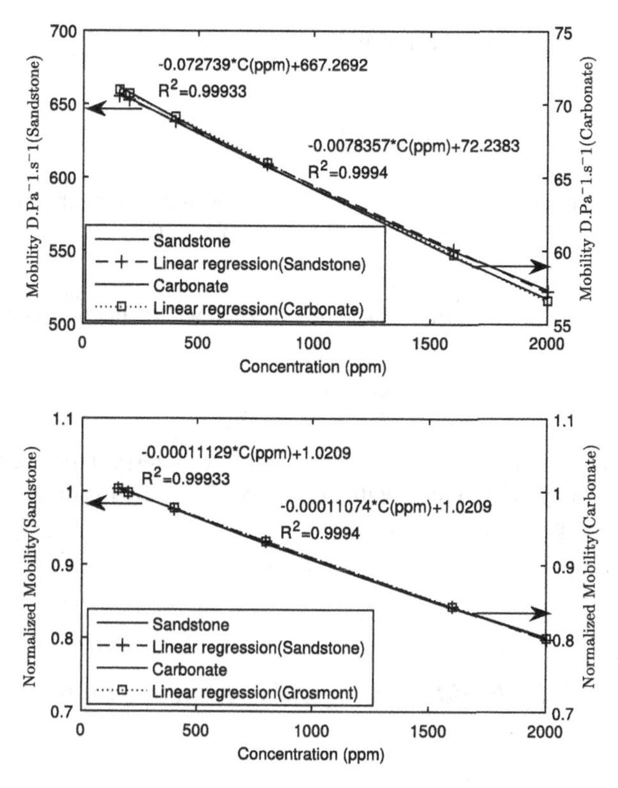

Fig. 6 Simulated Results of the Variation of the Mobility Based on the Polymer Concentration: (top) Absolute Mobility and (bottom) Normalized Mobility

simulation on the effective mobility variation are given in Fig. 6. As expected, the mobility seems to evolve inversely proportional to the concentration. Interestingly, while the viscosity exponentially depends on the concentration, the mobility seems to evolves linearly with the concentration. Furthermore, the normalized effective mobility leads to a master curve for both sandstone and carbonate rocks.

In order to interpret the linear evolution of the mobility with the concentration, let's assume that the fluid flow within the porous rocks follow a Poiseuille type flow. In this case, the flow rate can be related to the pressure gradient from Eq. (8) by

$$Q \propto \left(\frac{\Delta P}{\kappa}\right)^{1/n}, \tag{9}$$

where the flow consistency factor $\kappa \propto \exp(\alpha C)$. Using the definition of the effective mobility in Eq. (5) we have:

$$M_{eff} \propto \frac{LQ}{A\Delta P} \propto \frac{L}{A\Delta P}\left(\frac{\Delta P}{\kappa}\right)^{1/n}, \tag{10}$$

as ΔP, A and L are fixed, we can simplify

$$M_{eff} \propto \left(\frac{1}{\mu_C}\right)^{1/n} \propto \exp(-\alpha C/n). \tag{11}$$

On the other hand since $\varepsilon = \alpha C/n \approx 0.2 \ll 1$ even by taking the maximum concentration considered $C = 2000 \times 10^{-6}$, $\alpha = 100$, $n = 0.81$.

Therefore, by expanding Eq. (11) at the 1st order, we can deduce the following equation:

$$M_{eff} \propto 1 - \frac{\alpha C}{n} \tag{12}$$

which predicts a linear variation of the mobility with respect to the concentration, remarkably in agreement with the pore-scale simulation of the 2 samples depicted in Fig. (6).

4.2 Effect of the Fluid Rheological Parameters on the Mobility

We numerically investigate the sensitivity of the model to the rheological properties of the fluid, namely the flow behavior index n and the consistency factor K. We depict in Fig. (7) the impact of both n and K on the effective mobility (Fig. 7).

We observe that the mobility is reduced by an increase of both n and K. Even though, we have similar trends for both carbonate and sandstone on the fluid consistency factor(K), there is however a slight difference on the behavior index (n). In addition, the effect of K on the mobility seems to be more pronounced than that of n. It's worth noting the challenge posed experimentally to discriminate the effect of these two parameters contribution on the mobility. Hence, the possibility to perform such a sensitivity study numerically can help in the formulation of polymer solutions for EOR.

4.3 Effect of the Gradient of Pressure on the Mobility

We investigated the model sensitivity to different pressure gradient. In Fig. (8), we provide the results of simulations where the variation of the normalized flow rate and pressure gradient is found to follow a power-law, $Q/Q_0 \approx \Delta P^{1/\xi}$, where ξ is close to the behavior index. The non-linearity nature of the fluid flow within the porous

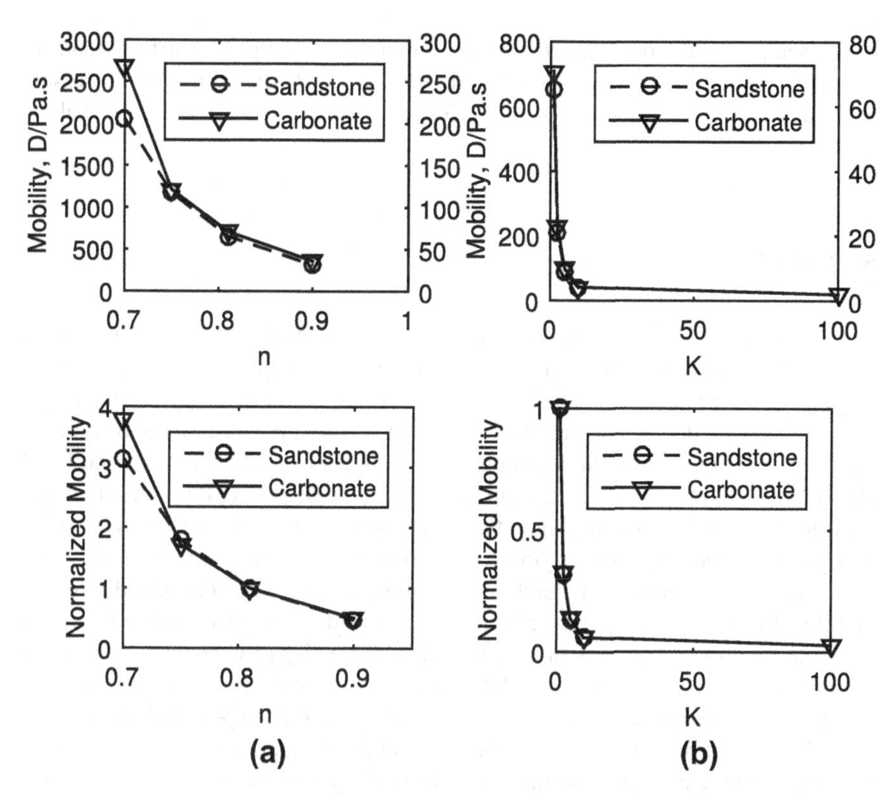

Fig. 7 The sensitivity of the effective mobility to (top) the flow behavior index n and (bottom) the consistency factor (K) at 200 ppm of the polymer concentration. The effective mobility is being normalized with that of water as detailed in Sect. 2

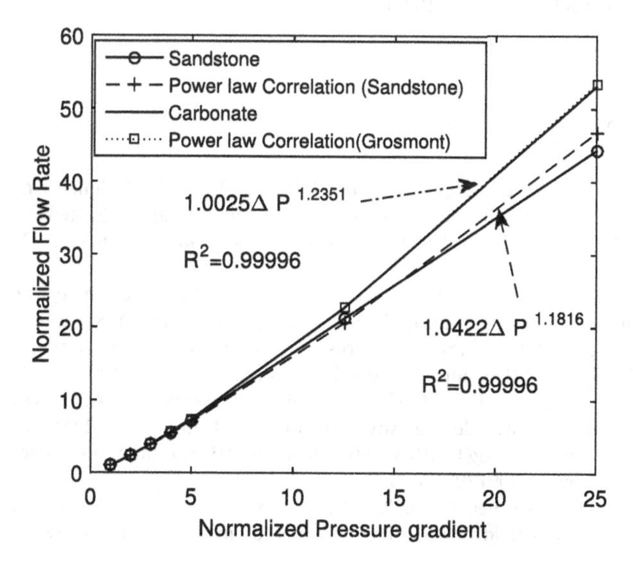

Fig. 8 Normalized flow rate and pressure gradient from pore-scale numerical simulations

media seems to be more marked for the carbonate, of complex structure, than for the sandstone sample. Interestingly, the dependency of Q on the power of pressure gradient is in agreement with the recent founding experimentally at Darcy scale in [4, 5] (Fig. 8).

5 Conclusion

In this work, we present a comprehensive numerical method based on the Finite Volume Method (FVM) and an adaptive meshing technique to describe the flow properties at pore-scale of a non-Newtonian fluid. The fluid rheology is modeled by incorporating its dependency to the concentration into a power-law viscosity fluid. Based on Newtonian fluid flow through rock samples from the literature, the FVM algorithm is validated against a Lattice Boltzmann Method (LBM). After implementing the non-Newtonian fluids, the model sensitivity to the rheology is tested by evaluating the effect of polymer concentration on the mobility as well as the relationship between flow rate and the pressure gradient. The normalized effective mobility function of the polymer concentration leads to a master curve while the flow rate function of the pressure gradient displays a disparity between the carbonate and sandstone. The effective mobility dependency on the polymer concentration follows a linear trend which theoretical interpretation has been provided. Besides the fluid rheological properties such as the flow behavior index and consistency factor lead to a master curve when normalizing the mobility for both the carbonate and the sandstone.

Acknowledgements The authors gratefully acknowledge the financial support from ADNOC under the Digital Rock Physics (DRP) project.

References

1. H. Andrä, N. Combaret, J. Dvorkin, E. Glatt, J. Han, M. Kabel, Y. Keehm, F. Krzikalla, M. Lee, C. Madonna, M. Marsh, T. Mukerji, E. Saenger, R. Sain, N. Saxena, S. Ricker, A. Wiegmann, X. Zhan, Digital rock physics benchmarks-part I: imaging and segmentation. Comput. Geosci. **50**, 25–32 (2013)
2. H. Andrä, N. Combaret, J. Dvorkin, E. Glatt, J. Han, M. Kabel, Y. Keehm, F. Krzikalla, M. Lee, C. Madonna, M. Marsh, T. Mukerji, E. Saenger, R. Sain, N. Saxena, S. Ricker, A. Wiegmann, X. Zhan, Digital rock physics benchmarks-part II: computing effective properties. Comput. Geosci. **50**, 33–43 (2013). https://doi.org/10.1016/j.cageo.2012.09.008
3. M.J. Blunt, B. Bijeljic, H. Dong, O. Gharbi, S. Iglauer, P. Mostaghimi, A. Paluszny, C. Pentland, Pore-scale imaging and modelling. Adv. Water Resour. **51**, 197–216 (2013) (35th Year Anniversary Issue). https://doi.org/10.1016/j.advwatres.2012.03.003, http://www.sciencedirect.com/science/article/pii/S0309170812000528
4. T. Chevalier, C. Chevalier, Clain, X., Dupla, J., Canou, J., Rodts, S., Coussot, P.: Darcys law for yield stress fluid flowing through a porous medium. J. Non-Newton. Fluid Mech.

195 (Supplement C), 57–66 (2013). https://doi.org/10.1016/j.jnnfm.2012.12.005, http://www.sciencedirect.com/science/article/pii/S0377025712002753
5. V. Di Federico, S. Longo, S.E. King, L. Chiapponi, D. Petrolo, V. Ciriello, Gravity-driven flow of herschelbulkley fluid in a fracture and in a 2D porous medium. J. Fluid Mech. **821**, 5984 (2017). https://doi.org/10.1017/jfm.2017.234
6. R. Guibert, M. Nazarova, P. Horgue, G. Hamon, P. Creux, G. Debenest, Computational permeability determination from pore-scale imaging: sample size, mesh and method sensitivities. Transp. Porous Media **107**(3), 641–656 (2015). https://doi.org/10.1007/s11242-015-0458-0
7. M.S. Jouini, N. Keskes, Numerical estimation of rock properties and textural facies classification of core samples using x-ray computed tomography images. Appl. Math. Modelling **41**, 562–581 (2017). https://doi.org/10.1016/j.apm.2016.09.021, http://www.sciencedirect.com/science/article/pii/S0307904X16304917
8. M.S. Jouini, S. Vega, A. Al-Ratrout, Numerical estimation of carbonate rock properties using multiscale images. Geophys. Prospect. **63**(2), 405–421 (2015). https://doi.org/10.1111/1365-2478.12156
9. A. Lavrov, Radial flow of non-newtonian power-law fluid in a rough-walled fracture: Effect of fluid rheology. Transp. Porous Media **105**(3), 559–570 (2014). https://doi.org/10.1007/s11242-014-0384-6
10. S. Longo, V. DiFederico, Unsteady flow of shear-thinning fluids in porous media with pressure-dependent properties. Transp. Porous Media **110**(3), 429–447 (2015). https://doi.org/10.1007/s11242-015-0565-y
11. M. Mackley, S. Butler, S. Huxley, N. Reis, A. Barbosa, M. Tembely, The observation and evaluation of extensional filament deformation and breakup profiles for non newtonian fluids using a high strain rate double piston apparatus. J. Non-Newton. Fluid Mech. **239**, 13–27 (2017). https://doi.org/10.1016/j.jnnfm.2016.11.009, http://www.sciencedirect.com/science/article/pii/S0377025716301483
12. C. Madonna, B.S. Almqvist, E.H. Saenger, Digital rock physics: numerical prediction of pressure-dependent ultrasonic velocities using micro-ct imaging. Geophys. J. Int. **189**(3), 1475–1482 (2012). https://doi.org/10.1111/j.1365-246X.2012.05437.x
13. P. Mostaghimi, M.J. Blunt, B. Bijeljic, Computations of absolute permeability on micro-ct images. Math. Geosci. **45**(1), 103–125 (2013). https://doi.org/10.1007/s11004-012-9431-4
14. T. Moussa, A.M. AlSumaiti, K. Rahimov, M.S. Jouini, Numerical simulation of non-newtonian fluid flow through a rock scanned with high resolution X-ray micro-CT, in *Lecture Notes in Engineering and Computer Science: Proceedings of The World Congress on Engineering 2017*, 5–7 July 2017, London, UK, pp. 958–960
15. Quadri, S.M.R., Shoaib, M., AlSumaiti, A.M., Alhassan, S.: Screening of polymers for EOR in high temperature, high salinity and carbonate reservoir conditions, in *International Petroleum Technology Conference*, 6-9 Dec 2015, Doha, Qatar. https://doi.org/10.2523/IPTC-18436-MS
16. T. Sochi, Non-newtonian flow in porous media. Polymer **51**(22), 5007–5023 (2010). https://doi.org/10.1016/j.polymer.2010.07.047, http://www.sciencedirect.com/science/article/pii/S0032386110006750
17. M. Tembely, D. Vadillo, M.R. Mackley, A. Soucemarianadin, The matching of a one-dimensional numerical simulation and experiment results for low viscosity newtonian and non-newtonian fluids during fast filament stretching and subsequent break-up. J. Rheol. **56**(1), 159–183 (2012). https://doi.org/10.1122/1.3669647

Study of Friction Model Effect on A Skew Hot Rolling Numerical Analysis

Alberto Murillo-Marrodán, Eduardo García and Fernando Cortés

Abstract In this chapter the existing friction conditions between the rolls and the workpiece of a hot skew roll piercing mill are evaluated. A modified model of this process without the inner plug has been simulated, using the Finite Element Method (FEM) and validated with experimental data extracted from the industrial process. Three friction laws have been considered for the simulation of the friction conditions between the rolls and the workpiece: Coulomb, Tresca and Norton. Then, their performance have been evaluated in terms of velocity, power consumption and sliding velocity at the interface. On the one hand, the inappropriateness of Coulomb law for this type of processes has been demonstrated. On the other hand, between Tresca and Norton laws, some differences are appreciable. Tresca law reproduces correctly the velocity of the process, but Norton law is more accurate regarding the estimation of frictional power losses. As hot rolling is a process with high energy consumption, Norton results to be the more complete law for the simulation of this kind of rolling processes.

Keywords FEM · Friction law · Friction modelling · Friction power
Rolling process · Sliding velocity

1 Introduction

Hot-rolled steel seamless tubes are demanded by petrochemical, power generation, mechanical and construction industries [1]. They are produced through the skew-roll piercing, which is the most established method and thus considered in this study.

A. Murillo-Marrodán (✉) · E. García · F. Cortés
Faculty of Engineering, University of Deusto, Avda Universidades 24, 48007 Bilbao, Spain
e-mail: alberto.murillo@deusto.es

E. García
e-mail: e.garcia@deusto.es

F. Cortés
e-mail: fernando.cortes@deusto.es

S.-I. Ao et al. (eds.), *Transactions on Engineering Technologies*,
https://doi.org/10.1007/978-981-13-0746-1_29

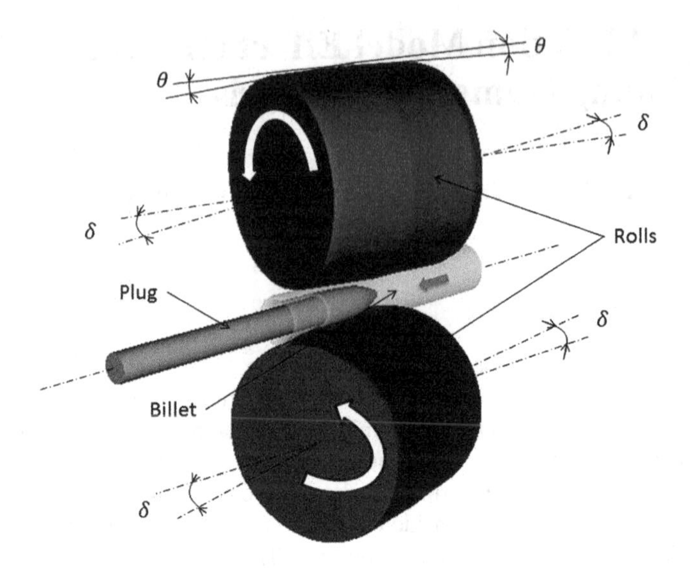

Fig. 1 Skew-roll piercing mill setup

Skew-roll piercing is the first step in the seamless tube forming process, the hollow part of the billet is generated in a mill that can be set up with different dispositions. There are mills with two or three director rolls, but in general two roll mills are more common in the industry [2]. The study considers the typical architecture of a Mannesmann mill, as shown in Fig. 1. It counts with two rolls, oriented with a feed angle δ, which determines the performance of the process. Their geometry is defined with the cross angle θ, which influences the conformation and final geometry of the tube. In addition, it includes the plug that creates the hollow part of the tube and two lateral Diescher discs, which contain the lateral deformation (not included in Fig. 1).

The present study belongs to a line of research of the authors [3] about this industrial process. According to the experiments and previous simulations, the contact between the rolls and the billet is responsible for almost the total energy consumed during the piercing process. This way, important process parameters such as the billet advance velocity and power consumed are strongly related to friction conditions in the contact region. In forging operations, the effect of friction on important process parameters has already been evaluated [4, 5]. Despite this fact, most of the skew-roll piercing studies hardly pay attention to the correct simulation of friction, which can lead to imprecisions.

The inherent complexity of the process hinders the required analysis of the friction phenomenon. In order to study the friction correctly the mill is simplified by removing the plug and discs, taking only into account the roll-billet interface. This leads to the analysis of a skew rolling process where conclusions can be extrapolated to the original process. In this paper, some widely extended friction laws like Coulomb or Tresca are studied through their application to the aforementioned process. In addition, Norton law is considered. It is a specific friction law valid for processes with visco-plastic material behaviour.

The friction laws considered for the simulation of the skew rolling process are the following:

- Coulomb's friction approach [6] is determined by a friction coefficient μ ($0 < \mu < 1$), which multiplies the normal stress σ_n, giving as a result

$$\tau = \mu \cdot \sigma_n. \tag{1}$$

  Zhao [7] used this friction law for the deformation analysis of the piercing process.
- Tresca friction model is used for shear stresses exceeding the shear flow stress of the material [8], thereby the shear stress is given by:

$$\tau = m \cdot k, \tag{2}$$

  where m stands for the friction factor with a value ranging between 0 and 1. The material shear yield strength k is dependent on the equivalent stress σ_{eq} and according to Von Mises is defined as

$$k = \frac{\sigma_{eq}}{\sqrt{3}}. \tag{3}$$

  Komori [9, 10] use this friction law for the piercing process whereas Ghiotti [11] for the rolling of the billet (without piercing) with coefficients m that range between 0.8 and 1 for the friction between the rolls and billet.
- Norton's or viscoplastic friction law assumes a viscous behaviour of the billet material close to the contact with the tool. A comparison between this friction law and the aforementioned showed better results in terms of torque and force for hot rolling [12]. The shear stress is given by

$$\tau(v) = -\alpha \cdot K \cdot |v_{rel}|^{pf-1} \cdot v_{rel}, \tag{4}$$

where α is the viscoplastic friction coefficient ($0 < \alpha < 1$), v_{rel} is the relative velocity at the interface, K is the material consistency and pf the sensitivity to sliding velocity. The term pf is given the same value as the strain rate sensitivity index of the rheology law, generally around 0.15 for hot steel forming [13]. The material consistency K is dependent on the material model, being in this case visco-plastic without considering elasticity. In this case it yields

$$K = \frac{\sigma_{eq}}{\left(\sqrt{3}\right)^{m+1} \cdot \dot{\bar{\varepsilon}}^{m}}, \tag{5}$$

where m is the sensitivity of material strain-rate $\dot{\bar{\varepsilon}}$, which takes also the value of the strain rate sensitivity index of the rheology law. The main contribution of this friction law is considering the existence of a thin viscoplastic interface layer making the friction forces dependent on the relative sliding velocity.

2 Numerical Model

In this section, the numerical model of the simplified skew-rolling mill is explained. It has been developed with the FEM software FORGE@, the principal elements of the model are the rolls and the billet but besides, it incorporates a guide and a simple press to replace the thrust bench action. The architecture is maintained as depicted in Fig. 1, the feed angle is set to 12° and the cross angle is 2° in agreement with the configuration used during the experimental tests. Rolls rotate at a constant velocity of 111 rpm and the billet is initially moved at a rate of 100 mm/s until it is grabbed by the rolls. The material of the billet is a low-carbon steel (0.12–0.17% of C) at an initial temperature of 1250 °C, considering the heat transfer by means of conduction and convection heat transfer coefficients *HTC* of 10,000 and 10 W/m^2K respectively. The behaviour of the material is assumed visco-plastic, omitting the elastic regime as deformation occurs under high temperature conditions. Hansel-Spittel law is used for the yield stress calculation, making it dependent on the temperature, deformation and strain rate.

The described model is used to test the three friction laws presented in the introductory section. For each law two friction coefficients have been considered, they are adjusted for each friction law for reproducing accurately the real velocity leading to a total of six study cases.

3 Experimental Tests

The tests performed for the acquisition of the experimental data are described in this section. A total number of 6 industrial tests have been performed where the power consumption and the process time have been measured.

The tests consist of the rolling of steel bars of 0.6 m length and 0.2 m of diameter showing an average duration of 0.92 s when the bar exits the mill. The process presents three different regimes: at the beginning the consumption increases (transient) until stabilizing, then it shows a stable value (steady phase) until the final part of the rolling (transient again), when the consumption decreases until the end of the process. The average power consumed during the steady phase results in 2467 kW.

In order to compare the simulated and real velocities of the billet, only the stable regime of the process is analysed. The value is obtained from the experimental data, giving an average result of 928.7 mm/s. In Table 1, the numerical and experimental results are compared in terms of velocity, while in Table 2 the comparison is based on average power consumptions.

Table 1 Velocities of the skew rolling process simulations

Case number	Friction law	Coefficient value (μ, m, α)	Velocity (mm/s)	Deviation (%)
1	Coulomb	0.3	903.1	−2.8
2	Coulomb	0.4	931.7	0.3
3	Tresca	0.7	925.2	−0.4
4	Tresca	0.8	931.3	0.3
5	Viscoplastic	0.7	917.1	−1.2
6	Viscoplastic	0.8	931.8	0.3

Table 2 Power consumption of the skew roll process simulations

Case number	Friction law	Coefficient value (μ, m, α)	Power consumed (kW)	Deviation (%)
1	Coulomb	0.3	2264.2	−8.8
2	Coulomb	0.4	2278.5	−8.3
3	Tresca	0.7	2323.2	−6.5
4	Tresca	0.8	2304.2	−7.2
5	Viscoplastic	0.7	2435.6	−1.9
6	Viscoplastic	0.8	2429.3	−2.2

4 Results and Discussion

In this section, the recorded experimental results of power consumption and velocity are compared to the results obtained from the numerical model presented in Sect. 2.

First, before analysing the process parameters, the friction laws are tested in terms of the contact conditions of the process according to the simulated results.

4.1 Contact Conditions

Through the analysis of the stress state of the contact between roll and workpiece, the values of pressure and equivalent stress are tested for the three models considered. The equivalent stress corresponds to the flow stress of the billet because its deformation occurs under a purely viscoplastic regime (without elastic deformation). Figure 2 shows the pressure at the contact and the equivalent stress in the same region for the case of Coulomb friction law.

According to the literature, Coulomb friction model is only valid when the mean contact pressure between the two contact bodies lies below the flow stress of the softer body. In Fig. 2, the pressure value overcomes the equivalent stress at the contact and thus Coulomb should not be used in this process. On the contrary, Tresca is valid under contact conditions in which the pressure is higher than the flow stress of the

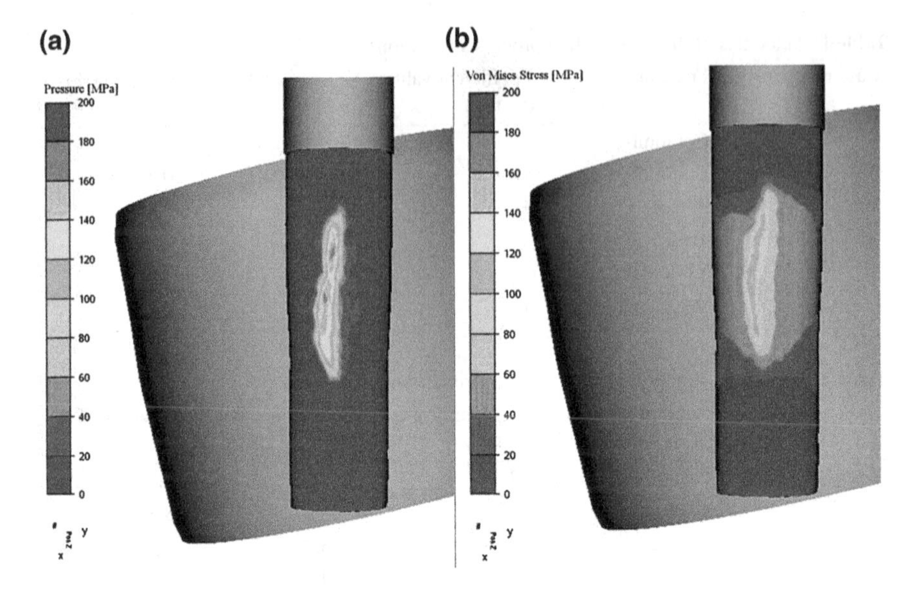

Fig. 2 **a** Pressure and **b** Equivalent stress at the interface with Coulomb friction law

softer body [8]. Regarding Norton law, it is valid for these specific contact conditions as is particularly oriented for materials under a plastic regime. Therefore, it would be more advisable to use Tresca or Norton instead of Coulomb friction law. Then, the three friction laws are compared in terms of operational process variables.

4.2 Process Parameters: Power and Advance Velocity

In this section, two process parameters are compared between the models and the real process namely power and advance velocity. For each friction law, the election of the coefficients is performed in a similar manner. As the aim is first to reproduce the process and secondly to assess the differences existing between friction laws, those coefficients that show a lower deviation from the experimental velocity have been selected and gathered in Table 1.

Hence, there are not noticeable differences in terms of working piece advance velocity between the simulated and real cases. However, this value is based on the velocity at which the material advances through the rolls, whereas the rotation is not measured.

In addition, another operational parameter that can help to identify the validity of friction laws is the power consumption. Table 2 shows that the mean power consumptions depend more on the friction law used than the election of the friction coefficient. This way, the deviation of the mean power with Coulomb is up to 8.8%, while in the cases of Tresca and Norton up to 7.2% and 2.2% respectively.

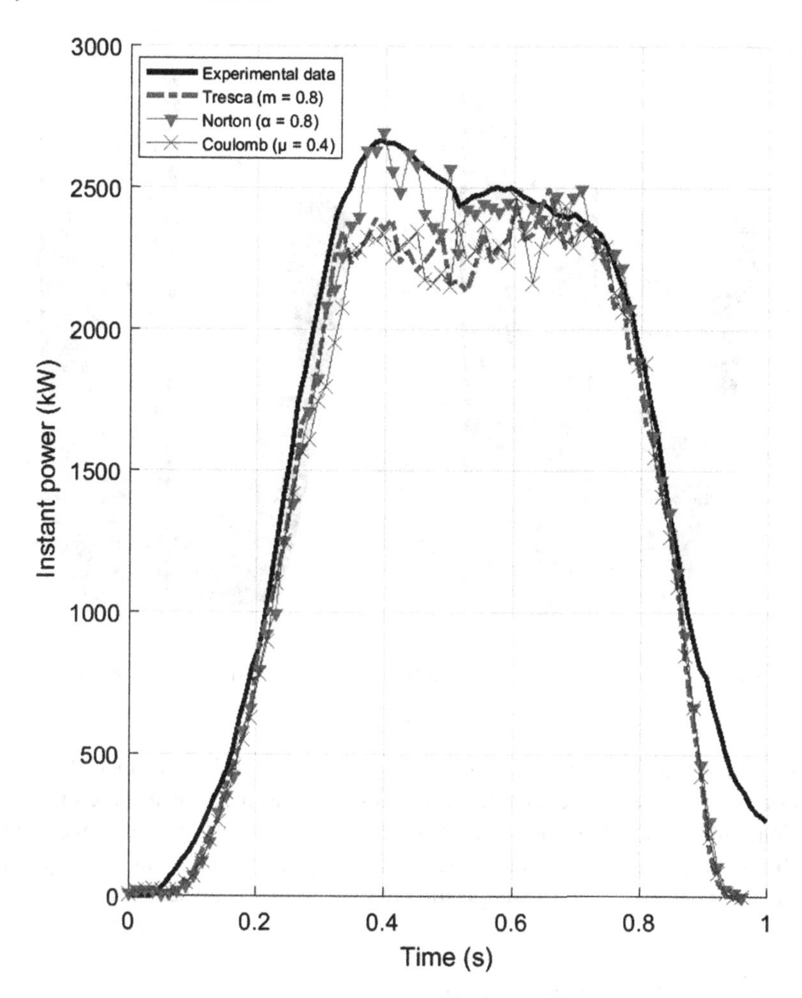

Fig. 3 Instant power evolution of the simulations in contrast with the experimental test

All cases show a good correlation with the measured mean power values according to the values presented in Table 2. Moreover, the reproduction of the instant power consumption is also precise according to the graphs shown in Fig. 3, fitting the experimental value in a correct way. During the final transient part of the process, the motor rotating at no-load explains the differences between real and simulated cases. However, the slope of the simulated curves must fit the real case.

As it can be observed in Fig. 3, power values are lower in the case of Coulomb and Tresca friction laws. Both show a similar evolution, with slightly better results in the case of Tresca. On the other hand, Norton friction law presents a lower deviation in terms of absolute power values and shape of the graph, mainly at the initial transient

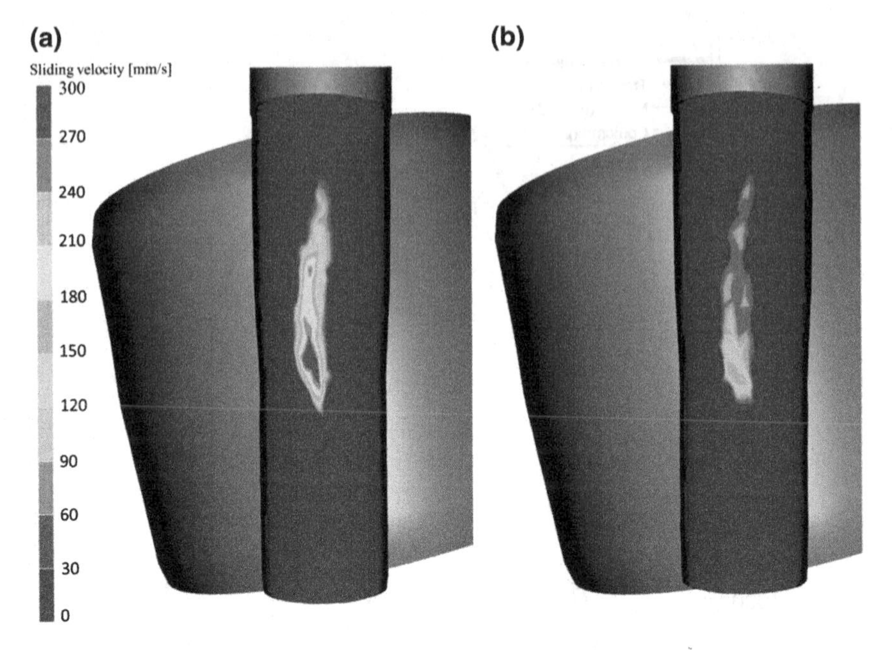

Fig. 4 Sliding velocity at the interface with **a** Norton (α = 0.7), **b** Tresca (m = 0.8) friction laws

stage, where the higher differences are shown. Besides, it is confirmed that all the cases reproduce the average advance process velocity as they finish at the same time.

Coulomb friction law shows the less precise values in terms of power and as has been shown in Fig. 2 is not advisable for the contact conditions of the process. Therefore, only Tresca and Viscoplastic laws are further analysed.

4.3 Friction Law Effect on Results

The election of the friction law as discussed earlier, presents a different impact on the material advance velocity and power consumed by the process. The main difference between Tresca and Norton friction models is the consideration of the sliding velocity at the interface. The implications over the process results are presented and discussed in this section.

In order to illustrate the differences between both models, the cases that show a highest deviation in terms of advance velocity are chosen, namely 4 (Tresca $m = 0.8$) and 5 (Viscoplastic $\alpha = 0.7$). The advance velocity of the material is measured at the tail of the billet, but along the contact with the rolls it is continuously fluctuating. It is illustrated in Fig. 4, in which the sliding velocity at the material-roll interface is shown.

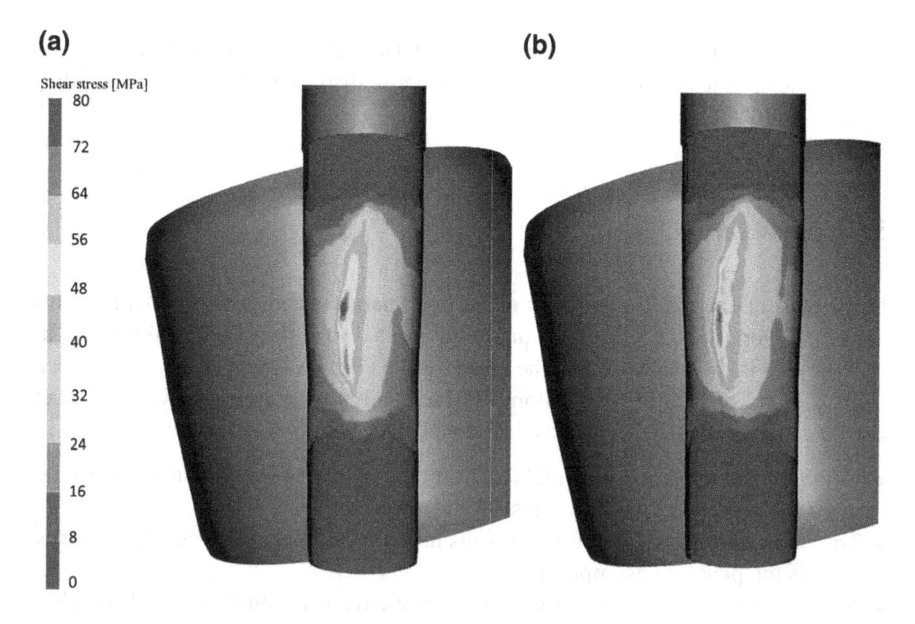

Fig. 5 Shear stress at the interface with **a** Norton ($\alpha = 0.7$), **b** Tresca ($m = 0.8$) friction laws

According to the results shown in Fig. 4, there is a higher value of relative velocity in the case in which Viscoplastic friction law is used. This is in agreement with the advance velocity values presented in Table 2.

Despite of the low percentage difference in terms of advance velocity shown in Table 1, high differences of relative velocity exist at the interface. It is caused by the characteristic position of the rolls, which leads to a higher velocity in the rotation (4500–4800 mm/s) than in the advance movement. Therefore, a small percentage difference in the advance of the material implies a high difference in the absolute value of relative velocity.

Another important aspect, is the relative velocity distribution at the interface. It is more homogeneous in the case of Tresca friction law, whereas Norton friction law shows marked differences. The relative velocity is lower in the region in which the material is initially compressed and higher in the region where the material abandons the mill.

The global strain rate value can be assumed similar in all the simulations (the section area reduction is conducted at a similar velocity). As the material model is similar in both simulations, it can be concluded that the power consumption deviations mostly correspond to a loss in terms of frictional sliding.

Figure 5 illustrates the shear stress at the interface, which is similar in both cases. The frictional power loss is calculated as the product of the shear stress and the relative velocity at the interface. Although both cases show a similar shear stress at the surface, the higher sliding velocity in the case of Norton friction law leads to a higher frictional power loss. It explains the higher accuracy for the instant and average power value given by Norton law over Tresca's approach. As a result, it can be concluded that this friction law shows more realistic results in terms of sliding velocity and frictional power loss.

5 Conclusions

In the present chapter, the problem of friction characterization of high temperature contacts (over 1250 °C) has been analysed applied to a skew rolling process. The specific aim is the evaluation of the effect of the friction law over the process performance and consumption. Coulomb, Tresca and Norton friction laws have been considered and after their analysis, it is concluded that

- Coulomb friction law is not valid for this process because the pressure value at the interface overcomes the flow stress of the material.
- Tresca friction law shows correct results in terms of process velocity but underestimates the power consumption.
- Norton law results are correct for both velocity and power consumed, with a slightly better accuracy on the power consumed.
- The sliding velocity consideration of Norton law, results to be the most realistic approach. It predicts a shear stress value similar to Tresca, while allowing a higher sliding velocity. Therefore, according to the experimental results, the frictional power losses are more realistic.

In short, despite the relatively low percentage difference between Tresca and Norton laws power consumption predictions, in terms of absolute values there is a considerable difference considering that rolling is a high energy consuming process.

References

1. M.S.J. Hashmi, Aspects of tube and pipe manufacturing processes: meter to nanometer diameter. J. Mater. Process. Technol. (2006)
2. H. Li, H. Yang, Z. Zhang, Hot Tube-Forming. *Comprehensive Materials Processing*, 2014. ISSN/ISBN: 9780080965338
3. A. Murillo-Marrodan, E. Garcia, F. Cortes, Friction modelling of a hot rolling process by means of the finite element method, in *Lecture Notes in Engineering and Computer Science: Proceedings of The World Congress on Engineering 2017*, 5–7 July 2017, London, U.K., pp. 965–969
4. R. Iamtanomchai, S. Bland, Study of wear and life enhancement of hot forging dies using finite element analysis, in *Lecture Notes in Engineering and Computer Science: Proceedings of The World Congress on Engineering 2015*, 1–3 July 2015, London, U.K., pp. 833–838
5. M.F. Erinosho, E.T. Akinlabi, Study of friction during forging operation, in *Lecture Notes in Engineering and Computer Science: Proceedings of The World Congress on Engineering 2016*, 29 June–1 July 2016, London, U.K., pp. 929–932
6. C.A. Coulomb, *Théorie des machines simples en ayant égard au frottement de leurs parties et à la roideur des cordages*, Bachelier, 1821

7. Y. Zhao, E. Yu, T. Yan, Deformation analysis of seamless steel tube in cross rolling piercing process, in *2010 International Conference On Computer Design and Applications*, vol. 3, 2010, pp. V3-320–V3-323. https://doi.org/10.1109/iccda.2010.5541258
8. H.S. Valberg, Applied *Metal Forming: Including FEM Analysis* (Cambridge University Press, 2010)
9. K. Komori, M. Suzuki, Simulation of deformation and temperature in press roll piercing. J. Mater. Process. Technol. **169**, 249–257 (2005). https://doi.org/10.1016/j.jmatprotec.2005.03.017
10. K. Komori, Simulation of Mannesmann piercing process by the three-dimensional rigid-plastic finite-element method. Int. J. Mech. Sci. **47**, 1838–1853 (2005). https://doi.org/10.1016/j.ijmecsci.2005.07.009
11. A. Ghiotti, S. Fanini, S. Bruschi, P.F. Bariani, Modelling of the Mannesmann effect. CIRP Ann.—Manuf. Technol. **58**, 255–258 (2009). https://doi.org/10.1016/j.cirp.2009.03.099
12. X. Duan, T. Sheppard, Three dimensional thermal mechanical coupled simulation during hot rolling of aluminium alloy 3003. Int. J. Mech. Sci. **44**, 2155–2172 (2002). https://doi.org/10.1016/S0020-7403(02)00164-9
13. Forge® Users' Manual, Transvalor

2. [illegible] Zhao [illegible] Deformation analysis of [illegible] steel tube [illegible] rolling [illegible] [illegible]

[illegible]

[illegible]

[illegible]

Residual Stress Simulations of Girth Welding in Subsea Pipelines

Bridget Kogo, Bin Wang, Luiz Wrobel and Mahmoud Chizari

Abstract Simulation of the welding of two dissimilar materials, stainless steel and mild steel, has been carried out using finite element together with experiments to validate the method and better understand the transient temperature profiles and the stress distribution in a cladded pipe. The results clearly show that the temperature distribution in the modelled pipe is a function of the thermal conductivity of each weld metal as well as the distance away from the heat source. The outcome of the study has been compared with previous findings.

Keywords Dissimilar material joint · FEA · Girth weld · Residual stress Subsea pipelines · Transient temperature response

1 Introduction

Welding of cylindrical objects is complex and poses a source of concern in the manufacturing processes. There are several benefits of welding as a joining technology which includes cost effectiveness, flexibility in design, enhanced structural integrity, and composite weight reduction. However, thermal stresses are usually initiated on the weld and the base metal [1–5]. Poorly welded joints result in leakages, pipe failures and bursts, which lead to possible environmental hazards, loss of lives and

B. Kogo (✉) · B. Wang · L. Wrobel · M. Chizari
Mechanical, Aerospace and Civil Engineering Department, College of Engineering, Design and Physical Sciences, Brunel University London, Uxbridge UB8 3PH, UK
e-mail: biddyagada@yahoo.com

B. Wang
e-mail: Bin.Wang@brunel.ac.uk

L. Wrobel
e-mail: Luiz.Wrobel@brunel.ac.uk

M. Chizari
School of Mechanical Engineering, Sharif University of Technology, Tehran, Iran
e-mail: mahmoudchizari@yahoo.com

S.-I. Ao et al. (eds.), *Transactions on Engineering Technologies*,
https://doi.org/10.1007/978-981-13-0746-1_30

properties. In this study, welding of dissimilar materials is carried out in-house using Gas Metal Arc Weld (GMAW). In parallel, a finite element analysis (FEA) on the pipes having different clad thicknesses of 2 mm and 12 mm has been carried out. For all models, the temperature versus distance profile obtained. The 12 mm cladded pipe results are discussed in this paper [6–8].

The welding process carrying out using an arc weld entails melting down the base metal and, in this research, it also includes melting down the clad metal. The result of this procedure generates a metallurgical structure positioning in situ the material which supplies superior tensile strength. The bulk of the material immediately after the fusion zone (FZ), which has its characteristics altered by the weld, is termed Heat Affected Zone (HAZ). The volume of material within the HAZ undergoes considerable change which could be advantageous to the weld joint, but in some circumstances, might not be beneficial. The aim of this paper it to closely look at the welding of dissimilar materials and to compare the results of the hardness profiles and tensile stress distribution across the welded zones.

2 Experimental Examination

The Gaussian transformation principle states that 'A Gaussian flat surface has a Gaussian curvature at each and every point of the magnitude of zero' [8]. Going by this principle, the surface of a cylinder can be said to be a Gaussian flat plane since it can be revolved from a piece of paper. Furthermore, the implication is that this can be done without stretching the plane, folding or tearing it. This means that the thermal distributions on the surface of a cylinder can also be appreciated by studying the thermal distribution on the surface of welded plates [6–8].

The series of welding experiment has been carried out using an arc weld entails melting down the base metal (parent plate) and also melting down the clad metal. The details has been explained in the previous studies carried out by the authors [6–8]. During the welding, filler metals are also melted and a solution formed by heating up the materials—holding them at that range of temperature long enough to permit diffusion of constituents into the molten solution, after which there is a rapid cooling down in order to maintain these constituents within the solution.

3 Finite Element Modelling

A fully cladded pipe has been modelled with an 8-node linear hexagonal element, generating a FEA mesh with 180,306 elements and 208,640 nodes. Linear hexahedral elements are recommended for their reduced computation time and ease of running analysis due to the structured grid which makes up the mesh, identical structured array and their uniform grid shape which eliminates skewness; however, a hexahedral

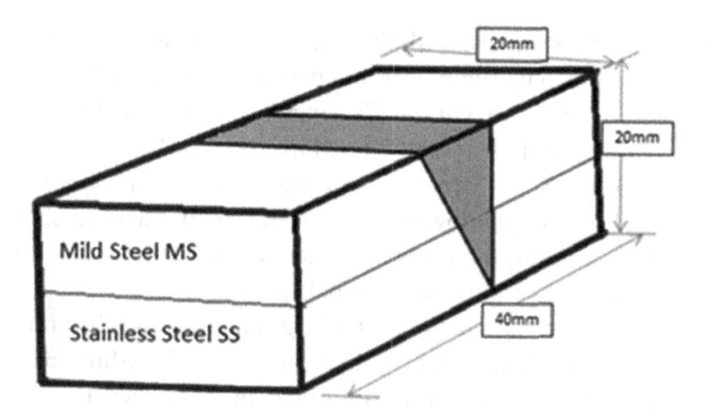

Fig. 1 A typical electronic microscopy test sample of a 12 mm cladded weld

mesh can also be unstructured depending on the manner in which element indexing is executed [8].

A thermal simulation for heat pass 1 in a full 3D pipe has been modeled [8] and the temperature contours representing each thermal distribution are recorded. Corresponding temperatures between 220 and 1,545 °C has been recorded on different zone [8].

4 Tensile Testing

A series of tensile tests on the specimens has been carried out. It was discovered that several factors such as temperature, strain rate and anisotropy affected the shape of the stress- strain curves. The parent metals have different elongation characteristics, and each exhibited this at different rates because of the applied stress under which it is stretched [8]. Similarly, the behaviour of the weld metal under the displacement time curve is also due to slip, which is caused by the elongation and failure of the different metals (mild steel and stainless steel) present within the weld samples, since they each have their original ultimate tensile stress (UTS). The volumetric change and yield strength as a result of martensitic transformation have influences on the welding residual stresses, increasing the magnitude of the residual stress in the weld zone as well as changing its sign [8].

5 Electron Microscopy Examination

A schematic of a typical sample used for microscopy is shown in Fig. 1. The sample contains a portion of the weld zone. The preparation of samples for Scanning Electron Microscope (SEM) carried out as following:

1. Two different weld samples, with the dimensions as shown in Fig. 1, were cut from the main weld. The samples of the parent materials: 12 mm stainless steel and 10 mm mild steel were cut into dimensions: 40 mm × 20 mm and 20 mm.
2. The parent material samples were formed into a mould using 5 spoonsful or 2.5 spoonful of Bakelite S and Struers mount press [9], to enable easy and controlled grinding [9]. Place sample down on the holder, pour Bakelite unto sample and start. After five minutes, mount press heats up for 3 min and cools down for 2 min. Use the electric scribing tool to label sample.
3. The samples were grinded using a grinding machine and different silicon carbide papers from 80, 120, 350, 800, 1200 and 1600. Polishing each time in the opposite direction to the scratches to eliminate the scratches. Polishing was done in alternating vertical and horizontal directions for each carbon paper change.
4. Polishing of samples with a polishing cloth with diamond paste.

Both hand polishing and machine were carried out. The sample preparation is like that for SEM and is explained below: Polishing weld samples with of varying degree of polish paper (carbide). At each stage, the sample was washed clean with water and its surface preserved (to avoid oxidation) with ethanol or methanol before drying. In the case of washing with methanol, protective breathing, eyes and hand clothing such as gloves and eye protection were worn for safety. The sample is then examined under an electron microscope to see if the required microstructure has been achieved, otherwise polishing continues.

After a mirror surface is achieved, the diamond paste and Colloidal Silica Suspension (also called Oxidizing Polishing cloth) was used. Rinsing the surface with nitric acid to clean surface and especially preserve from corrosion.

6 Results and Discussion

6.1 Micrograph Examination

The polishing made the heat affected zone and the weld zone visible as shown in the Fig. 2; however, the unique features in the weld zone and heat affected zones were only appreciated after etching as shown in Figs. 2 and 3.

The experiment was carried out in agreement with the standard for grinding and polishing stainless steel cladding with mild steel [10, 11]. These cladded samples were examined with the aid of an optical microscope to observe the microstructural evolutions within the heat affected zone. The heat affected zone is as shown in Fig. 2. The HAZ is the boundary or zone surrounding the welded zone. This area is of paramount interest in this research because of the grain size formed as well as the constituent elements that make up that zone, most especially because the large grains formed in this zone as a result of austenitic cooling of the martensitic grains get oxidized when the pipe is laid on the sea bed or in deep offshore operations. As this layer gets eroded, they expose the layer of the cladded pipe beneath resulting

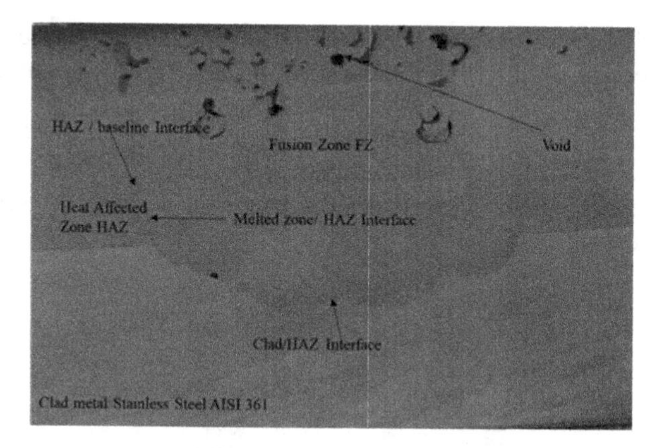

Fig. 2 Micrograph of FZ, HAZ and stainless steel-clad metal

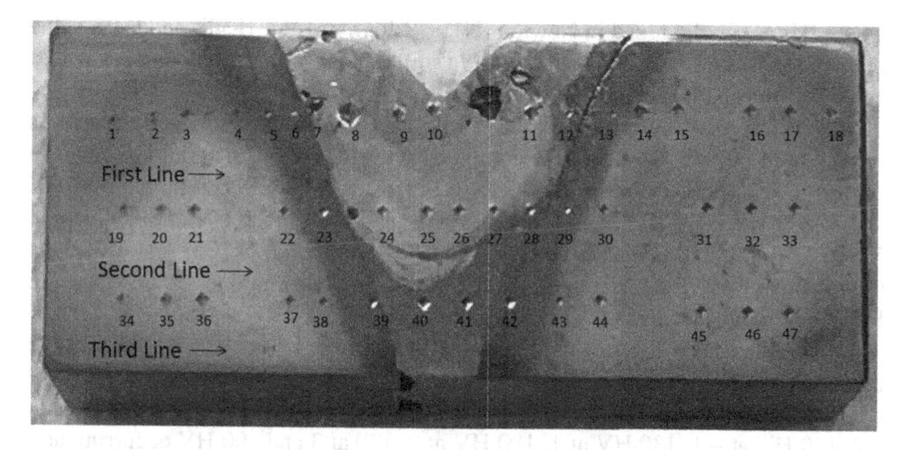

Fig. 3 Pattern and order for diamond stud imprints across HAZ and fusion zones. NB: Etching enables a distinct view of features

in pitting. Pitting, if not handled properly as result of the pressure of fluid within the pipe and the forces acting on the pipe from surrounding environment, the ocean current also contributing, could lead to a leak, which could result in a burst until there is complete failure of the pipeline.

6.2 Vickers Hardness Test

Using the Vickers hardness machine, a diamond stud pattern was created with the aid of a pyramidal diamond indenter following strictly the pattern in Fig. 3 across the HAZ, the FZ and the region close to the welded zones of the welded metals. These were repeated in the second and third lines as seen both in the welded sample and the chart. Alternatively, there is a vertical array of similar types of pattern across the HAZ.

The length of the diagonals of the diamond stud was measured both in the vertical (D1) and horizontal (D2) axis and the average reading recorded. This average value of the distance was inputted into a standard equation known as the Vickers hardness equation as shown below.

$$Hv = \frac{\text{Applied load (kgf)}}{\text{Surface area of impression (mm)}} \tag{1}$$

In the case where an indenter is used as in this research;

$$HV = [2Fsin(\vartheta/2)]/D2 \tag{2}$$

where F is the applied load in Newton (N) or Kilogram-force (KgF) and has a value of 20 N, D is the mean diagonal length in millimeters (mm), θ is the angle with a value of 136° and HV is the Vickers' hardness measured in HV.

It is observed that, for the first line of all the 12 mm samples 1–3, the hardness is very high in both HAZ and FZ compared with the parent material as shown in Fig. 4a–c. The peak values of hardness in Fig. 4 (a) from left to right are 330 HV at −4, 100 HV at 1; (b) 270 HV at 4, 190 HV at 5 and (c) 180 HV at 6, 290 HV at 5, 195 HV at 4, 210 HV at 3, 290 HV at 1, 200 HV at 2 and 210 HV at 4. For the second line of the 12 mm samples 1 and 2 of the Fig. 4, the hardness is higher in the HAZ than in other regions.

Of significance is the fact that hardness is also high in the FZ and HAZ of the third line of 12 mm samples 1 and 2 in Fig. 4a–b. This can be seen in graphs (a–b) readings which are, respectively from left to right, (a) 100 HV at −4, 190 HV at −2, 170 HV at −1, 130 HV at 1, 180 HV at 2, 170 at 3 and 100 HV occurring at 4 likewise; (b) 100 HV at −4, 130 HV at −2, 160 HV at 1, 150 HV at 2, 170 HV at 3 and 100 HV at 4.

From the charts in Fig. 4, it is noted that there is a unique trend in the increased hardness profile across the FZ and HAZ of 3rd line in the 12 mm samples 1 and 2.

From the Vickers hardness test, it is obvious that the weld hardness is 30–70% greater than the parents' metal. This is due to the very high rate of Martensite formation during rapid cooling of the melt pool. Throughout the weld process, there is continuous reheating taking place as the weld touch passes to and from the weld metals. The average hardness of the dilution zone is comparable to that of the clad.

From the hardness profile in samples 1, 2 and 3, it is obvious that the hardness of the HAZ varies linearly from the clad/HAZ interface to the HAZ/baseline interface with values 200–330 Hv accordingly. The reason for the direct variation of hardness in the HAZ is the difference in heating temperature in the HAZ resulting in the variation in grain growth.

The result of the 12 mm samples 1, 2 and 3 in Fig. 4a–c is the formation of coarse grain because of tall peak temperatures leading to coarser microstructure formed

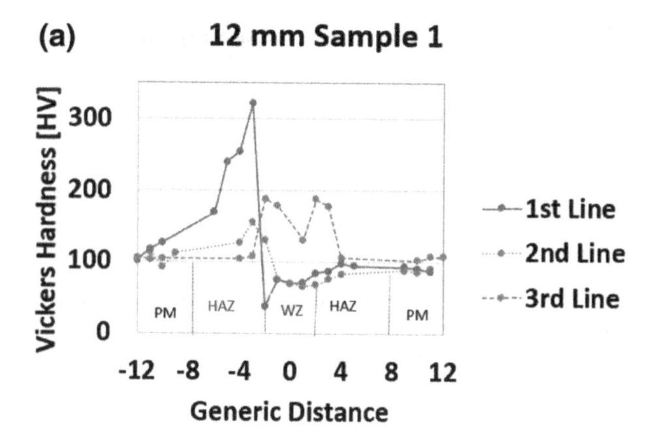

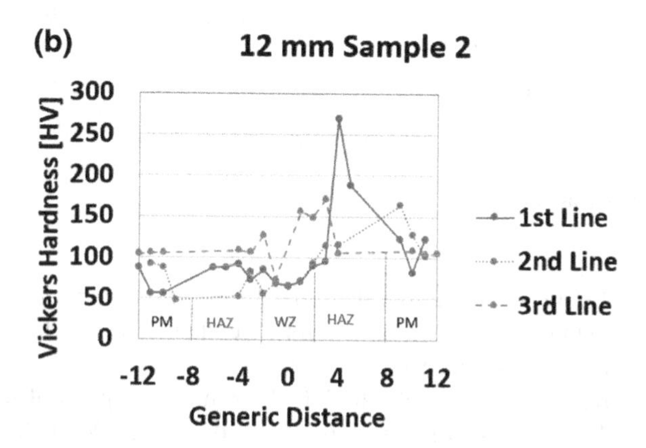

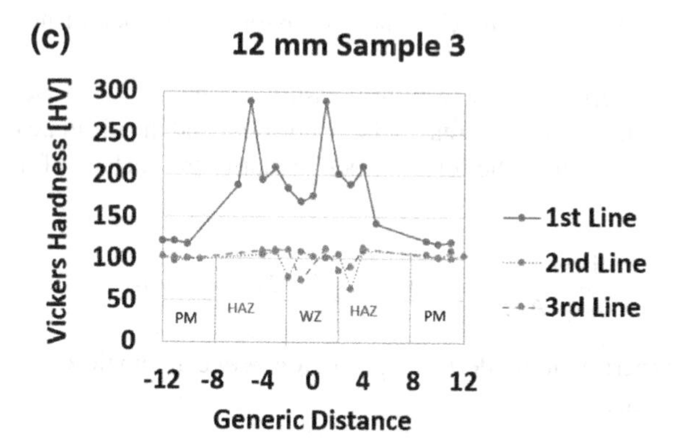

Fig. 4 **a–c** Hardness profile of three samples of 12 mm cladded weld

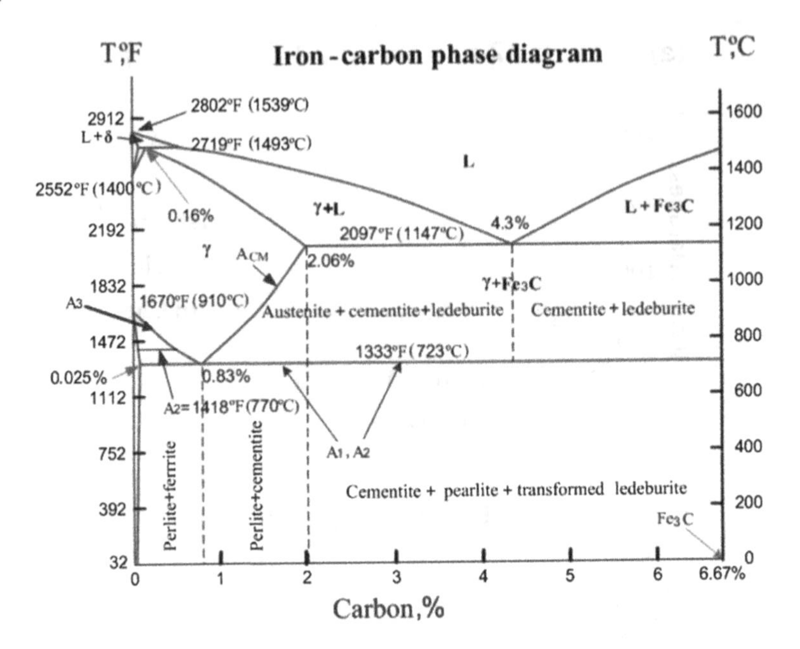

Fig. 5 Iron-Carbon phase diagram

close to the Clad/HAZ interface. On the other hand, finer grain sizes are formed because of decreasing temperatures away from the clad/HAZ interface. Overall, a finer grain size is harder than a coarse grain size.

The HAZ increases proportionally to 330 HV, which is typical of the hardness observed during heat treatment ranging from 710 to 170 °C. When A1 is attained in this temperature range, there is a sharp fall in the hardness [12, 13] at the end of the HAZ which implies that there are no γ transformations occurring, as shown in Fig. 5.

Mechanical properties are relevant to pressure vessels [14]. An increase in temperature or in irradiation dose increases the yield stress and the ultimate tensile stress. From Vickers hardness test, the force perpendicular to the surface of the indenter is given as:

$$H_V = \frac{\text{Load}}{\text{Contatc Area}} = 0.927\,\text{P} = 0.927\,\text{P} \times 2.96\,\sigma_y = 2.74\,\sigma_y \tag{3}$$

The Busby experimentally derived correlation between tensile strength and hardness is expressed as:

$$\sigma_y = 0.364\,H_V \tag{4}$$

where H_V and σ_y are measured in kg/mm^2 or

$$\sigma_y = 3.55\ H_V \quad (5)$$

where H_V is in kg/mm^2 and σ_y in MPa.

Gusev et al. [15] derived the relationship between yield stress and micro-hardness as:

$$\Delta\sigma_{0.2} = k_\Delta \cdot \Delta H_\mu \quad (6)$$

$$\Delta\sigma_{0.2} = 2.96 \cdot \Delta H_\mu \quad (7)$$

where k = 2.96 and $\Delta\sigma_{0.2}$ is the difference in property correlation.

The above equation was derived for both stainless steels irradiated and non–irradiated at elevated temperatures of 300 °C, whereas the hardness was carried out at room temperature. The drawback of this system of correlation is that it requires a material property correction factor due to the difference in the non-temperature dependent property. This material property correction factor is expressed as:

$$k\_\Delta = 3.03 \quad (8)$$

This applies to all steel types irrespective of the make-up of the steel.

The advantage of this experimentally derived correlation is that both hardness tests and tensile tests were carried out at room temperature, sodium transformed surfaces were removed before the micro-hardness test and as many indentations as could be were punched unto the metallic surfaces. Experimental studies carried out on larger scale did not factor in the slightest change in composition and possible deposition of ferrite onto the surfaces. Another possibility is also that the brittle nature of the stainless steel at elevated hardening levels could be due to martensitic distortion while carrying out micro hardness dimensions of the low alloy steel [15]. This universal correlation (k_Δ) enables the determination of yield stress form the micro hardness value hence improving labour efficiency.

Ductile to Brittle Transition Temperature (DBTT) varies in dissimilar materials, some being more severe than others, which can be accounted for via a temperature sensitive deformation process. The procedure and behaviour of a Body Centred Cubic (BCC) lattice is triggered by temperature and responds to reshuffling of the dislocation core just before slip. This could result in challenges for ferritic steel in the building of ships. Neutron radiation also influences DBTT, which deforms the internal lattice hence reducing ductility and increasing DBTT.

6.3 Hardness Versus Yield Stress

Figure 6a–c further show a linear relationship exists between the yield stress and the hardness of the weld samples which further confirm, for the 12 mm stainless

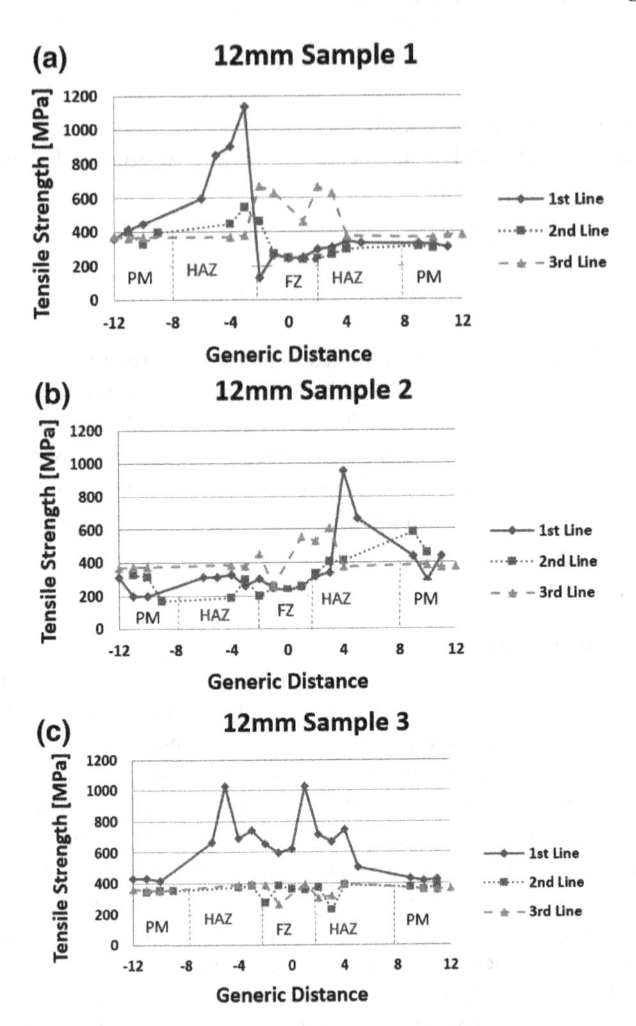

Fig. 6 Plots of yield stress versus hardness for **a** sample 1; **b** sample 2; **c** sample 3. The hardness unit is MPa

steel clad, that the value of hardness increases with the decrease in temperature and applied load. Hardness versus yield stress summarized in Table 1.

There is a similar trend in the increased tensile profile across the FZ and HAZ on the third line in 12 mm samples 1 and 2 as shown in Fig. 6(a and b), respectively. According to graphs, the peak tensile strength values for the graph (a) is 668 MPa at −2 and 664.26 MPa at 2 whereas for the graphs (b) it is 452.59 MPa at −2, 553.83 MPa at 1 and 606.16 MPa at 3, respectively.

The strength is also high in the FZ and HAZ of the second line for the12 mm samples 1 and 2 graphs (a–b). For the 12 mm samples 1 and 2 graphs (a–b) respectively, the strength is higher in the HAZ than in other regions.

Table 1 Hardness versus Yield stress

Specimen	Hv (MPa)	Yield stress (MPa) = (3.55) HV
12 mm sample 1	320	1136
12 mm sample 2	269	954.95
12 mm sample 3	288	1022.4

Table 2 Elements in Weld Steel of 12 mm SS/MS Clad

Element	Fe	C	Cr	N	Mn	Si	Ca	Mo
SS	71.30	5.08	13.41	7.05	1.51	0.41	0.08	1.16
MS	92.02	5.51	0.61	–	1.13	0.41	0.07	0.25

It is observed that for the first line of all the 12 mm samples 1–3, the yield strength and consequently the Ultimate Tensile Strength (UTS) is very high at both HAZ and FZ compared with the parent materials, as shown in Fig. 6(a–c) respectively.

High thermal gradients were experienced during the Butt welding procedure leading to residual stress and discrepancy in hardness [16–20]. Because of the high concentration of thermal stress in the clad, the presence of residual stresses usually affects the inherent resistance to corrosion and fatigue cracks. To improve the mechanical properties of the clad/base metal interface, as well as reduce the residual stresses generated, post-heat treatments are carried out.

6.4 Microstructures

There is an element of carbon in the stainless steel which is evident from the result of the SEM analysis of the weld, with Table 2 displaying the elements present in both stainless steel and mild steel. The Fe content is higher in the mild steel than in the stainless steel. The Nickel content is very high in the stainless steel and absent in the mild steel. Likewise, the Molybdenum content is higher in the stainless steel than in the mild steel. The Cr content is very high in the stainless steel compared with the mild steel, and further displayed in the spectrum [8] for stainless steel.

Since the weakest point of the weld is the clad/HAZ interface due to inconsistent fusion and reheating [17], during Butt welding, there are high thermal gradients experienced during the procedure leading to residual stress and discrepancy in hardness. The presence of residual stresses as a result of high concentration of thermal stress in the clad usually affects the inherent resistance to corrosion and fatigue cracks. In order to enhance the mechanical properties of the clad/base metal interface, as well as reduce the residual stresses generated, post-heat treatments are usually carried out.

The presence of Nickel and Manganese in steel decreases the eutectoid temperature lowering the kinetic barrier whereas Tungsten raises the kinetic barriers. The presence of Manganese increases hardness in steel and likewise Molybdenum.

Within the transition zone next to the 12 mm weld metal, the stainless-steel part of the microstructures contains acicular ferrites, which are formed when the cooling rate is high in a melting metal surface or material boundaries [8]. Different ferrites are formed starting from the grain boundary. Such ferrites include plate and lath Martensite, Widmanstatten ferrite, and grain boundary ferrite.

6.5 Weld Direction—Nomenclature

The usual concept of 90°, 180°, 270° and 360° has been used in a clockwise manner to describe the direction of the weld, as well as the 3 o'clock, 6 o'clock, 9 o'clock and 12 o'clock convention. However, there is a new system of nomenclature known as the 45°, 135°, 225° and 315° reference system, which is obtained by simply rotating the cross-section of the pipe through an angle of 45° in the clockwise direction [8].

The above style of representation of a welding direction is known as 1:30, 4:30, 7:30 and 10:30 h face of a clock using a temporal connotation. Representing the four positions of interest on the pipe circumference onto a plate, following the Gaussian transformation principle [8], implies that different weld directions can also be represented on a plane surface as shown on a 2D plate [8].

6.6 Thermal Analysis

The effect of the clad on the weld is such that the clad has effectively reduced the operating temperature thereby limiting the thermal conductivity of the welded path. The reduction in thermal conductivity enhances the insulating effect of the cladding. Since the thermal diffusivity varies directly with the density and specific heat of the material, it implies that the thermal diffusivity reduces as the thickness of the insulating material increases. Therefore, the material density is directly related to the insulation performance.

Bearing in mind that the temperature imparts directly on the toughness, modulus of elasticity, ultimate tensile strength and yield stress, this means that an increased operating temperature will also impact upon these properties of the cladded pipes.

6.7 Stress Analysis

It was also observed from the residual axial stress of the cladded pipe that, close to the weld vicinity, compressive and tensile stress fields are present in and near the section of the weld both on the external and internal surfaces of the pipe. This is as a result of the varying temperature profiles on the inner and outer surfaces of the pipe. By virtue of the thickness of the cylinder wall and because of the proximity to the weld line, tensile and compressive residual stress fields are generated due to shrinkage occurring within the weld pipe [5, 8, 18–21].

The differences in the values of the residual stresses are a result of the different material properties such as yield strength for the base and filler metals, weld geometry and heat source parameters. There have been volumetric change and yield strength as a result of martensitic transformation, which have effects on welding residual stress, by increasing the magnitude of the residual stress in the weld zone as well as changing its sign. The simulated results show that the volumetric change and the yield strength change due to the martensitic transformation have influences on the welding residual stress [8].

An axial inclination of the constraint free end of the pipe usually occurs when the weldment cools down. The thickness of the pipe is considered for the radial shrinkage and measured for four different increments.

Similarly, the shrinkage is measured and plotted against the normalized distance from the weld path for different increments of the axial length. The axial shrinkage at lower increments is slightly different from those at higher increments, because there are high thermal gradients experienced during Butt welding leading to residual stress and discrepancy in hardness, hence a creep effect is observed at higher increments [8].

7 Conclusion

It is significant to note that there exists a linear relationship between the tensile strength and the hardness of the weld and consequently, the ultimate tensile strength. The EBSD and XRD analysis were discussed in line with this. From the hardness profiles, the following can be deduced:

- The hardness in the FZ and HAZ is 30–70% more than that in the parent material.
- A linear relationship exists between the yield stress and the hardness of the weld samples, which further confirm for the 12 mm stainless steel clad that the value of hardness increases with the decrease in temperature and applied load.
- There is a similar trend in the increased tensile profile across the fusion zone and HAZ.
- The hardness of the HAZ varies linearly from the clad/HAZ interface to the HAZ/baseline interface with values 200 to 330 HV accordingly.
- The reason for the direct variation of hardness in the HAZ is the difference in heating temperature in the HAZ resulting in variation in grain growth.

Acknowledgements The authors want to express their gratitude to IAENG and WCE for the great privilege of being awarded Best Student Paper Publication at the WCE 2017 Conference and for the opportunity to publish this paper as an IAENG publication.

The authors want to express their gratitude to Brunel University London for the facilities provided and conducive research environment.

The first author also thanks The Petroleum Technology Development Fund (PTDF) for their funding and support through which this research has been made possible.

References

1. S. Lampman, *Weld Integrity and Performance* (ASM International, Ohio, 2001)
2. J.A. Goldak, *Computational Welding Mechanics* (Springer, New York, 2005)
3. A.G. Youtsos, *Residual Stress and its Effects on Fatigue and Fracture* (Springer, Dordrecht, The Netherlands, 2006)
4. E. Muhammad, *Analysis of Residual Stresses and Distortions in Circumferentially Welded Thin-Walled Cylinders* (National University of Science and Technology, Parkistan, 2008)
5. N.U. Dar, E.M. Qureshi, M.M.I. Hammouda, Analysis of weld-induced residual stresses and distortions in thin-walled cylinders. J. Mech. Sci. Technol. **23**, 1118–1131 (2009)
6. B. Kogo, B. Wang, L. Wrobel, M. Chizari, Analysis of girth welded joints of dissimilar metals in clad pipes: experimental and numerical analysis, in *Proceedings of the Twenty-seventh (2017) International Ocean and Polar Engineering Conference San Francisco*, CA, USA, 25–30 June 2017, pp. 1–5. San-Francisco, CA, USA: Copyright © 2017 by the International Society of Offshore and Polar Engineers (ISOPE) (2017)
7. B. Kogo, B. Wang, L. Wrobel, M. Chizari, Thermal analysis of girth welded joints of dissimilar metals in pipes with varying clad thicknesses, in *Proceedings of the ASME 2017 Pressure Vessels and Piping Conference PVP2017* (ASME PVP, Waikoloa, Hawaii, USA, 16–20 July 2017), pp. 1–9
8. B. Kogo, B. Wang, L. Wrobel, M. Chizari, Residual stress simulations of girth welding in subsea pipelines, in *Lecture Notes Engineering and Computer Science: Proceedings of the 25th World Congress on Engineering (WCE 2017)*, 5–7 July 2017, London, UK, pp. 861–866
9. Struers, Grinding and Polishing, Struers Ensuring Certainity (2018), www.struers.com. Accessed 16 Jan 2018
10. Struers, Metallography of Welds, Struers (2018), www.struers.com. Accessed 17 Jan 2018
11. G.F. Vander Voort, *Metallography of Welds. Advanced Materials and Processing*, June 2011, www.asminternational.org
12. D.S. Sun, Q. Liu, M. Brandt, M. Janardhana, G. Clark, Microstructure and mechanical properties of laser cladding repair of AISI 4340 steel, in *International Congress of the Aeronautical Sciences*, 2012, pp. 1–9
13. S.A. South Africa, Technologies, Iron-Carbon Phase Diagram (2012), SubsTech, http://www.substech.com/dokuwiki/doku.php?id=iron-carbon_phase_diagram. Accessed 07 Sept 2016
14. J.T. Busby, M.C. Hash, G.S. Was, The relationship between hardness and yield stress in irradiated austenitic and ferritic steels. J. Nucl. Mater. 267–278 (2004)
15. M.N. Gusev, O.P. Maksimkin, O.V. Tivanova, N.S. Silnaygina, F.A. Garner, Correlation of yield stress and microhardness in 08Cr16Ni11Mo3 stainless steel irradiated to high dose in the BN-350 Fast Reactor. J. Nucl. Mater. 258–262 (2006)
16. C.A. Sila, J. Teixeira de Assis, S. Phillippov, J.P. Farias, Residua stress, mirostructure adn hardness of thin-walled low-carbon steel pipes welded manually. Mater. Res. 1–11 (2016)
17. W. Suder, A. Steuwer, T. Pirling, Welding process impact on residual stress distortion. Sci. Technol. Weld. Join. 1–21 (2009)
18. G. Benghalia, J. Wood, Material and residual stress considerations associated with the autofrettage of weld clad components. Int. J. Press. Vessel. Pip. 1–13 (2016)

19. D. Dean, M. Hidekazu, Prediction of welding residual stress in multi-pass butt-welded modified 9Cr-1Mo steel pipe considering phase transformation effects. Comput. Mater. Sci. **37**, 209–219 (2006)
20. N. Mathiazhagan, K. Senthil, V. Balasubramanian, V.C. Sathish Gandhi, Performance study of medium carbon steel and austenitic stainless-steel joints: friction welding. Oxid. Commun. 2123–2134 (2015)
21. P.K. Sinha, R. Islam, C. Prasad, M. Kaleem, Analysis of residual stresses and distortions in girth-welded Carbon steel pipe. Int. J. Recent Technol. Eng. **2**, 192–199 (2013)

Index

S.-I. Ao et al. (eds.), *Transactions on Engineering Technologies*,
https://doi.org/10.1007/978-981-13-0746-1

Printed in the United States
By Bookmasters